INSIDE THE VACUUM TUBE

Other Books
by
JOHN F. RIDER

VACUUM-TUBE VOLTMETERS

AUTOMATIC RECORD CHANGERS AND RECORDERS

THE METER AT WORK

FREQUENCY MODULATION

SERVICING BY SIGNAL TRACING

THE OSCILLATOR AT WORK

THE CATHODE-RAY TUBE AT WORK

SERVICING SUPERHETERODYNES

PERPETUAL TROUBLE SHOOTER'S MANUAL
VOLUMES I TO V ABRIDGED
VOLUME VI TO VOLUME XIV INCLUSIVE

ALIGNING PHILCO RECEIVERS, VOL. I AND VOL. II

AUTOMATIC FREQUENCY CONTROL SYSTEMS

SERVICING RECEIVERS BY MEANS OF RESISTANCE MEASUREMENT

AN HOUR A DAY WITH RIDER

ON:

RESONANCE AND ALIGNMENT

AUTOMATIC VOLUME CONTROL

ALTERNATING CURRENTS IN RADIO RECEIVERS

D-C VOLTAGE DISTRIBUTION IN RADIO RECEIVERS

INSIDE THE VACUUM TUBE

BY

JOHN F. RIDER
Lt. Col. U. S. Signal Corps, (Ret.)

Illustrated by
Baxter Rowe

JOHN F. RIDER PUBLISHER, INC.
404 Fourth Ave. New York 16, N. Y.

Printed in the United States of America

Dedicated to

JANET

1330

who still is an inspiration
to her father

AUTHOR'S FOREWORD

OF ALL THE INVENTIONS that have ever made scientific history, the discovery by Lee DeForest that a few turns of wire in between the two elements of a diode vacuum tube were able to control the flow of electrons, can without doubt be classed as among the foremost in its benefits to mankind. Not only has the three-element vacuum tube been the means of providing entertainment and education to millions of people throughout the world, but it has proved to be the keystone in the arch upon which has been built innumerable branches of scientific thought.

Considering the effect of these few turns of wire—the grid in a triode—on the transmission and reception of intelligence through the medium of radio communication alone, to put it mildly, this discovery was tremendous. Of course, the discovery of the high-frequency alternator by Alexanderson and the arc transmitter by Poulson made possible the exchange of intelligence by radio, yet this means of communication was limited in that it did not possess the versatility of electronic apparatus typified by the vacuum tube. The advent of the triode vacuum tube as an ancestor for the vast number of different types that we know today, was really a beginning of a new era in world thought. Where would television, facsimile, radio broadcasting, and radar be without it? The answer is obvious. . . .

In view of all this, it was felt that this book was a necessity and it has but one purpose: to present a solid, elementary concept of the theory and operation of the basic types of vacuum tubes as a foundation upon which can be built a more advanced knowledge of tubes in general. Here then are the elements—the rest is up to the reader. . . .

We have set certain boundaries in this book so that its scope may be considered limited to some degree, but after all—and we repeat—this is an *elementary* explanation of vacuum tube behavior and operation. We have kept the essential mathematics in as simple a form as possible and have included only the minimum amount. In the final chapter, we have omitted mention of certain types of tubes, for instance the lighthouse tube, the magnetron, and the klystron, because even a superficial description of their functioning would involve matters which were considered to be beyond the limits which had been set. So also did we omit the subject of reactive loads in our discussion of this important phase of tube functioning, because its inclusion would take us over those boundaries of elementary study.

The reader will find several innovations in this book. Following the explanation of the electron theory, is a new presentation of text concerning the vacuum tube: a discussion of electrostatic fields. It is our feeling that by understanding the distribution and behavior of the fields within a tube, the reader will gain a better picture of why amplification is accomplished within a tube and how the grids and plate are inter-related. Throughout the book, which covers diodes, triodes, tetrodes, and pentodes, the aim is to present a clear physical picture of exactly what is occurring in a vacuum tube, inclusive of the development of characteristic curves of all kinds, load lines, and dynamic transfer characteristics, discussion on power amplifiers, the use of pentodes as triodes, the cathode follower, etc.

One of the problems in book making has been to make illustrations easily accessible to the text describing or discussing them. We have, therefore, had certain diagrams and graphs printed on both sides of the page so that they can be consulted with a minimum of bother to the reader while reading the text referring to them.

Another innovation in the publishing of radio texts is the use of line anaglyphs that provide three-dimensional pictures which up to now have been presented in one plane only. We have used but three of these stereoscopic illustrations, because we believe that if the reader once gets the idea of a field, for instance, from an anaglyph, he can imagine how it appears under other conditions from a drawing in one plane.

We wish to express our gratitude to the RCA Manufacturing Co. for their cooperation in providing us with the hitherto unpublished data on the 6SJ7 pentode with low screen-grid voltages, among others. Also we wish to thank the General Electric Co. and other firms for their kind cooperation. Our gratitude is expressed to Mr. Robert Lorenzen and Mr. G. C. Baxter Rowe for their contribution to this text and their conscientious criticism, and to Mr. Louis Prior for his valuable suggestions and graphical drawings.

JOHN F. RIDER

September, 1945

VIEWING THE ANAGLYPHS

The three two-colored anaglyphs facing pages 9, 37, and 122, must be viewed through the red and blue spectacles that will be found in an envelope fastened to the inside back cover. Hold the spectacles close to the eyes with the blue filter over the right eye and the red over the left eye. For a person with normal eyesight, the stereoscopic effect will be obtained with the page held at ordinary reading distance; this distance may vary for persons whose vision is not normal and uncorrected.

In the case of anyone who may be color blind, he may be unable to see the stereoscopic effect, depending on the degree and nature of the visual defect. In such event, looking at an anaglyph with one eye through one of the filters will enable him to see the illustration in at least two dimensions.

TABLE OF CONTENTS

INSIDE THE VACUUM TUBE

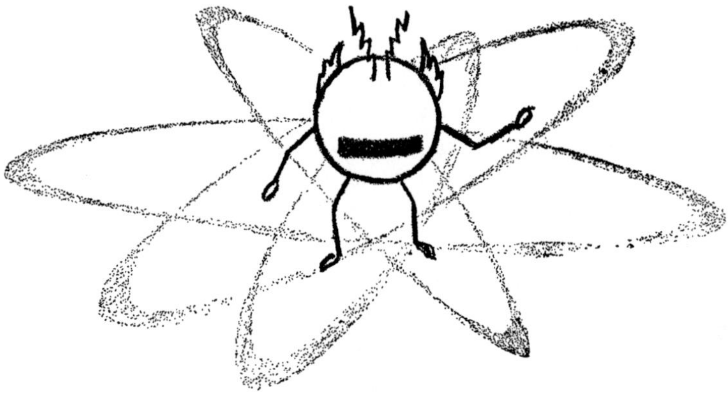

Chapter 1

INTRODUCING THE ELECTRON

THIS IS A TALE about the second smallest thing in the world. For centuries men sought it, yet they knew not what they were seeking. Over two thousand years ago thinkers started guessing what it was and strangely enough, some of those ancients' guesses were surprisingly accurate. They felt there had to be "something" on which to build their conceptions of the how and why of things, but what this something was—there was the rub—that they did not know.

It was not until men had discovered that research and experiment going hand-in-hand with logical thinking were the keys to the age-old puzzle, that any progress was made in the tracking down of this famous and elusive unknown. Then for a century or so, guesses became better and better—closer and closer to the truth—and it was finally decided that men were looking for something common to everything—a universal building block. Chemists and physicists guessed and experimented and, just about a half century ago, their efforts were at long last rewarded: they discovered the Electron! As far as we know, it is the *second* smallest thing in the world.

It was discovered more than fifty years ago, but has never been seen. Science knows its weight, its size, its electrical character, how and when it moves and how fast it travels—yet we have never seen it. Strange, isn't it? We know that it has an enormous capacity for doing many jobs which we are sure it has been doing since the beginning of everything, but the important fact is today we are learning how to make it do the many jobs we require done. One of the most important of these jobs for the benefit of hundreds of millions of people is its lightning-like flight in a device known as a vacuum tube.

In order to secure a proper understanding of the vacuum tube, that most valuable of devices which has made possible the many forms of electrical transmission of intelligence, television, and radar, it is essential that you have a solid foundation in your knowledge of the electron. Not that we have any intention of considering the highly technical aspects of this subject, for, after all, this is an elementary book. But elementary as it may be, it is our personal feeling, based upon study and experiment, that every student of the vacuum tube should know something about the electron and its behavior.

The reason for this is not difficult to understand, if you appreciate the fact that the operation of the vacuum tube is based upon the actions of the electron. Realizing that many different types of vacuum tubes are utilized for many different applications and that the presence of the electron within the vacuum tube is due to different conditions, it would seem quite impossible to understand the manner in which these various kinds of tubes function to accomplish their purposes unless you understood the behavior of the electron. Under the circumstances, there seems to be no more fit starting point than a discussion of the electron.

The Electron

If we choose to say that electricity is "something" which is capable of doing work, then that "something" which can do this work for us

is carried from place to place by a very tiny particle known as the *electron.* In other words, the electron is either a carrier of a certain amount of electricity or it is a certain amount of electricity. Whichever viewpoint you wish to adopt is immaterial, because both lead to the same conclusions. For the sake of simplicity, our preference is for the latter; so in all future discussions in this book, reference to the electron signifies a definite amount of electricity.

As to the amount of electricity involved when one speaks about a single electron, it is an extremely small figure, as witnessed by the fact that 1 ampere of electric current means the flow of about 6,000,000,000,000,000,000 electrons past a point in a wire in one second. But this extremely small amount of electricity in an electron is by no means a limiting factor in the discussion of the electron, for when we describe the behavior of a single electron, it is understood that this description applies to *all* of the electrons which may be under the same influence as the single one mentioned, so that in reality we are discussing a reasonable amount of electricity.

This is made possible by the fact that scientists agree that there is only one kind of electron and that when we speak about an electron we are talking about the same thing no matter what its source or condition. For example, it is an accepted fact that *electric current* in a wire, whether alternating, direct, or pulsating, whether started by batteries, dynamos, generators, or some other device, and operating a refrigerator, locomotive, airplane, or flashlight, is an electron stream moving in a definitely directed manner through the electrical system containing the device. *All the electrons are the same,* each identical to the other, even when there are many billions upon billions of them moving through the circuit every second.

Now it would appear from what we have said that this electron is something tangible—something physical. Although no one has ever seen the electron—and it is doubtful if anyone ever will due to its extremely small size—it is quite in order to form a mental picture of it and to accept its existence as a particle. Some scientists may not agree with this point of view, but since it makes understanding of the subject so much easier, we use it. Since science goes so far as to speak about the radius and mass of an electron, we feel fairly safe in forming the conclusion that its shape is round. And since we have occasion to speak freely about its movement under conditions which might be difficult to understand, it might be well to attempt to create for you, some idea of its extreme smallness, which will explain why it can behave and move as we shall describe. Perhaps the comparison in itself is enough to tax your imagination, but even if you do not grasp its true size, it will not interfere with your comprehending the contents of this book.

To give you some idea of the smallness of the electron, imagine the electron compared with a 1 5/16-inch pingpong ball. The electron is as small compared to this 1 5/16-inch ball, as the 1 5/16-inch ball is compared to the orbit of the earth, which is 186,000,000 miles in diameter. If you want to compare the two in inches and like to calculate, multiply the

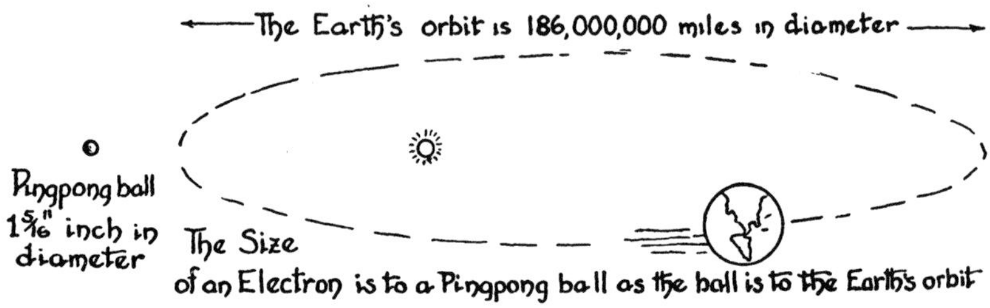

orbit of the earth in miles by 63,360, which is the number of inches in a land mile. If you can use another example, the electron is about $\frac{1}{13{,}500{,}000{,}000{,}000}$ inch in diameter.

The simple statement that an electron is a tiny invisible particle representative of the basic elemental charge of electricity, that it is round and can be imagined to be a tiny ball, and that electric current consists of a tremendous number of these tiny balls in directed motion, is not sufficient. In view of the manner in which these electrons appear in the operation of different types of vacuum tubes, we must say more. We must talk about where they come from and the conditions under which they are made available for use in vacuum tubes. Before so doing, we must add that there are some who do not agree that the electron is a particle, for there are conditions under which it does not behave as such. But, since this is an elementary book and the world has not found fault with an imaginary picture of the electron as a tiny ball, and because it simplifies comprehension, we shall treat it as a particle and disregard those conditions when it does not behave that way.

The Atomic Theory

Electrons are in everything. While we may speak about electrons in connection with specific things such as solids, liquids, or gases, it should be understood that electrons already exist not only in those things, but in everything. This short paragraph says a great deal, but deserves elaboration.

Matter is anything – water of a river flowing past the steel and concrete of a city – the birds flying over it and the trees where they nest – glass of a bottle and the paper and ink of a book

All of this can be explained by saying a few words about what this world is made of and the theory which science has established concerning the structure of things. If we attempt to find one word which identifies everything in the world, within it, above it, and upon it, it is *Matter*. By Matter is meant anything and everything which has weight and occupies space.

But being so broad in definition, Matter must, of necessity, have certain sub-classifications. In other words, Matter exists in three major states: solids, liquids, and gases.

Matter can be in any one of three states—for instance.

WATER can be SOLID LIQUID or GAS

In the *solid* groups would be included any thing from the earth itself to the finest pulverized powder, for while a mass of powder may not behave as a solid, a single grain of that powder is a solid. Among the *liquids*, there is everything which flows freely, from water to molten steel or molten rock, the lava which has engulfed many cities during volcanic eruptions. Among the *gases* there is everything from air, which is a mixture of many different gases, to that which is given off by a flower and sensed by the action of the membranes of the nose and nerves of the brain as a fragrance. To attempt to list all of the examples of Matter is of course impossible, as you can readily appreciate. The limits we have expressed in the preceding paragraph are by no means specific, but were selected in preference to technical definitions as expressed by Webster.

Matter Can Be Transformed

All forms of Matter, whether solids, liquids, or gases are subject to transformation of state. By this is meant that a solid upon being heated, may change into a liquid as for example iron, and, if the temperature is raised to sufficient height, becomes a gas. In the same manner, a liquid that is sufficiently cooled becomes a solid. For example, at a sufficiently low temperature water freezes into ice, whereas, if it is heated it becomes a gas called steam. A gas, on the other hand, if cooled sufficiently and also compressed, passes through the stage of becoming a liquid and then a solid. For example, air can be liquified by reducing its temperature simultaneously with an increase in compression. But when this liquid air is allowed to stand in the open air so that its temperature is raised, it again returns to its original gaseous state and mixes with the surrounding air.

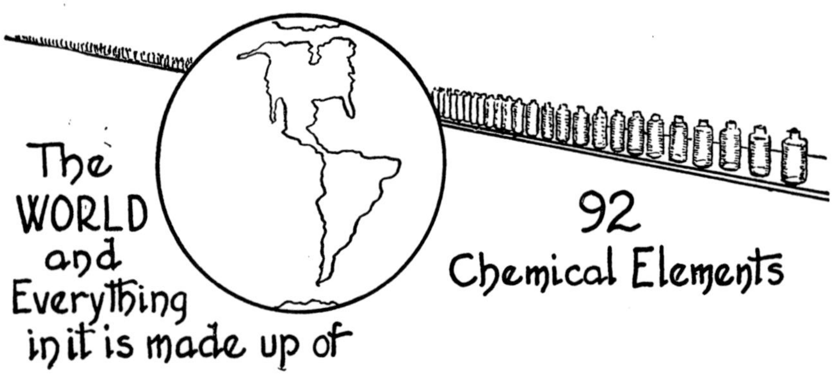

Perhaps it should have been mentioned before, that the chemist, while recognizing the three states of matter, breaks them down into even a finer degree—into *elements*. It has been found that at least 92 different substances exist on the earth that are pure unto themselves. These are known as *elements*. In other words, the finest possible division of an element, even by chemical means, results in the same substance with the same characteristics without the presence of any other substance. The smallest possible particle of such an element is known as an *atom* of that substance or element.

The Atom

These atoms are extremely tiny particles which are too small to be seen, but which are accepted as existing and of such dimensions that

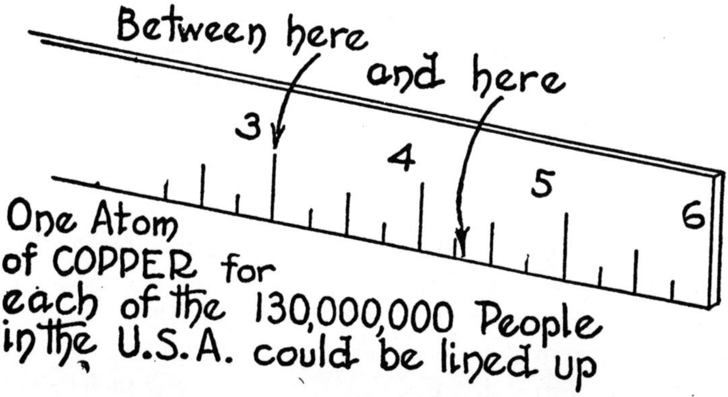

tremendously great numbers of them would fit upon the point of a pin. (About 100,000,000 atoms can be placed side by side in a length of 1 inch.) Just as there are 92 different kinds of chemical elements, so there are 92 different kinds of atoms, each element having its own kind of atom. Thus the smallest possible subdivision of copper, which is a chemical element of the metal variety, would be an *atom of copper*. On the other hand, the smallest possible subdivision of carbon, which is not a metal, would be an atom of carbon. And the smallest possible subdivision of oxygen would be an atom of oxygen. Now many, many different substances other than just the elements exist on the earth. The chemist designates those as *compounds* which have two or more elements combined *chemically*. An example of a compound is water, which is a chemical combination of two atoms of hydrogen and one atom of oxygen. Another grouping of elements or of elements and compounds, is called a *mixture*. Here the elements or compounds are merely mixed *physically* and do not combine chemically. For example, the air we breathe is a mixture and is composed of hydrogen, nitrogen, oxygen, neon, and several other gases.

The total of the different kinds of chemical compounds which may be formed from the basic chemical elements is tremendous. As you look around, you see far more examples of compounds than you see of elements. For example, in the manufacture of vacuum tubes use is made of many compounds as well as elements and it is surprising to note the number of elements which are employed; particularly those of the metal, rare earths, and gas variety. Rare earths and metals come under the general classification of solids.

Recognizing that a chemical compound is the result of a combination of various chemical elements, it stands to reason that the smallest possible subdivision of a compound which still retains the characteristics of that substance must be something other than the atom. For, if the subdivision is carried down to the point where the compound is subdivided into its constituent atoms, it no longer is the same substance. It is here that we encounter a new term which helps us to develop the construction of matter.

The Molecule

Whereas the smallest subdivision of an element is the atom, the smallest subdivision of a chemical compound is the *molecule.* For every kind of chemical compound which exists there is that kind of a molecule. For example, the smallest subdivision of water, as water, is a molecule of water; for illuminating gas it is a molecule of this gas; for steel it is a molecule of steel; and so on. Generally speaking the size of the molecule is larger than the atom, for the molecule is a structure of atoms. There was a time when it was said that the molecule, like the atom, was invisible. This is still true in almost all instances, but not in all, for recent reports about the new electron microscope, a device capable of tremendous magnification, state that large molecules of certain materials have been seen. The molecule, however, is still an extremely small particle of matter and is generally considered to be invisible.

Thus we have reached the second building block in the constructional pattern of matter, first the atom and then the molecule. Lest you build up the wrong impression concerning the molecule, we wish to add that molecules are not associated only with compounds; they are to be found as elements as well, but such molecules of elements differ from the molecules of compounds in that the molecule of the element is made up entirely of atoms of that one element, whereas molecules of a compound are made up of different kinds of atoms which, in certain combinations, comprise the compound. As to the number of atoms which form a molecule of an element it is safe to say that it would be one or more, although this is not of great significance to us. In the case of compounds, the number varies with the compound; some molecules are

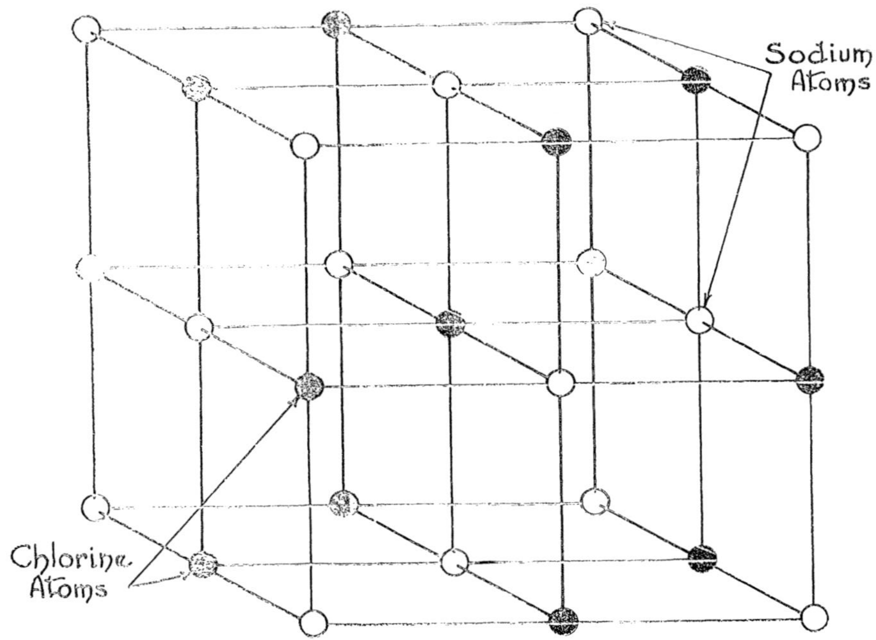

PLATE I. The arrangement of chlorine and sodium atoms in a crystal of rock salt.

very complex and others are very simple, as, for example, water, which has two atoms of hydrogen and one atom of oxygen in each molecule.

In some instances the arrangement of atoms is quite complex. In a crystal of common rock salt, the sodium and chlorine atoms are grouped as shown in the anaglyph, Plate 1. The sodium atom at the center of the cube is equidistant from six of the chlorine atoms and each chlorine atom is likewise equidistant from six sodium atoms. This interlocking formation is carried throughout the entire piece of rock salt. This also illustrates the fact that the atoms of a substance are not packed tightly together, but that they are separated by relatively great distances.

The Atom Too Is A Structure

Referring once more to the atom, while it is a fundamental building block of all matter, it too, is a structure. In other words, it is not an indivisible particle. If, as we have stated, there are 92 different kinds of atoms, one for each kind of element, how do they differ? What is the difference between say an atom of copper and an atom of iron? The answer may surprise you, for *the difference between atoms is the amount of electricity within the atom and the arrangement of the various particles of electricity which comprise the atom!* If from this statement you conclude that every material thing in this world is made up of electricity, then you are correct, for such is the case. In other words, you may recall comment to the effect that such and such a device makes electricity. That is not so! . . . *Electricity is never made or created—it already exists in everything!* Those devices which are classified as being producers of electricity, like batteries or generators, are simply devices which can make electricity (for example electrons) *move* in an electrical system. *They never produce electricity!*

As to the atom being a structure, it is a very interesting one. Although there is not entire agreement among all scientists concerning the make-up of the atom, and of the specific behaviour of those things which make it up, certain ideas have been accepted.

Concerning the structure of the atom, the most commonly accepted idea is that it resembles a miniature solar system, something like our Sun and the different planets which revolve around it. In accordance with this concept, the atom contains a center part or *nucleus*, which

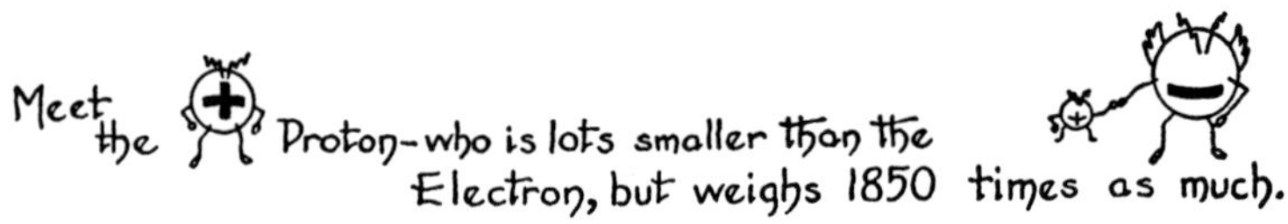

would be the equivalent of the Sun. This nucleus in the simplest of atoms, the Hydrogen atom, consists of a particle of positive electricity called a *proton*. Around the proton and at a great distance from it, considering the size of the particle involved, rotates a single electron which is a particle of negative electricity. The amount of electricity represented by the proton and the electron is equal in magnitude but opposite in sign. However, the *mass* (weight) of the atom is represented essentially by the proton, and is about 1850 times greater than

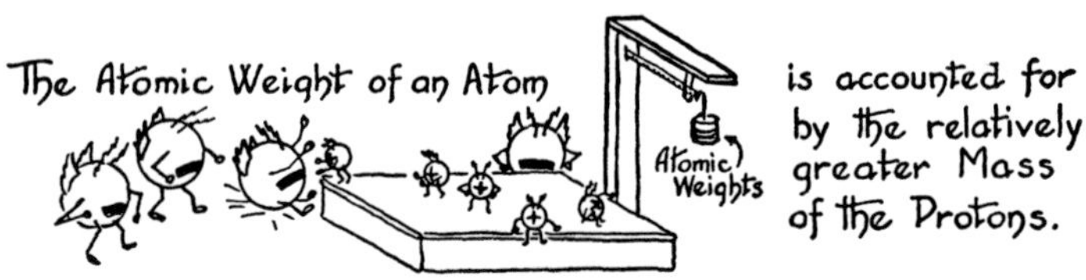

that of the electron, although the diameter of the electron is many times larger than that of the proton. It is interesting to note that as far as we know the proton is the smallest thing in the world.

In more complicated atoms the nucleus was originally supposed to consist of all of the protons (positive charges) and some of the electrons, with as many electrons remaining outside of the nucleus as would be required to balance the excess number of protons within the nucleus. These electrons outside of the nucleus were identified as *planetary electrons* for they revolved around the nucleus in various orbits. Thus, if some element (manganese) was supposed to contain 55 protons and 55 electrons and of these 55 electrons, 30 were within the nucleus, there would be 25 planetary electrons. As to the exact nature of these orbits and the number of electrons within each orbit, that is still a subject of discussion and dissension, but it is generally admitted that planetary electrons exist beyond which we need not go in this discussion.

The *atomic number* is the number of planetary electrons which revolve around the nucleus. Thus, for copper, the atomic number is 29, which means that this atom has 29 electrons revolving at high speed in its planetary system, with great spaces between the respective electrons.

The Modern Idea

Today's theory is sometimes modified although it does not materially alter the picture. Whereas the old ideas of the structure of the nucleus of an atom evolved by Bohr considered individual protons and electrons, the modern conception recognizes the existence of the protons and electrons within the nucleus, but it states that the electrons within the nucleus are, to put it in simplest terms, in affinity with an equal number of protons in the form of pairs, each pair consisting of a proton and an electron. This new kind of particle is called a *neutron* and

since each of these particles has within it a single plus charge and a single minus charge, electrically speaking, the net charge of the neutron is zero. In other words, the neutron is electrically neutral.

This does not of course disturb the electrical balance of the entire atom, for if an atom (cobalt) has 59 protons and 59 electrons, of which number of electrons 27 are planetary, the 32 electrons within the nucleus are in affinity with 32 of the total 59 protons so as to form 32 neutrons. The remaining 27 protons remain within the nucleus to balance the electrical charge of the 27 planetary electrons.

Actually, other particles have also been identified as being within the atom, but we need not discuss them for they have no immediate bearing upon our subject and it is entirely correct technically, considering the scope of this book, to view the atom as we have, inclusive of the neutron. Even if we had omitted the reference to the neutron, it would not have altered the presentation of facts associated with the operation of vacuum tubes.

Continuing with the atom, we are approaching that portion of this discussion which is of greatest significance, as far as vacuum tubes are concerned. As is readily evident, the major difference between atoms is in the distribution of the electrons and the total number of protons.

Therein lies the difference between the nature and behaviour of the various kinds of elements. From the electrical viewpoint, there is an-

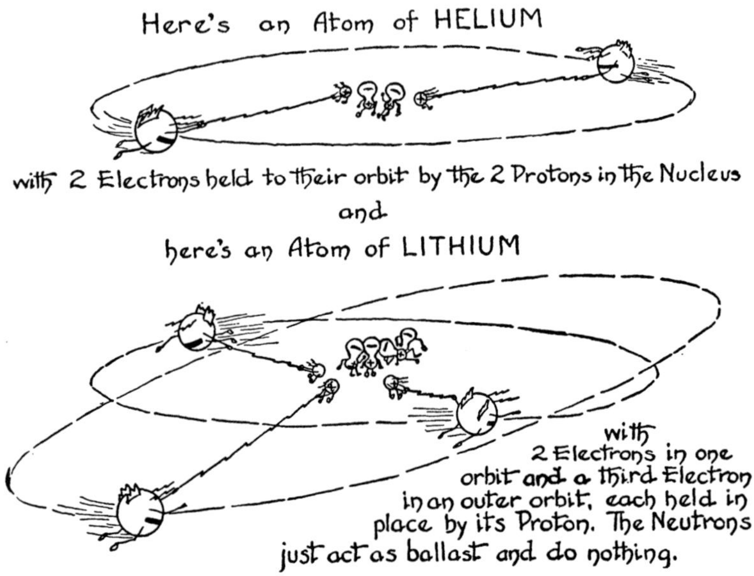

other difference between atoms which is associated with the behaviour of the planetary electrons.

While it is true that the protons within the nucleus display a powerful attracting force upon the planetary electrons and the electrons have a like attraction for the protons, this force is not the same in all atoms. In some atoms, the outer electrons, that is, those which revolve in the outermost orbits, are often under the influence of some external force, such as collision between atoms or extreme agitation of the atoms, and will fly out of their orbit or be knocked out of it and so leave the atom. When this happens, the atom, which is shy one or more electrons, now has a preponderance of positive charges and will therefore exert an attracting force upon other electrons. Then again, as a result of some action, one or more electrons which are not attached to any one atom, having been freed from some other atom, may become attached to an atom which was originally electrically neutral, thus making it more negative and giving it a tendency to repel other free electrons. These excess electrons may then, due to some agitation or collision, be knocked off the atom and will join some other atom so that a transfer or movement of electrons is continually taking place between the atoms of a substance especially in metals. The freedom with which such inter-atomic movement of electrons takes place is an indication of the suitability of that substance as a conductor of electric current. The greater the numbers of such free electrons in a substance, the better that substance is as a conductor of electricity. Free electrons are extremely abundant in metals of all kinds.

Conductors and Insulators

The difference between an electrical *conductor* and an *insulator* is, of course, a matter of chemical constituency, but from the atomic viewpoint, it is merely that the atoms comprising a conductor part with one or more electrons readily, whereas the atoms of an insulator hold on to their electrons with extreme tenacity. When an insulating material is said to "break down," it simply means that the electrical stress upon the atoms is so great as to tear electrons from the atoms, and create a directed movement of a stream of electrons, thus constituting a flow of electric current.

In this connection it might be well to make specific reference to gases as well as metallic conductors, for both are of importance in vacuum tube operation. The atoms of certain gases are of such a type that an easy liberation of electrons from the atoms of the gas occurs. This, as you will see later, plays a dominant role in the operation of some vacuum tubes.

Ions and Ionization

Certain identifying terms are applied to atoms which are not electrically neutral due to a change in the number of planetary electrons.

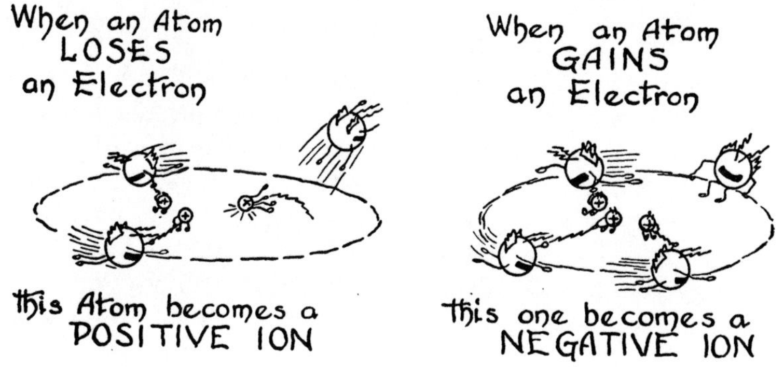

An atom which has an excess of positive electricity within it, due to loss of one or more planetary electrons, is called a *positive ion;* whereas an atom which has an excess of electrons, and which, therefore, is electrically negative, is identified as a *negative ion.* The process of producing ions is known as *ionization.*

In all of this discussion you cannot help but note that everything relating to the electrical is in terms of motion of the electron, that is, the movement of the negative particle of electricity. The positive portion of the atom does not, under most conditions, move, other than as a part of the complete atom. This is important in connection with what follows, for later in this text reference will be made to negative and positive polarities in electrical circuits, at which time it should be understood that such designations represent a redistribution of electrons and not a movement of the positive particles of the atoms.

Referring again to the electrons in an atom, another important consideration must be mentioned because of later comment. We are referring specifically to atoms of gases or vapors. Anything which is

A Ball at rest on a table has no Kinetic Energy | When it is thrown, it has Kinetic Energy because of its Motion | When it stops, its Kinetic Energy changes into Heat & Sound Energy

in motion is capable of doing work. For example, if you throw a ball and it strikes a window, it will break that window. In breaking the window the ball is *doing work;* the same amount of work which you performed upon the ball by getting it to move. When you caused that ball to move you gave to it a certain amount of energy—*kinetic energy*—the *energy of motion.* When that ball struck the window, it *gave up* the energy you imparted to it. In the same way an electron in motion is capable of doing work, for, being in motion it possesses kinetic energy.

If, for some reason, electrons are caused to be freed from atoms of certain gases or vapors and these electrons move between these atoms, they may strike or collide with the atoms. When they are moving slowly and strike atoms, they may become attached to these atoms, or may bounce away from them without any effect. However, if they strike the atoms at high speed, they may give up all or part of the energy they possess to the atoms they strike. When this happens the atoms radiate light and the gas is said to glow. As a rule, the ionization of a gas atom is accompanied by such a glow of light and this is quite commonplace in vacuum tubes which contain a gas. The exact color of

the glow is dependent upon the kind of gas within the tube. Although somewhat removed from the subject, neon advertising signs are typical of such action, wherein light is radiated by the atoms of the gas due to such bombardment of the gas atoms by high-speed electrons.

To summarize the foregoing we may state that electrons move within the atom and under proper conditions, move freely between atoms. Further, since everything upon this earth consists of atoms, electrons are available in everything. Whether or not they can move freely is another matter, depending upon the nature of the substance, that is, the kind of atom. And last, although the particles within the nucleus of the atom are supposedly packed tightly together (about this no one is certain), there is plenty of room between the planetary electrons and the nucleus. In other words, there are great spaces within the atom.

How about the complete atom? How does that behave? This is largely a matter of conjecture. Scientists say that the atoms of a substance are always in a state of agitation and this applies to all substances. They move about and collide with each other and with free electrons. The greater the number of free electrons which have been liberated from atoms, the greater the number of these collisions. Since the individual atoms are in a state of agitation, it stands to reason that they are *not solidly packed in any substance.*

With this much behind you, you are apt to ask a question, particularly if you recall the previous comment that the electron in motion represents electric current. If this is the case and electrons are continually moving between atoms of a substance, whether it is a gas, a solid, or a liquid, why does not electric current flow at all times? This is simply explained. Unless an electric force is applied to the system, the movement of the electrons between the atoms is random and as many move in one direction as in another. The result is that

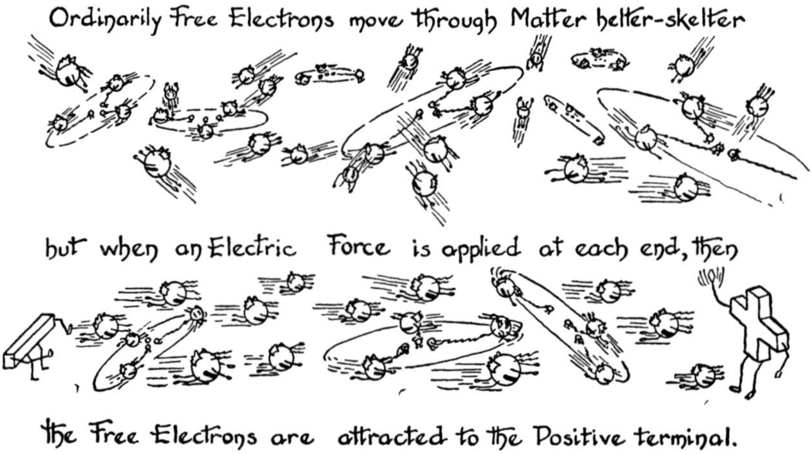

the electrical effect of one electron moving in one direction is offset by another moving in the opposite direction, so that the net electrical effect is zero. However, when an electrical force is applied, the direction of movement of the electrons between atoms is controlled, thus resulting in a *directed* stream of electrons, and electric current is said to be flowing.

Atomic Spacing

Now, as to the molecules. Are the atoms packed solid in the molecules? The answer is in the negative. Whether the molecules are those of an element or of a compound, although a certain affinity exists between the atoms in order to make up the individual molecules, the molecule is another entity in which there is plenty of room. In other words, five, or twenty, atoms may make up a molecule, but whatever the sum, the atoms are not packed solid. There is plenty of space between them. In addition, each of the molecules is in a state of agitation, and undergoes a to-and-fro motion.

This action can be compared to an imaginary box without sides, and which has its boundaries established by the location of a ball at each corner. Without any suspension, each of these balls is moving around a mean position which represents the limits of the box. Within this imaginary space are mounted a few other balls, again without any visible suspension other than some kind of an affinity for each other and for the balls at the imaginary corners. The space between the balls is many times the dimensions of the balls. Each ball and the entire box is in a state of agitation—and there you have your molecule. Not that molecules are square in shape for they may assume various shapes, depending entirely upon the nature of the substance.

The atoms within a molecule behave as if they were by themselves, that is, with respect to the movement of electrons between atoms and the movement of atoms which are not attached to molecules.

As to the packing of molecules, there again, there is plenty of space between molecules. In fact, the state of the substance identifies the

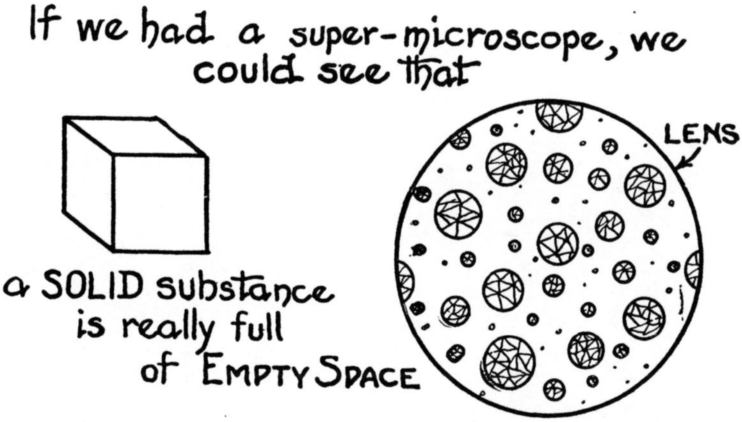

relative space between the molecules. In gases, for example, the separation depends entirely upon the pressure applied to the gas, for a gas will normally expand in accordance with the size of the container. Such expansion is nothing more than increased separation between the molecules. If the gas is removed from one container and forced into a smaller one, the molecules are crowded more closely together. Such freedom of movement exists among the molecules of a gas that the volume of gas will assume the shape of the container employed to store it.

In liquids, like gases, there is some separation between the molecules. There is not as much freedom of motion as with gases, for while a liquid will change its shape to fit that part of the container with which it is in contact, it will not expand to fill the container.

In solids, which normally hold their shape, unless they are forcibly compressed, the molecules are quite close together. But even here they are not solidly packed, but have a definite freedom of motion, and are in a state of agitation.

So much for the subject of electrons as it relates to the electronic theory of electricity and the atomic theory of matter.

Chapter 2

ELECTRON EMISSION

In Chapter 1 we discussed the electron concept of electricity in general terms, namely, that an electric current is a directed stream of tiny particles of negative electricity, identified as electrons. Also that these electrons are one of the fundamental building blocks of all forms of matter and are to be found in everything. In this second chapter we shall consider another phenomenon associated with electrons, namely, electron emission. To state this differently, *electron emission is the liberation of electrons from substances under particular conditions.* This action is of vital importance in the operation of every type of vacuum tube.

In contrast to the general discussion of electricity and the reference to all kinds of materials, electron emission, as related to vacuum tubes, narrows the kinds of matter which need be considered to just one kind,—solids—and among solids specifically to metals, for it is with metals that we have to work. Concerning metals, which are generally good conductors, you will recall that we mentioned the profusion of free electrons which rove haphazardly through the material. Also, that collisions between atoms, which normally are in an agitated state, results in the dislodgement of still more electrons. All in all, such materials contain an abundance of these roving negative particles.

Recognizing the manner in which these electrons behave in such metals, one is tempted to ask a very natural question. If these electrons move about so readily, why don't they move right out of the substance? What keeps the free electrons within the boundaries of the metals, say a length of wire? Why don't they break right through the surface and get into free space?

Well, the answer is that some do just that, but they are so few in number under normal conditions that they are negligible. In fact, they are so few that the generally accepted description of what happens is that there is no emission of electrons from metals unless certain special conditions are deliberately established. As to why these electrons do not leave the metal, the answer was advanced first by O. W. Richardson, in 1901. He assumed a certain state of affairs which has subsequently been proved to be right.

Imagine, if you will, a number of individuals within a circle formed by a number of men standing shoulder to shoulder or holding hands. The men who form the ring are determined that the individuals inside will not get out, and those who are encircled make but a feeble effort to get out. Although they are milling around, those of the inner group who come in contact with this human barrier bounce back or are deflected in different directions and remain inside the circle. As a result of the slowness of their movements, the kinetic energy possessed by those within does not enable them to overcome the work being done by the encirclers to keep those who are inside within the circle.

But suddenly those inside the circle speed up their movements. Instead of moving slowly towards the encircling ring of people, they start running rapidly and thereby gain kinetic energy (the energy of motion). When they reach the barrier the energy they possess is converted into work, and if it is greater than the work being done by those trying to keep them inside, they break through the human barrier and are free.

In a general way that is what happens inside of a metal. At the surface of the metal a sort of barrier exists, the exact nature of which has not as yet been fully determined, but which tends to keep the electrons within the confines of the metal. Inside the metal some of the free electrons move very slowly and others move more rapidly. Some move with sufficient speed so that the energy they possess, is sufficient to permit them to break through the barrier but, as has been stated, these are extremely few in number.

Assume now that something is done to the metal, which causes increased speed of movement of all the electrons. As the electron speed of movement increases, the kinetic energy acquired by the electrons also increases until finally great numbers of them break through the surface and escape from the metal. This is the phenomenon of *electron emission.*

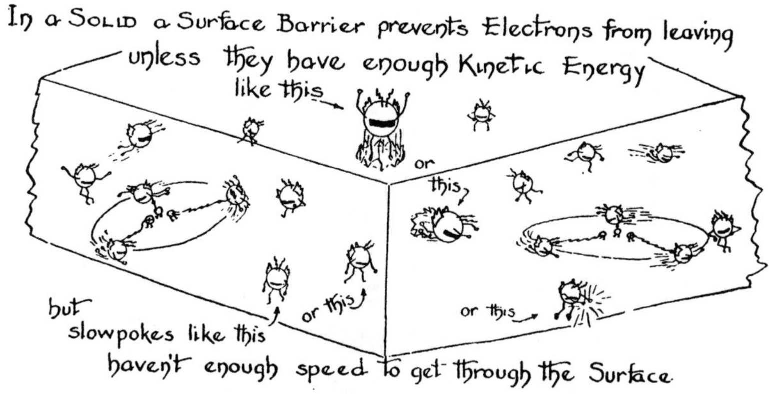

Electron Emission Caused by Energy Conversion

Naturally, we are interested in what is done to the metal to cause such emission of electrons. What will speed up the motion of the electrons so that the work they can do is sufficient to cause them to break through the surface? Numerous conditions will create such a state, for there are a number of ways in which additional energy can be imparted to the electrons within the metal. For example, there is the application of heat. Heat is a form of energy, and if the temperature of a piece of metal is raised by any one of a number of heating processes, some of the energy in the heat is transferred to the electrons and they are made to move faster than their normal speed of motion. This is the equivalent of conversion of heat energy into kinetic energy. When electrons are freed by heating a metal it is called *thermionic emission.*

Then there is the application of light. Light, which is an electromagnetic radiation, is a form of energy. Although the exact mechanism whereby electrons are liberated from substances when electromagnetic waves of the proper frequency impinge upon them has not as yet been determined, it is a well-established fact that electron emission does take place. Very many different electrical systems are in use today which depend upon this phenomenon for operation, and these will be discussed later. But for the present, we can accept the statement that the energy contained in the light rays which strike the substance under consideration is in some way transferred to the electrons

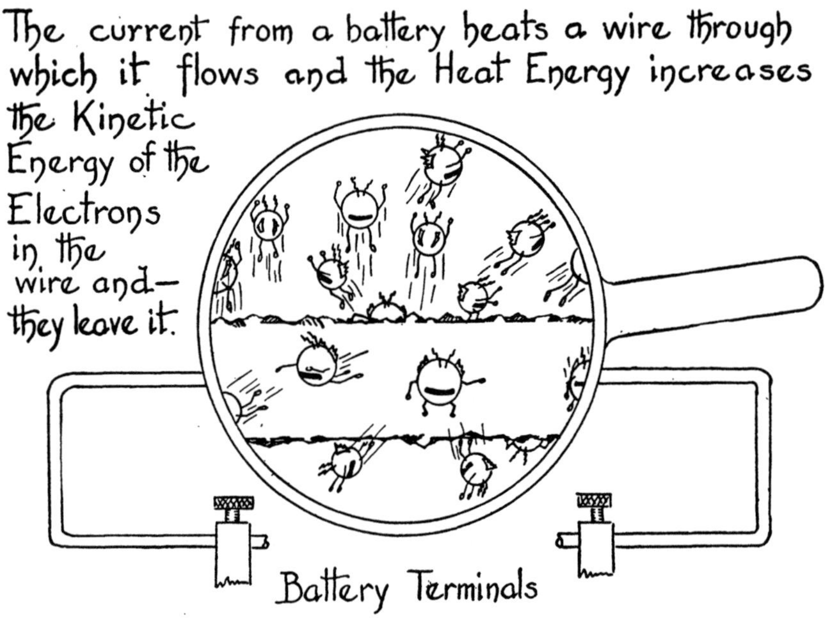

in the substance, and that their motion is accelerated sufficiently so that they are liberated from the substance. This action is called *photoelectric emission.*

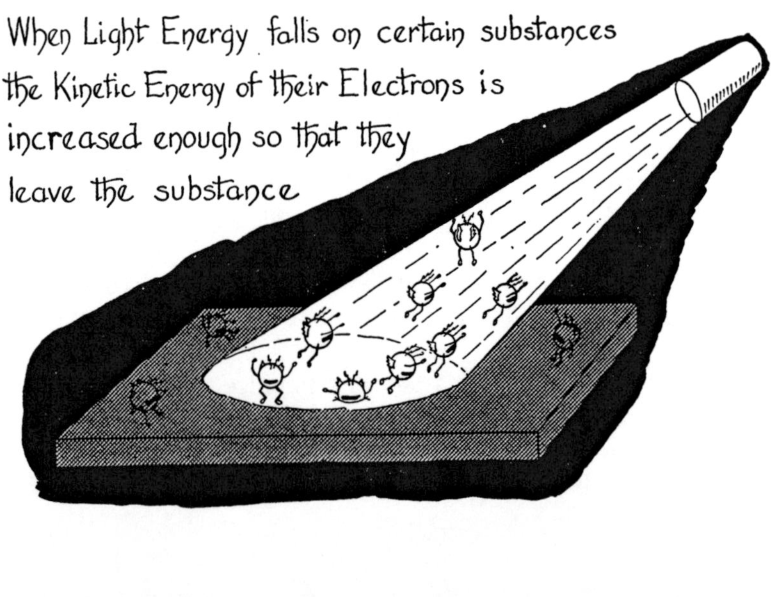

A third type of emission is called *secondary emission.* This is the liberation of electrons from a substance as a result of the bombardment of the substance with fast-moving electrons. In other words, the

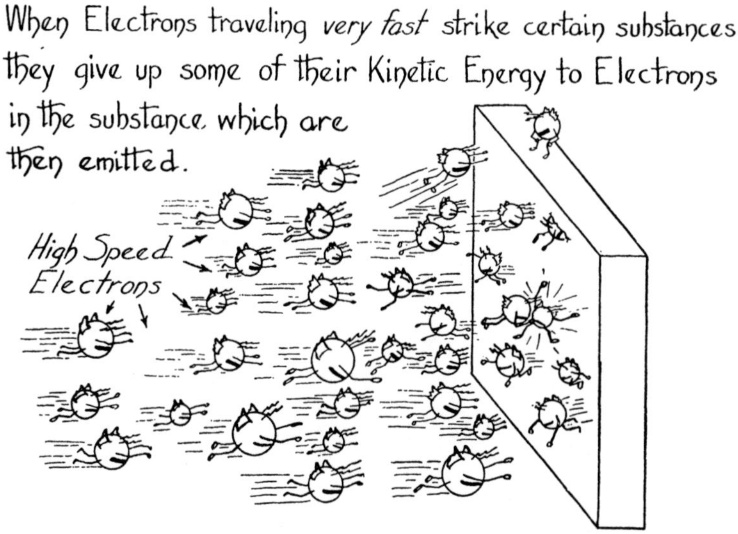

energy contained in these high-speed electronic bullets is given up to the electrons and atoms which they strike, and in so doing increases the kinetic energy of the electrons in the metal and thereby enables them to gather sufficient speed so that they can break through the surface of the metal and be liberated. You will see later in this text that the existence of this condition is undesirable in certain vacuum tubes and provision has been made to prevent it. On the other hand this is a deliberately created condition in other tubes. Naturally, the high-speed electrons, which are used as the bullets, are also secured from some source by the process of emission. In secondary emission electron emission is also attained by means of the transfer of energy.

A fourth method is identified as *cold-cathode emission.* In this the electrons are literally pulled out of a substance by the attracting power of an extremely strong electric attracting force. Since the electron is a negative particle, the existence of a point which is positive with respect to the electron source (the cathode), will tend to draw electrons towards it. If this attracting source is sufficiently powerful it will pull electrons out of a metal, without any heat being applied or the application of any other action which would tend to free the electrons from the metal. Due to the extremely high voltages involved, such cold-cathode forms of emission are not commonly used, therefore we shall not discuss such tubes in this book.

When a strong positive field is near a metal plate, it attracts the Electrons and pulls them from the plate

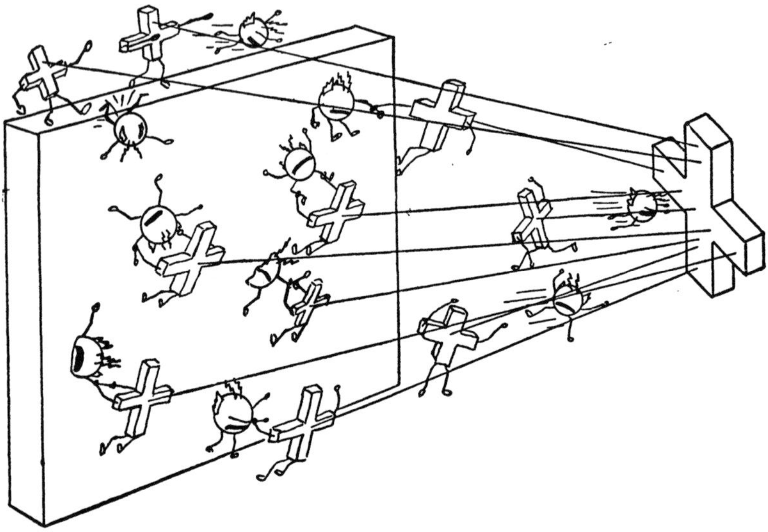

Thermionic Emission

As we stated earlier in this chapter, of the various methods of obtaining electron emission from a metal that which employs heat is of the greatest interest to us. While it is true that some types of vacuum tubes are of the photoelectric variety and others of the secondary-emission variety, they are relatively few in comparison with those which depend upon thermionic emission for their operation. Under the circumstances, we shall discuss thermionic emission in this chapter and reserve further comment upon the other systems until we reach that chapter in which special-purpose tubes will be described. As to the tubes, which are of the thermionic variety—they employ heat as the means of adding energy to the electrons in the metal so that they will be liberated—every vacuum tube which is spoken of as using a heater or filament or indirectly-heated cathode comes within this category.

It might be noted in passing that the discovery of the emission of electrons from heated bodies was not born of the vacuum tube we know today. Investigators back in the 18th century noted that open areas adjacent to heated bodies were better conductors of electricity than those which were not heated. Just why this was so they did not know, for the electron theory was non-existent at that time, but the results were observable. In the 19th century experimenters established that an insulated body located near an incandescent metallic body would acquire an electric charge, thus indicating the presence of some action associated with the state of heat of the incandescent body.

Then the action came to the attention of Edison when he was working with his incandescent light and also to the attention of Fleming. Edison noted that if a metal plate was located within his incandescent lamp and this plate was joined by a wire to the positive side of the filament, an electric current would be indicated upon a meter connected into this circuit. Just what was taking place he did not know, until in 1899, J. J. Thomson described it as the flow of negatively charged particles between the filament and the metal plate. Since the plate was connected to the positive end of the filament, the filament was negative with respect to the plate, hence electrons liberated from the filament

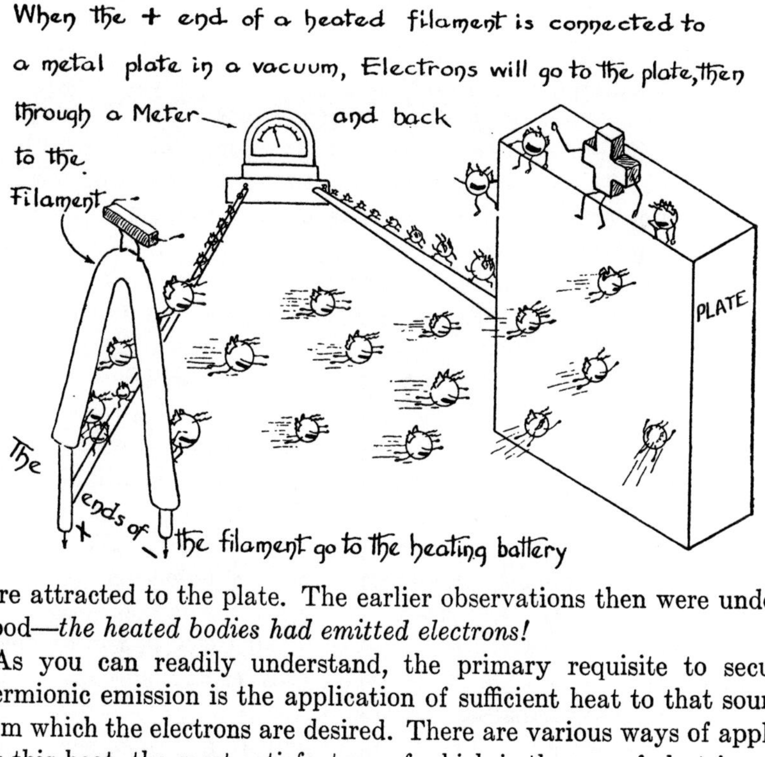

were attracted to the plate. The earlier observations then were understood—*the heated bodies had emitted electrons!*

As you can readily understand, the primary requisite to secure thermionic emission is the application of sufficient heat to that source from which the electrons are desired. There are various ways of applying this heat, the most satisfactory of which is the use of electric current. In modern tubes, two methods of employing electric current to heat the electron emitter are in use. One is the *direct method,* wherein the electric current, which may be either a-c or d-c, is caused to flow through a *filament* of wire and heat it to either a cherry red or a white heat, depending upon the nature of the filament.

The Indirectly Heated Cathode

The second method, and the one which is most common today, employs an *indirectly-heated cathode.* The *cathode* is the element which is the electron emitter, and this is heated to the electron emitting temperature, which is usually a cherry-red heat, as a result of the heat from the *heater.* The electric current flows through the heater. The cathode surrounds the heater and the heat energy created in the heater is conveyed by conduction to the cathode by means of an electrically insulating sheath. This is somewhat similar to what happens in an ordinary soldering iron or household iron. The heating element is inside and is electrically insulated from the outside surface. The outer sheath of metal to which the soldering tip is attached in the soldering iron becomes hot as the result of the heat from the heating element. Incidentally, such heated surfaces, like the soldering iron, the curling iron, the heating elements in toasters, all emit electrons but in such small quantities as to be entirely impractical of use.

The amount of electron emission that can be obtained from an incandescent filament or cathode depends upon a number of factors. Essentially it depends upon the temperature of the emitter which, in ordinary vacuum tube operation, is determined by the current through the filament or the heater, depending upon the type of tube, and also upon the nature of the emitter.

Considering these separately, it would seem that the higher the temperature of the emitter, the higher the emission. That is correct, but for a limiting agency: the effect of excessive temperature upon the emitting material. These substances are designed for operation within certain temperature limits in order to obtain maximum operating life. Excessive temperature causes an extremely rapid deterioration of the material, and thereby results in an extremely great reduction in operating life if not immediate irreparable damage. This applies equally to the filament, heater, and cathode, hence to all types of tubes which use heated electron emitters. This topic is further discussed under the heading of "Plate Dissipation" in Chapter 13.

The abundance with which electrons are emitted also is influenced by the chemical nature of the emitter. Certain materials, such as oxide-coated filaments and cathodes, are more prolific emitters than either tungsten or thoriated tungsten. During the early days of vacuum tube development, filaments and cathodes coated with the oxides of alkaline earths were used only in special tubes and were very highly prized. Today they are used in the majority of receiving tubes and in most of the low-powered transmitting tubes. These electron-emitting surfaces are in reality surfaces baked upon a core material.

Types of Electron Emitters

In the early days the core material of such oxide-coated filaments, made famous by the Western Electric "E" and "J" tubes, was either platinum or an alloy of platinum and iridium. Today it is frequently an alloy of iron, cobalt, nickel, and titanium, which is commercially known as Konel. This core is covered by a coating of a paste of strontium and barium, the oxides of which are formed during manufacture. Such tubes are operated at a cherry-red temperature, as are also the oxide-coated cathodes.

The reasons why such oxide-coated filaments are more prolific emitters of electrons than filaments not coated is still a matter of disagreement. Van der Bijl says that such a coated filament enables electrons to break through the barrier at a lower speed than they could in the uncoated filament.

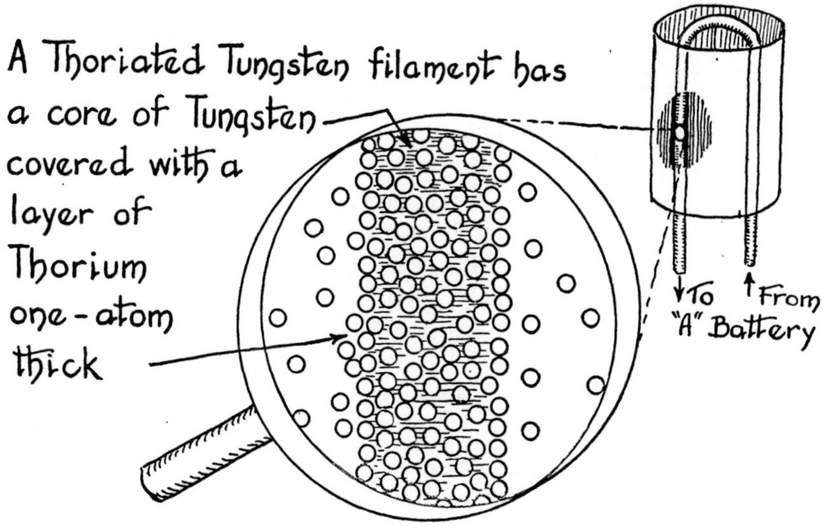

Thoriated-tungsten filaments as electron emitters largely supplanted tungsten filaments in receiving and low-powered transmitting tubes. This type of filament affords the advantage of greater electron emission than is obtained from a plain tungsten filament when operated at the same temperature. The basic purpose of the thorium oxide, which is added to the tungsten during manufacture, is that it reduces the force present at the surface which tends to keep the electrons *inside* the material. The thorium appears as a layer one atom thick at the surface of the tungsten and gradually evaporates during use. As it evaporates it is replenished from within the filament wire. In other words, when the thorium is mixed with the tungsten it is not chemically combined with the tungsten. The thorium atoms remain free to move, so that under normal conditions a monatomic layer of thorium exists at the surface of the filament.

If, by chance, the single layer of thorium atoms is burned away by excessive filament temperature, the electron emission is very materially reduced, since the absence of the layer of thorium changes the surface work function to that of the tungsten only, and this is higher than that of the mixture of tungsten and thorium. When this happens the tube must be *reactivated*. This operation, consists of *flashing* the filament at approximately three times its normal voltage for about 10 or 15 seconds and then operating the filament at approximately 1.5 times its normal voltage for about 2 hours. Both operations are carried out without any other voltages applied to the tube elements. This procedure restores the layer of thorium.

Of course, if the thorium within the tungsten filament has been exhausted due to long use, then such reactivation is of little value and the emission is reduced to that of the pure tungsten, which, considering the basic design of the tube, is insufficient, and the tube must then be replaced.

Tungsten as material for an electron emitter is not as good from the emission standpoint as either the thoriated tungsten or the oxide coatings, but mechanically, it is the strongest of the three types. Because it is capable of high overloads, it still finds application in high power transmitting tubes, but does not justify such detailed discussion as was accorded to the other two types of emitters. It is of interest, however, to note that it is frequently used as the heater in indirectly heated tubes.

As a final thought concerning the three types of electron emitters, it might be well to mention a detail relating to operation. Since both the oxide coated filament and cathode as well as thoriated tungsten filaments depend upon the action of the coating for their efficient operation, the general structure of the filament remains intact for practically the full span of efficient operating life. In the case of the tungsten filament, however, a gradual evaporation of the tungsten accompanies its use, so that the filament becomes thinner and thinner, and consequently its resistance rises and its current carrying capacity decreases. In order to maintain the longest possible life, the voltage across the terminals is usually held constant, rather than keeping the current through the filament constant. Accordingly filament circuit indicators used in high-power transmitting systems employing tungsten filament tubes are voltmeters instead of current meters.

So much for the present about the subject of thermionic emission. What we have said is by no means a complete explanation of the subject. Much has been left unsaid, some of it because there will be further discussion later in this text, and the remainder because it is too detailed for inclusion in a book of such elementary character.

Chapter 3

MOVEMENT OF CHARGES

THE EDISON EFFECT was mentioned in Chapter 2 without any special comments that the conditions of operation associated with it were of extreme importance in describing the basic manner in which vacuum tubes work. Yet that is actually the case. The Edison effect is the movement of electrons from the vicinity of the filament to the metal plate which is joined to the positive end of the filament. The current indicated upon the meter in the circuit is greater for this method of connection than when the metal plate is connected to the negative end of the filament. Here is a clear example of the motion of charges between bodies. This example of the movement of the electron, is not a complete description of the movement of charges, for it is possible to have both positive and negative charges in motion in tubes of special types as you will see later. However this brief comment on electron motion serves as a satisfactory starting point.

But we must do one thing more before we delve into this discussion. For the sake of clarity, it is necessary to set up certain distinguishing features concerning charges and the charged body, for otherwise some confusion may result.

As you go along you will find reference to charges in motion. Now, in accordance with what we have told you so far in this text, there are two elemental charges, the two basic particles of electricity: the negative charge or the electron, and the positive charge or the proton. Of these two, we have mentioned that only one is mobile, that is, moves from place to place. This is the electron. Either the nucleus or the proton we assume to be immobile.

We have also spoken of the existence of two states in electrical circuits—of something being positive and of something being negative, two terms which appear quite frequently in all electrical discussions. These two states will hereafter be viewed as representative of condi-

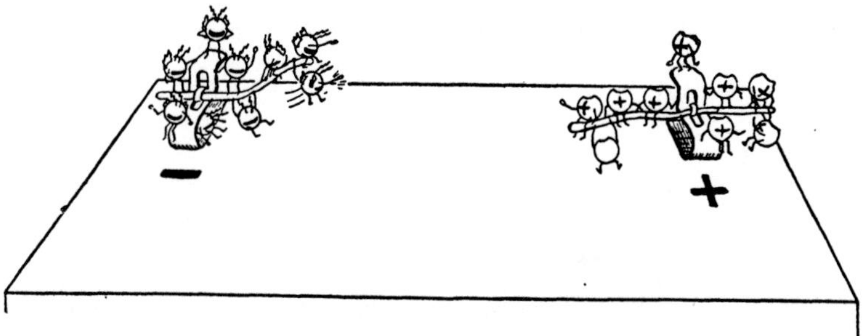

The NEGATIVE point in any electrical system has the greatest gathering of Electrons and the POSITIVE point has a corresponding surplus of Positive Ions.

tions associated with the presence or absence of electrons. *That point or place in any electrical system where there is a surplus of electrons with respect to the electrons at another point is referred to as being a negative point in the system, and that place where there is a shortage or deficiency of electrons with respect to the electrons at another point is a positive point in the system.* Thus we establish the two polarities in electrical circuits in terms of surplus and deficiency of stationary electric charges, each being mentioned with respect to the other, for in an electrical circuit there may be both a deficiency as well as a surplus of electrons at different points.

The reason for speaking about stationary electric charges at this point, rather than charges in motion (electric current), is because the conditions created in a vacuum tube so that it is capable of functioning, are examples of charged bodies, which subject receives more attention later. Between these charged bodies is a movement of negative electric charges in most of the tube types, and the movement of both negative and positive in a few of the tube types.

(A disciple of absolute technical accuracy might claim that the analysis here presented is not scientifically rigorous, since we have neglected to take into account the relation between charge, capacitance, and potential. Disregard of this relation is deliberate, for it was felt that the rigor that would have resulted by its inclusion was of less importance than the aid in visualization that resulted from its omission.)

The Positive Ion

There need be little difficulty in visualizing the existence of a surplus of electrons at a certain point, for we have asked you to visualize the electron as a tiny ball. The visualization of the positive state, that is a shortage of electrons at a point which is supposed to be positive, might appear somewhat more difficult. Actually this need not be so, for as we stated in this text, the removal of an electron from an atom leaves that atom with a net surplus of an elemental positive charge.

Such an atom was given the name of *positive ion*. That same name applies if the atom has lost more than one electron, and difference in its state, either with one, or with more than one electron shy from its structure, is the magnitude of positive charge which that atom will display. Thus, if one electron has left an atom, we can say that the positive ion which is the result, has a positive charge equal to the one surplus proton in the structure. Alternatively, we can say that the aforementioned positive ion has a positive charge which is equal in magnitude to the negative charge of the electron which has left the atom. When we say a point in an electrical system is shy a number of electrons, and hence is positive, we mean that at that point, an accumulation of positive ions exists.

For example, if by some means we create a shortage of electrons upon a piece of metal—and if we say that this metal therefore becomes positively charged and that this charge is uniform across its surface, a visual picture that we can form of this condition is that the surface of this metal is covered with a number of such positive ions. We can actually visualize that these ions have come from places within the metal and have formed such a surface layer. It is possible that the actual condition is not exactly as we have outlined, but it is convenient to accept something of the order of what we have described. Actually, the positive charge which is displayed by the metal is not

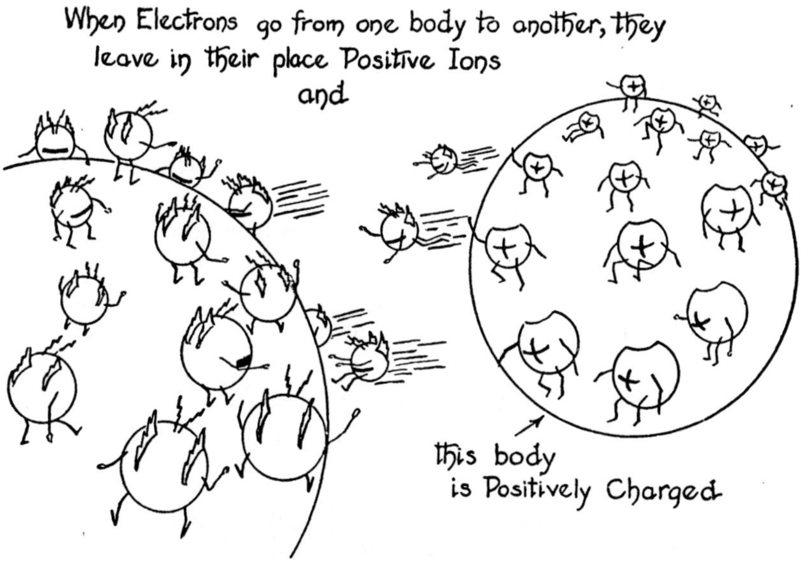

altered whether or not our conception is correct. If, during other operations, this positively charged plate is again made neutral, this is accomplished by supplying electrons *to* the plate, and not by the movement of positive ions from the plate to some other place in the system.

The Negative Ion

In line with this discussion of charges, one more point must be stressed. You will remember that we discussed the possibility that an atom under various conditions may gain, although only temporarily,

one or more surplus electrons, and thus become a *negative ion.* What we failed to mention at that time is that this condition is seldom associated with the atoms of metal. However, such a condition is not of any importance, for if we had the negative ion as a carrier of an electrical charge, it would be doing nothing more than could be done by the electron itself, which is far more mobile than the negative ion. In gases and liquids however, the negative ion occurs sufficiently often to justify its recognition as another example of a carrier of a negative charge.

Now, using this summary of the properties of charges as our basis, we can make further progress in our effort to clarify the problem of charges and charged bodies. Actually, the smallest particles of electricity in order of size are the proton and the electron, so that there are only these two elemental charges. All things larger than this can very readily be classified as charged bodies. A distinction should be made between electrons and ions, which are *charges,* and a *charged body,* which is the material object upon which the charges exist.

Laws of Attraction and Repulsion

Referring again to what Edison found in his incandescent lamp, we now are in a position to consider the motion of the electrons in that tube. Why did the electrons move to the metal plate which Edison connected to the positive end of his filament? The reason is expressed in the most basic laws of electricity, the laws of attraction and repulsion, which state that

1. *Like charges repel one another.*
2. *Unlike charges attract one another.*

This means that electrons have no appeal for other electrons; positive ions do not like other positive ions, and negative ions also do not like other negative ions. On the other hand, electrons like positive

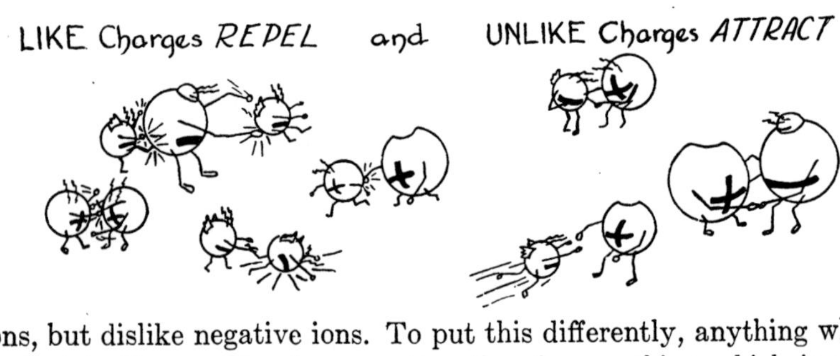

ions, but dislike negative ions. To put this differently, anything which is electrically negative has an attraction for anything which is electrically positive, and vice-versa, but things electrically positive do not like other things which are electrically positive, and the same applies for two things which are electrically negative. Thus, in Edison's tube the metal plate which he connected to the positive end of the filament, became a positively charged body and therefore attracted the electrons towards it.

Electrostatic Fields (Zone of Influence)

Now, if you ask *why* this attraction or repulsion takes place, we cannot give you this answer. All we know is that scientists have established these laws of attraction and repulsion and we accept them as fact. However, some explanation does exist, developed no doubt as a convenience, which describes some of the conditions involved in this process of attraction and repulsion, but it still does not answer the question "why." That some sort of invisible influence does exist between charges has been definitely established, and the fact that this influence is a force, is also conclusively proven since it can cause objects to move.

This something, which we have referred to as an influence, has been named an *electrostatic field* and it is inseparably associated with the

When an Atom *loses* an Electron, it develops a charge equal to that of the Electron but opposite in its directive force.

When an Atom *gains* an Electron, it gains a charge equal to that of the Electron and of the same directive force.

elemental charges and ions, and hence it is associated with all objects larger in dimensions than these elemental charges—for all things are composed of elemental charges. Where such an elemental charge exists, such a field also exists, for the two are inseparable. In the normal atom where there are equal numbers of unlike elemental charges, the field of each charge exists, but it does not manifest itself. Everywhere around that atom the fields of the unlike charges are counteracting each other insofar as influence upon something else outside of the atom is concerned; hence the net effect is that there is no *external* electrostatic field. But, if an electron is taken from that atom so that it becomes a positive ion, then it will be surrounded by an electrostatic field, having a magnitude representative of that one surplus positive charge which that positive ion contains. The same is true if an electron is added to an atom, thus making it a negative ion. This too will then display an electrostatic field, equal in magnitude to that of the surplus electron.

According to previous statements the charge upon a proton is equal to that upon the electron except that it is of the opposite kind. It therefore is evident that the strength of the electrostatic field of a positive ion is equal to that of a negative ion, provided that the surplus of positive charges upon the former is the same amount as the surplus of negative charges upon the latter. And as far as these fields are concerned, while it is true that each elemental charge has its own electrostatic field, when a number of these charges are accumulated upon a spherical surface, the nature of this field behaves as if there was but one charge equal in magnitude to the entire number of the charges upon that body and all located at the center of the sphere.

Radial and Parallel Electrostatic Fields

As to the nature of this electrostatic field, we find it valuable to consider it from two viewpoints: the *radial field* and the *parallel field.* This is necessary because those conditions which exist for radial fields are not used in connection with the conditions assumed to exist in vacuum tubes. Inasmuch as our primary interest lies in the vacuum tube, we shall devote just sufficient time to the radial field as is necessary to give you some familiarity with the subject. To present this picture, we shall concern ourselves with the two elemental charges: the proton and the electron.

Since pictorial representation is valuable in creating familiarity

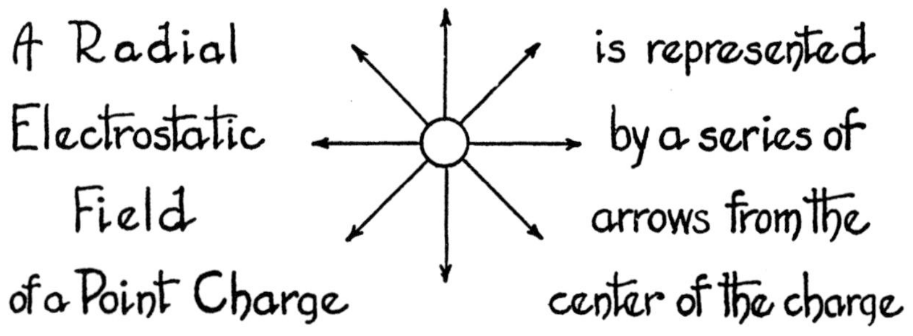

with a subject, scientists have seen fit to illustrate the electrostatic field by means of a number of lines, each line representing the direction of the force which would be exerted on a test charge at any point along the line of force.

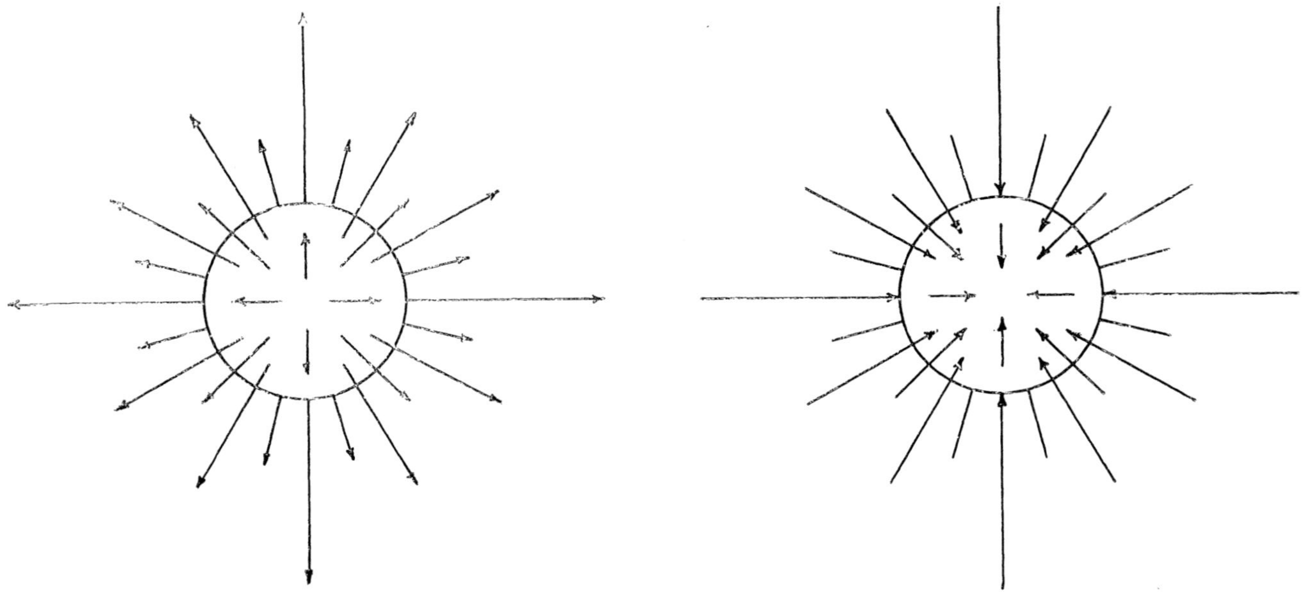

PLATE 2. The arrows indicate that the lines of force are away from the electron and towards the proton; the direction of the force exerted upon a negative test charge is indicated by the arrows. The lines of force radiate in all directions.

In the case of a point charge, the lines of force radiate in all directions. In other words, *point charges have a radial field.* These lines of force are the same for both kinds of charges, and differ only in their direction. The single plane of the paper ordinarily permits showing the lines of force in one plane only, whereas these lines really radiate in all directions, just as if you had stuck a number of long pins into an orange pointing all of them towards its center and using as many as are needed to cover the entire surface of the fruit.

This is illustrated in the anaglyph of Plate 2. Of course, only a comparatively few lines are shown in the three-dimensional drawing for the sake of clarity.

We mention the fact that the entire surface of the orange is covered so as to explain that there is no space between these lines of force. Nowhere within this field would you be able to find any space where no lines of force exist. All the pins are pointed toward the center of the orange to illustrate that the lines of force originate at the center, since a spherical body whose surface is covered with charges behaves as if all the charges were concentrated at the center of the sphere.

Direction of Lines of Force

It is of interest to consider the direction which the lines of force emanating from a charged body will take when they act upon another charged body. Here we find a peculiar condition—one which seems to have been established as the result of a bad guess by Benjamin Franklin almost 200 years ago.

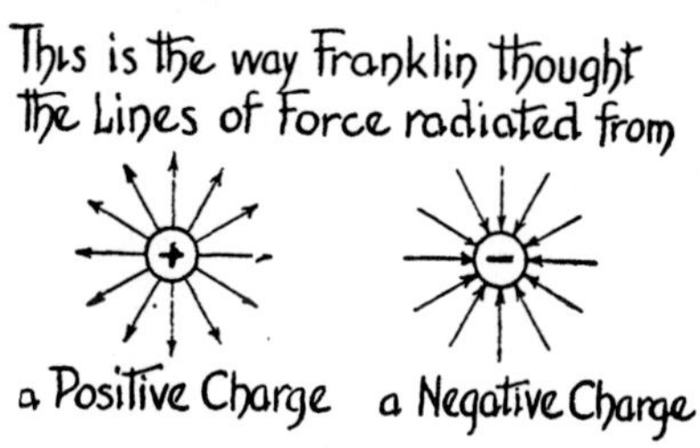

During the days of Franklin, Dufay, and others who were conducting electrostatic experiments with sealing wax, cat's fur, pith balls, glass rods, and silk, a great controversy arose concerning the nature and behaviour of electricity. Without going into details, although it was upon this work that the basic electrical laws previously mentioned were founded, Franklin's guess was that positive electricity was transferred to the sealing wax by the cat's fur and that positive electricity was removed from the glass rod rubbed with silk. Today we know that this guess was wrong. Until the birth of the electron concept the belief was that the flow of electric current was from positive to negative. The result has been, even in such basic things as representations of the direction of the lines of force which emanate from the elemental charge, that the positive charge is shown as being mobile.

Since the electron is the essential item in the operation of the vacuum tube and in the majority of vacuum tubes the only mobile kind of electricity is negative, we shall consider all things in terms of the modern conception of electron movement as being *from minus to plus.*

Under the circumstances, our illustration of the direction of electrostatic lines of force will assume the motion of the negative charge,

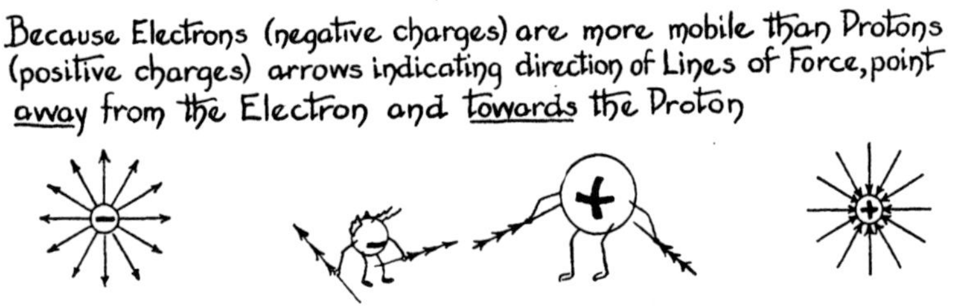

therefore our test charge whenever mentioned, will be negative. This means that if an arrowhead is used as an indicator of the direction of a line of force of an electron, it will point *away from the electron.*

In the case of the positive charge, the arrowheads indicating the direction of the force exerted by the electrostatic field is towards the positive charge, since the field would cause the negatively charged body to move towards the positively charged body along a line of force.

When we associate a negative charge and a positive charge, according to our convention, the direction of the lines of force of the electron is toward the positive charge and the direction of the lines of force of the positive charge are in towards itself, thus indicating that the negative charge would move to the positive charge. (See Plate 2.)

Forces Present In The Radial Electrostatic Field

Continuing with the discussion of the electrostatic field between point charges, we shall next consider the force which exists between such point charges.

Experiments performed by Coulomb towards the latter part of the 18th century, established the law that the force exerted by electrostatic fields varies inversely as the square of the distance between the point charges responsible for the fields. This applies to attraction as well as repulsion and means that if two similar point charges are 1 inch apart and exert a certain repelling force upon each other, increasing the separation to 2 inches; twice what it was before, the force now active between the two decreases to 1/4 of what it was before. If the spacing is increased to 3 times what it was, then the force decreases to 1/9 of what it was originally. Conversely, if the separation between charges is reduced, the force increases inversely as the square of the distance. Thus, if the separation is reduced to 1/2 of what it was, the force increases 4 times; if it is reduced to 1/3, the force increases to 9 times, etc. This relationship is expressed by the equation

$$f = \frac{q \times q'}{d^2}$$

where f = force

q and q' = magnitude of charges

d = separation between charges

Assuming unit charges for q and q' in air, then the force is

$$f = \frac{1}{d^2}$$

While this relationship is often applied in various electrical arrangements, it is not employed in the same way in vacuum-tube theory. Due to the structure of the vacuum tube, a different relationship between charged bodies has been found more convenient. The use of the inverse-square law and radial electrostatic fields does not lend itself properly. This, however, does not imply that any discrepancy exists in the inverse-square law or its application to numerous other subjects.

Electrostatic Field Between Parallel Plates

The conditions existing within a vacuum tube, as far as electrostatic fields are concerned, differ from those used in the discussion just completed. The electrostatic fields used when considering the vacuum tube are those between charged bodies in the form of parallel plates of infinite size. If ever you have seen a vacuum tube you know that this structure applies more directly to what you find in a vacuum tube than electrodes or elements of very small dimensions.

In days of old, friction between two bodies was the method whereby electrons were removed from one body so that it would be positively charged and added to another body so that it would be negatively charged. In vacuum-tube systems we operate on a different principle.

We use batteries or other kinds of voltage sources to attain the same end. This is made possible by virtue of the fact *that the plus and minus terminals of a voltage source, during the time that a voltage difference is present between the terminals, are electrically charged bodies due to a deficiency and a surplus of electrons which exists upon these respective terminals.* Thus, if we connect the plus or positive end of say, a battery to one metal plate and the negative end to another plate, the former will become positively charged and the latter will become negatively charged. The metal plates joined to these terminals assume the same electrical relationship as the two terminals of the battery.

The positive terminal of the battery has a deficiency of electrons, that is, a preponderance of positive ions, equal to the number of surplus electrons upon the negative terminal. When the uncharged metal

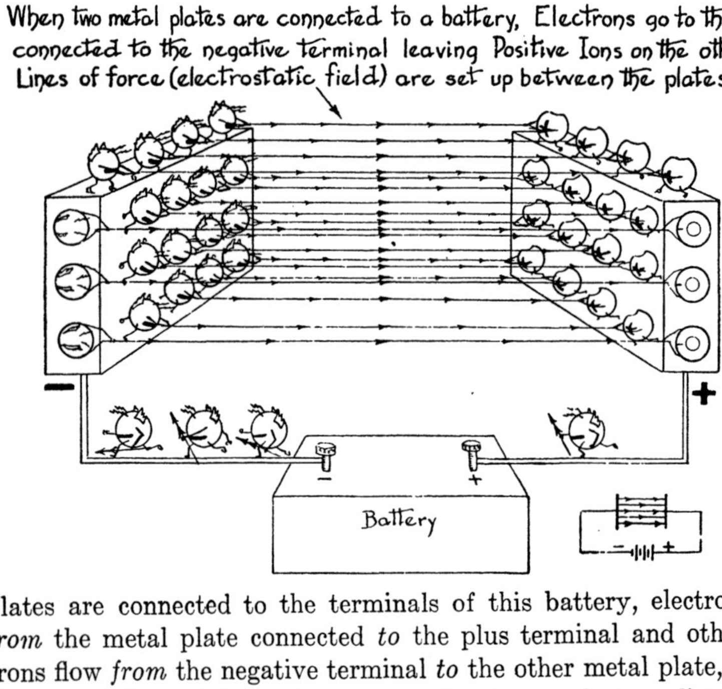

plates are connected to the terminals of this battery, electrons flow *from* the metal plate connected *to* the plus terminal and other electrons flow *from* the negative terminal *to* the other metal plate, so that these extensions of the voltage source develop an electron distribution, shortage and surplus, exactly equal to that which exists at the terminals of the voltage source. The charges upon these plates are assumed to be uniformly distributed upon the respective surfaces.

If the positive terminal of the battery is 100 volts plus with respect to the negative terminal, then the plate connected to the positive terminal becomes 100 volts plus with respect to the plate connected to the negative terminal. In like manner several more plates or other structural forms may be connected to intermediate points between the positive terminal and the negative terminal. These will become electrically charged in exactly the same way as the first two plates we discussed. In each case the number of electrons added or removed is determined by the condition of the voltage terminal to which the plates are electrically connected. These plates are either positive or negative with respect to each other depending upon their connection to the voltage source and by an amount which is determined by the distribution of electrons.

Distribution of Charges

Thus, if we have three metal structures joined to a voltage source, the one which is connected to the most negative terminal has the greatest surplus of electrons upon it. That plate which is connected

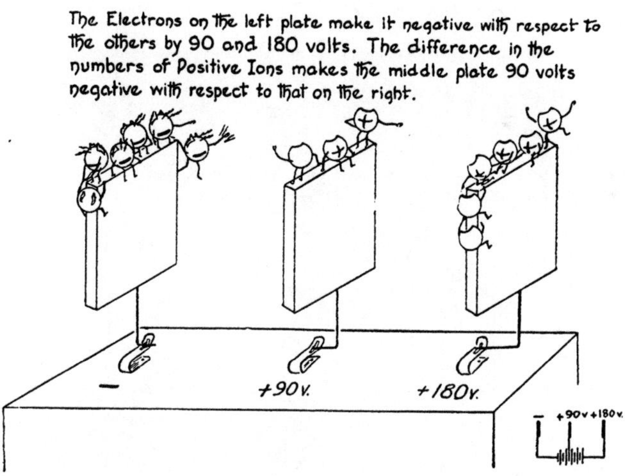

to an intermediate terminal, which happens to be 90-volts plus, then has the deficiency of electrons which would charge that plate 90-volts plus with respect to the negative plate. If the third plate is connected to the plus 180-volt terminal of the voltage source, that terminal has a deficiency of electrons (a preponderance of positive ions) equal to the surplus upon the negative plate, but twice as many positive ions as the plate which is joined to the 90-volt terminal of the voltage source.

Suppose for a moment that we consider only the two plates which bear a preponderance of positive charges with respect to the negative plate. You can readily see that the plate connected to the 90-volt junction can be viewed as being negative with respect to the plate connected to the 180-volt junction of the voltage source. The reason is that the plate connected to the 90-volt junction has fewer positive ions upon it than the plate connected to the 180-volt junction. In this example one plate (90 v) is less positive than the other (180 v). The plate which is less positive (90 v), can be viewed as being negative with respect to the other (180 v) although both (90 v and 180 v) are positively charged with respect to the third (—) plate.

In this way there can be any assembly of elements or plates or electrodes, all of which are joined to different points of a voltage source and all of which become charged and can therefore exert a force upon an electron test charge. The direction of this force is dependent upon the polarity of the charges upon the plates. *A plate which is positively*

The intervening plate need not be solid to have a charge; it can be a grid-like structure

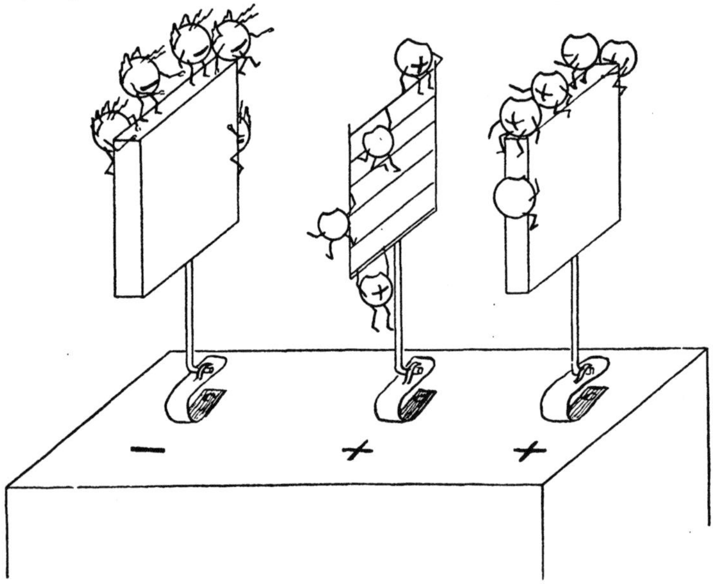

charged will have a field which will act in such direction as to attract a negative charge and repel a positive charge, while a plate which is negatively charged will attract a positive charge and repel a negative charge.

Charged bodies need not be solid plates. They can take various shapes as for example, a wire bent back upon itself to form a ladder-like structure. The active surface upon which the charge exists upon such a structure is the wire forming the ladder. If an electrostatic field exists between this ladder-like structure and a solid plate, the lines of force are found to extend from the wire rungs to the solid plate. Lines of force also emanate from the edges where the wire supports the mesh, but in this discussion we will neglect their presence. If such a screen or any form of structure which does not present a solid surface is located between two solid plates or two other charged bodies, the center grid-like structure will act as a partial shield between the solid plates, so that only part of the electrostatic field of the two outer plates penetrates through the open spaces between the wires. Naturally the closer these horizontal wires are to each other, so that there are less open spaces, the more completely will the screen structure act as a shield located within the field of the other two plates and which will affect the nature of the field existing between the two solid plates. An idea of the electrostatic field which would exist in such an assembly is shown in a subsequent chapter.

The Force Existing Between Two Parallel Charged Plates

As we stated earlier in this text, the inverse-square law is not used when examining the force existing in the electrostatic field of structures which are the equivalent of vacuum tube electrodes. Because of conditions within the tube, it has been found preferable to consider the field between two parallel plates (theoretically of infinite size, which stipulation we can neglect) as being uniform throughout the space between the two plates. Lines of force shown between such plates, one of which is positive and the other negative, would appear as parallel lines starting from one plate and ending upon the other. The direction of the lines of force are given as they would act upon a negative charge: namely, *from* the negative plate *to* the positive plate.

At first thought, the reference to a uniform field existing between two parallel plates of infinite size may be confusing; however, a little further consideration will clear away any misunderstanding that may exist.

If a voltage is applied between two parallel plates which are separated by a fixed distance and an electron is inserted in the space between the plates, the force which will be exerted upon this test charge will be the same no matter where the test charge is located between the two plates. Although actually only one force acts, for purposes of illustration we may consider that two forces are acting upon this test charge. As has been shown in the illustrations, the electrostatic lines of force which fill the space between the plates are in such a direction as to move the electron *to* the positively charged plate, but this movement of the electron towards this plate is not due solely to the attraction of the positively charged plate. A distinct contribution to this action is made by the negatively charged plate as well. During the time that the *positively* charged plate is *attracting* the test charge, the *negatively* charged plate is *repelling* the charge towards the positively charged plate. Hence two forces are acting upon the test charge (electron).

Now if we imagine the electron initially located midway between the two charged plates, equal forces of attraction and repulsion will move

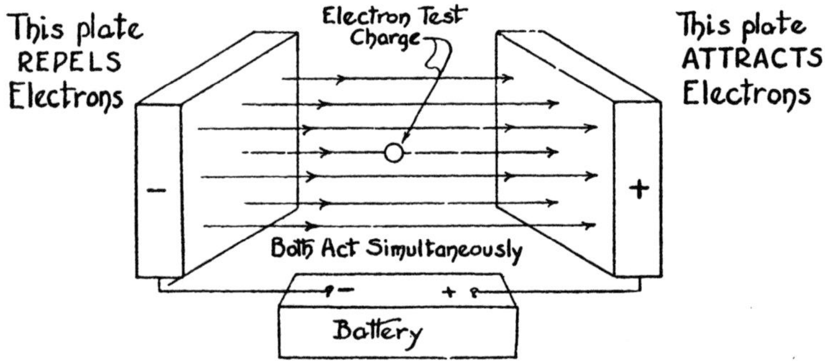

The uniform field is due to the existence of two forces acting on the electron and the fact that the sum of these forces remains the same no matter where the electron may be located between the two fixed charged plates.

the electron in a single direction, i.e. towards the positively charged plate. The total force moving the electron will be the sum of these two individual forces, both, you will remember, acting in the same direction.

If the location of the electron is changed so that it no longer is located midway between the two plates, but instead is located nearer the negatively charged plate, the total force acting upon the electron is still the same as before. Although it is true that the magnitude of the *attracting* force has *decreased* due to the greater separation between the electron and the positive plate, the magnitude of the repelling force due to the closer proximity of the electron to the negative plate, has been correspondingly increased, thus maintaining the total force constant.

Reversing the location of the electron, that is, locating it nearer the positive plate and farther from the negative plate, maintains the total force the same as it was before. Now the reduction of the force acting upon the electron by the negatively charged plate is compensated for by the proper corresponding increase in the attracting force by the positively charged plate.

Thus it is clearly evident that the electrostatic field acting upon the electron anywhere between the two plates is uniform throughout the space between the charged surfaces. Accordingly, it is possible to say that the magnitude of the force which acts upon the electron and tends to move it towards the positive plate, is the same everywhere within the space between the two plates. This holds true for any one set of conditions which establish the electrostatic field and for any one set of plates which are separated by a certain distance.

Relationship of Force, Voltage, and Plate Separation

The relationship between the force f upon a unit charge, the applied voltage E, and the distance d between the two plates, is expressed by the equation

$$f = \frac{E}{d} \tag{1}$$

where f is in dynes, a unit of force equal to approximately 1/445,000 pound or 1/28,000 ounce; E is in statvolts (1 statvolt equals 300 practical volts), and d is in centimeters (2.54 centimeters equal 1 inch).

The term statvolts not being in general use, it is much more convenient when talking about voltages applied to tubes, to refer to conventional or practical units. This can be done by a modification of equation (1). Since the term E is equal to 300 practical volts, the equation can be changed to

$$f = \frac{E/300}{d} \tag{2}$$

and the quantity $E/300$ can be expressed in volts. A much more convenient form for this equation is

$$f = \frac{E}{300\,d} \tag{3}$$

wherein E is in conventional or practical volts and f and d remain just as they were in equation (1). Accordingly, future applications of the basic equation (1) will be in the form represented in (3) and the term statvolts will be replaced by the normal volts as used every day.

From equation (3) we learn that the force f is proportional to the voltage between the plates and inversely proportional to the distance or separation between the plates.

Suppose for the sake of illustration that we imagine two parallel plates 10 centimeters apart with 300 volts applied between them. According to equation (3), the force upon a unit charge at any point between the two plates is

$$f = \frac{E}{300\,d} = \frac{300}{300 \times 10} = \frac{300}{3000} = 0.1 \text{ dyne}$$

and according to the relationship previously expressed, reducing the separation between the plates to 5 centimeters, increases the force to 0.2 dyne; decreasing the separation to 1 centimeter, increases the force to 1.0 dyne. This is indicated in the illustration. It is also clearly

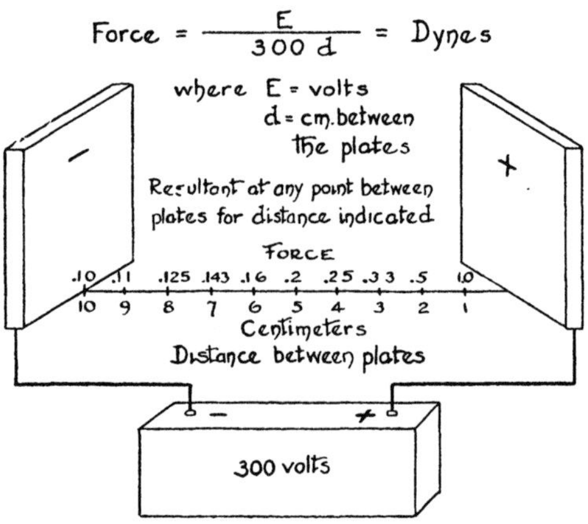

evident that increasing the separation between the plates to 20 centimeters would decrease the force to 0.05 dyne, assuming, of course, that the voltage between the plates remains 300 volts.

Continuing with our example of two plates separated by 10 centimeters and 300 volts applied to the plates, let us further imagine a unit negative charge located somewhere in the space between the two plates. Also that the movement of the unit charge due to this force is towards the positive plate. According to equation (3), this unit charge will move towards the positive plate under the influence of a force equal to 0.1 dyne.

Let us decide that we consider this force insufficient. How can we increase it? Equation (3) shows that this can be done in two ways:

by a reduction in the separation between the plates or by an increase in the voltage applied to the two plates. Of course, a third method is a combination of both the others. Let us decide upon the voltage method, leaving the separation constant at 10 centimeters.

For instance, we desire a force of 0.19 dyne rather than 0.1 dyne to act upon the unit charge while the plates are separated by 10 centimeters. What is the voltage necessary between the plates? Transposition of equation (3) to solve for E results in

$$\begin{aligned} E &= f \times 300\, d \\ &= .19 \times 3000 = 570 \text{ volts} \end{aligned} \tag{4}$$

Since we were already applying 300 volts between the plates, we would have to increase this voltage by 270 volts. In the same way, if we desired to decrease the force upon the unit charge to 0.01 dyne, it would require a voltage of

$$E = 0.01 \times 300 \times 10 = 30 \text{ volts}$$

This is a reduction of 270 volts from the original value.

Controlling the Force by a Third Charged Body

Such control is not as satisfactory as another means which is available. This is the insertion of another charged structure between the test charge and the positive plate. Assume that the formation of this

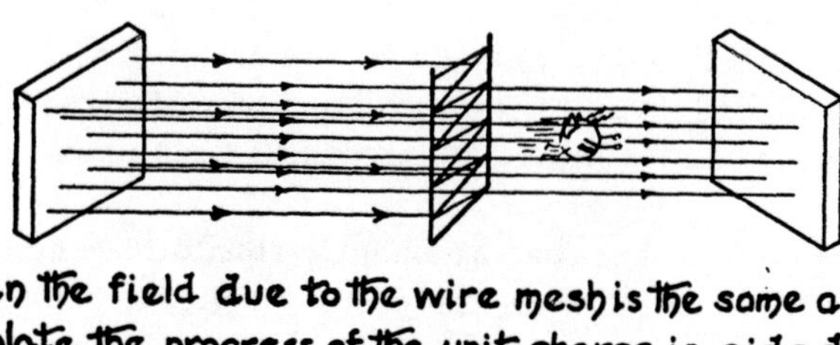

When the field due to the wire mesh is the same as that of the plate, the progress of the unit charge is aided, but—

When the field due to the wire mesh is opposite to that of the plate, the progress of the unit charge is hindered.

structure is such that it will not prevent the movement of the unit charge towards the positive plate, as, for example, if it were a wire mesh so that there are spaces within the structure. Its electrostatic field can then either aid or oppose the existing electrostatic field due to the voltage between the two outside plates and in that way speed up or slow down the movement of the unit charge towards the positive plate. In other words, it can act as a control surface by controlling the effectiveness of the electrostatic field at that point where the unit test charge is located. This comes about as the result of the location of this intervening plate with respect to the location of the positive plate.

As you probably realize, if you are at all familiar with the vacuum tube, this third charged surface resembles the control grid in the normal triode vacuum tube, *for basically the control grid functions as an electrostatic field control.* Let us see how this third charged surface functions. In the explanation which will follow we are limiting our discussion in accordance with the elementary nature of this book, and hence omit some of the special considerations which are technical in nature and not essential for this text.

Let us now recall that when the plates were 10 centimeters apart the first change in attracting force which we desired upon the unit test charge, was to be accomplished by raising the voltage upon the positive plate from 300 to 570 volts. This increased the force upon the unit charge located between the plates from 0.1 to 0.19 dyne.

The second suggested change in force upon this unit test charge was a reduction from 0.1 to 0.01 dyne or a total swing of 0.18 dyne from the minimum of 0.01 dyne to the maximum of 0.19 dyne. This, you will remember, called for a total swing of 540 volts upon the positive plate.

Suppose for the moment that we imagine this third charged surface is located 7 centimeters in from the positive plate, thereby making it 3 centimeters from the negative plate. Knowing the required values of force upon the unit charge and also the change in voltage upon the positive plate, we can now imagine that the positive plate is disconnected and that a unit charge is located between the remaining plate and the grid-like structure. Now we must establish what voltages are required upon this intermediate or control body to fulfill our requirements. The minimum force is 0.01 dyne, hence according to the equation for voltage and with 3 centimeters as the value of d in the equation, the voltage on the control electrode will be

$$E = 0.01 \times 300 \times 3 = 9 \text{ volts}$$

For the maximum force of 0.19 dyne upon the unit charge, the voltage on the control electrode will be

$$E = 0.19 \times 300 \times 3 = 171 \text{ volts}$$

The swing in voltage between the minimum and maximum voltage limits is 162 volts upon this third electrode, whereas the swing in

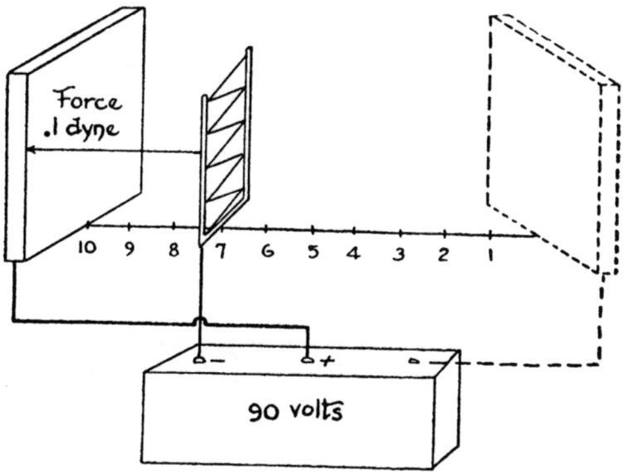

voltage required upon the original plates, which were separated by 10 centimeters in order to create the same change in the force upon the unit charge, is 540 volts. Thus we see that this third electrode is about 3.3 times as effective as the positive plate in controlling the force upon the test charge.

By varying the separation between the original positively charged plate and the intervening plate, it is possible to arrange for any predetermined ratio of effectiveness in the force exerted between the two plates by the electrostatic fields due to the charges on these plates. You can see from the example given that by applying a negative charge to this intervening plate it is possible to offset the force of the positive charge upon the positive plate at any point which is so located that the intervening control surface is between it and the positively charged plate. In like manner a positive charge on the intervening surface can aid the existing field.

Suppose that we wanted the intervening plate to exercise an even greater control over the test charge than was the case in the preceding example. As has already been pointed out, this may be accomplished by moving the intervening plate still closer to the left-hand plate. Let us assume that this separation is 1 centimeter and that we still wish to maintain the force exerted upon a unit test charge at 0.1 dyne. Applying the equation, we obtain

$$E = 300\, f\, d = 300 \times 0.1 \times 1.0 = 30 \text{ volts}$$

From this example it will be observed that when the intervening plate is only 1 centimeter away from the left-hand plate, only 30 volts need be applied between them. This is in contrast to the 300 volts

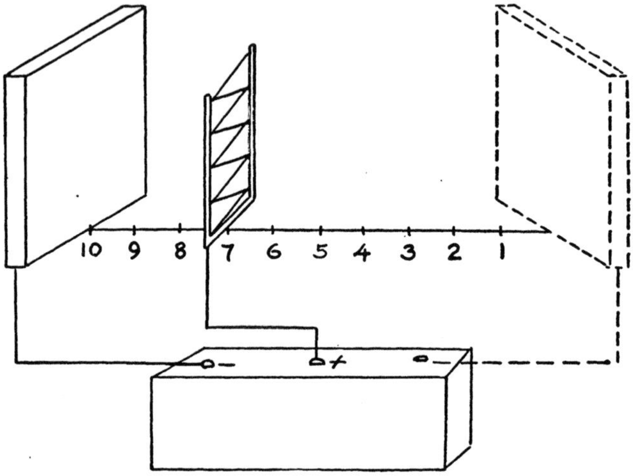

Locating a charged body closer to the test charge requires a much lower voltage to exert a certain constant force upon an electron than another charged body further removed.

needed between the plates when they are separated by 10 centimeters. In other words, the plate separated by 1 centimeter is 10 times as effective as when the plates are separated by 10 centimeters.

In the same way it is possible to visualize more than two charged surfaces acting upon a unit charge. This is done in practice by making all the surfaces intervening between the highly charged positive plate and the location of the unit charge, of such a type that they can be charged and have electrostatic fields, yet not interfere with the movement towards the positive plate of the unit charge being acted upon. Their specific purpose will be considered later on.

Admittedly, many things related to charges and charged bodies have not been told, nor will they be discussed even in the remainder of this book. What we have described is in accordance with the technical level of this book and we hope will serve as an aid in understanding the discussion which follows.

Chapter 4

SPACE CHARGE AND PLATE CURRENT

IN AN EARLIER CHAPTER we discussed what happens in a metal when it is heated to a sufficiently high temperature so that it is capable of emitting electrons. What we did not then mention, but which we intend to cover in this chapter, is what happens to the electrons after they leave the emitter.

The emission of electrons in a vacuum tube is from that element which is referred to as the *filament* or *cathode.* The electrons tear through the surface of the emitter material and become free in space. By space, in this instance, we mean the space within the glass or metal envelope which houses the various components of the tube. Incidentally, the fact that vacuum tubes are housed within such evacuated envelopes is of great importance, for were such emitters heated to their proper electron-emitting temperature in open air they would almost immediately burn up.

Now let us imagine that the emitter is contained within a glass envelope from which all air and gases have been removed, at least to such an extent that the remainder is entirely negligible. In other words, it is a *vacuum tube*. Further, let us imagine that we have control over the emitter, which we shall hereafter identify as the cathode, so that we can vary its temperature, and hence the electron emission, at will. The question then arises, "Where do these electrons go after they have left the cathode?"

In answering this question we shall speak about that phenomenon known as *space charge* which has a tremendous influence upon the actual operation of the vacuum tube, particularly that action known as the flow of plate current through the tube.

Electrons Leave the Cathode

Let us suppose that we raise the temperature of the cathode very gradually and, when we reach the electron-emitting value, that we can follow the first group of electrons which are emitted from the full length of the emitting surface. We shall imagine these electrons leaving the cathode in a sort of a layer, although actually emission is anything but uniform because of the different speeds of the emitted electrons. However, for illustrative purposes we shall assume such uniformity. Thus, these electrons, in the form of a layer one electron thick, are traveling away from the cathode.

This movement of electrons away from the cathode deserves some comment in view of the fact that for many mathematical computations relating to vacuum tubes it is customary, in order to avoid unnecessary complication, to consider the electron as being emitted with zero velocity. This is also done later in this text. However, our statement that the electron leaves the cathode with a finite velocity does not create any real conflict between these two viewpoints, because first, we are not concerned with the mathematics of the tube and hence are not required to avoid the complications introduced by the velocity, and second, the electrons actually do leave the cathode with different finite velocities.

The speed with which these electrons move away from the cathode again requires us to make an assumption for the sake of simplicity. We shall assume that the electrons are moving with a uniform speed when they leave the cathode, although actually, both high-speed and low-speed electrons are emitted. This depends upon the amount of kinetic energy imparted to the electron within the emitter and upon how much of this energy the electron retains after it has given up that amount which is required to overcome the surface forces tending to keep it within the boundaries of the emitter material. The greater the kinetic energy retained by the electron after it has overcome the surface force, the greater its speed of movement away from the cathode. The phenomenon associated with such high-speed electrons will receive special mention later.

In the meantime, let us forget about these high-speed electrons and consider only that group which leaves the emitter at a comparatively slow speed.

Electrons in Space

Since the temperature of the cathode is such that emission can continue, other electrons will follow the first group, and still other electrons will follow the second group until tremendous numbers of electrons have left the cathode and are out in space. In our discussion we are assuming an orderly emission and an orderly motion of these electrons, whereas, in fact, the motion is definitely haphazard.

The first group of electrons have nothing in front of them to retard them, whereas they have the second group in back of them, the third group in back of the second, the fourth group in back of the third, etc.

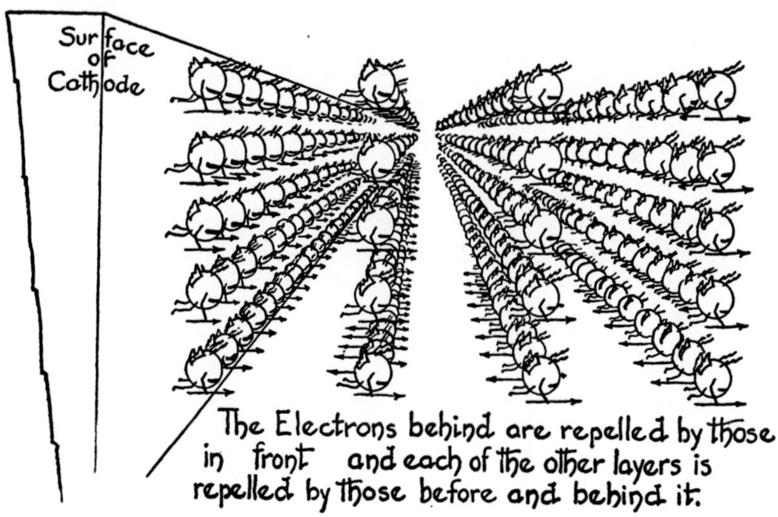

Since the basic law pertaining to charged bodies states that like charges repel one another, the electrons in back of the first group are experiencing a repelling force due to the electrons in front of them and the direction of this force is such as to tend to retard the motion of the electrons in their advance away from the cathode.

So we have layer after layer of electrons moving from the cathode because of their velocity of emission. We ultimately arrive at a distribution of these electron layers wherein those nearest the cathode are crowded closely together and those further away towards the boundaries of the glass envelope, are further and further apart. In other words, the electrons occupying the space within the tube are non-uniformly distributed, *being relatively dense in the vicinity of the cathode and less and less crowded as the distance from the emitter is*

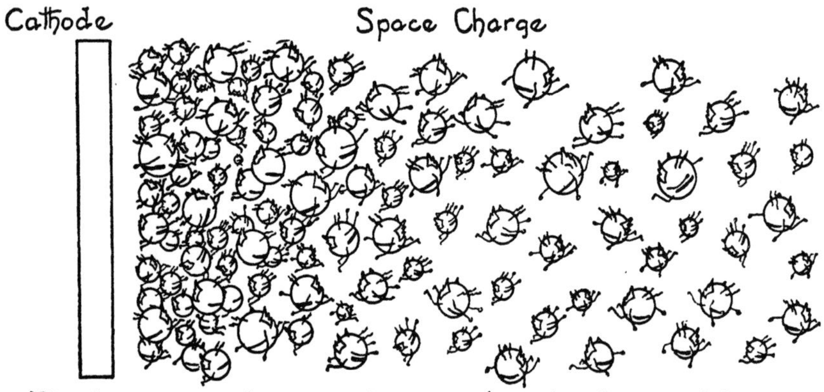

Electrons near the Cathode are close together and less crowded further away

increased. You must realize that our simplified representation is necessarily crude because of the difficulty of showing such exact graduations that do exist in the layer-like arrangements of the electrons.

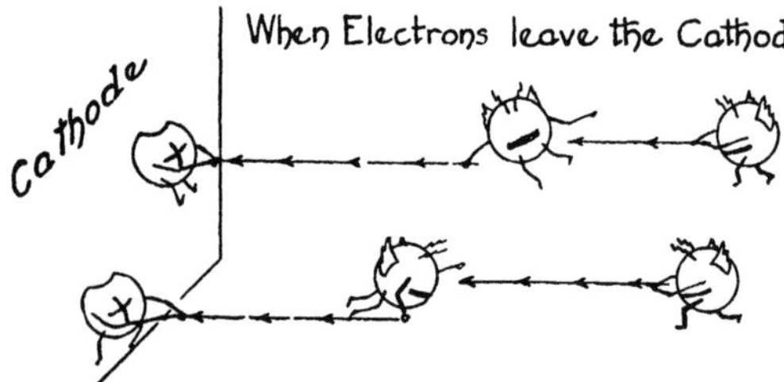

When Electrons leave the Cathode their speed is reduced by the field of the Positive Ions they leave in the Cathode and by the opposing field of the Electrons in the Space Charge

Actually, these arrangements are haphazard and the distribution is not uniform, for the reasons mentioned above.

There is another factor contributing to a maximum density of the electrons at the cathode. This is the fact that as the cathode loses an electron, the cathode becomes positive to the extent of that charge, and hence tends to exert a force upon that electron pulling it back to the cathode. While it may not accomplish this, due to the initial velocity of the electron when it left the cathode, as well as other factors which may exist and which will be discussed later, it can, however, slow down the electron speed. Cumulative action of this kind and also the repelling force acting upon a newly emitted electron by those already out, tends to cause an accumulation of electrons outside the cathode and near it.

Space Charge

Now, each of these electrons is an elemental charge of electricity and since they are out in space, the very descriptive name of *space charge* is given to them. Furthermore, you have learned that each elemental charge of electricity, positive or negative, is inseparably associated with an electrostatic field. When an accumulation of charges exists, the electrostatic field is assumed to be that of the entire group of charges, so that *the space charge which surrounds the cathode in all directions has associated with it an electrostatic field.* This field has

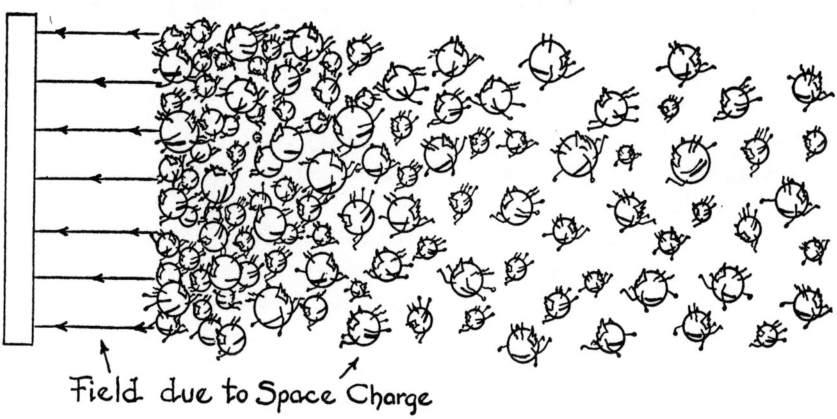

a direction such that it would tend to repel a negative test charge in the vicinity of the field.

The nature of this field is similar to the electrostatic fields previously discussed in connection with vacuum tubes. In other words, if we imagine the emitter to be a surface, then the lines of force emanating from the space charge would be parallel straight lines between the space charge and the emitting surface. We make specific mention of this in order to preclude any possible misconception of radial fields emanating from the individual electrons in the space charge.

What we said should give you an idea of what happens to the electrons after they leave the emitting element, although you still do not have a complete picture because all of the facts have not been stated. These will now follow for this sample tube containing just the emitting cathode. Naturally, you must realize that this is not a typical tube, for there is more than one element within even the simplest of tubes, the diode. However, the existence of the space charge as described will be found in all vacuum tubes and plays a tremendously important role in them. It is further true that in conventional tubes you will not be interested in the spreading out of this space charge to occupy all of the space within the tube, but rather in the concentrated space charge between certain elements.

Returning to the space charge in our sample tube, let us examine some of the significant details associated with it. We know that the space charge is due to the emitted electrons. Since this is the case it is only natural to wonder about a few things associated with this emission. For example, what is the relationship between the space charge and the electron emission? There must be some controlling influence, otherwise the space charge density will continue increasing —and what is its limit?

Density of Space Charge

There is such a relationship—in fact a very important one. While it may appear at first glance as if the density of the space charge would increase as long as there is emission, a limit actually exists in the space-charge density for each value of cathode temperature. In other words, for any given value of cathode temperature, which represents a certain rate of electron emission, a state of equilibrium (known as *emission saturation*) is established whereby the maximum density of the space charge is fixed. The fact that electrons are being emitted does not increase this density, because the electrons emitted from the cathode are *repelled back* to it. To express this differently, for every electron which gets into the space charge after this state of balance or equilibrium has been reached, the space charge returns an electron to the cathode.

The processes involved in this action are a subject of conjecture. For every value of cathode temperature a certain magnitude of thermal agitation occurs within the emitter material, and also a definite average speed for the emitted electrons. The higher the temperature of the cathode, the greater is the internal thermal agitation, and the higher is the average speed of the emitted electrons. Equilibrium at any temperature is reached when the number of electrons which have formed the space charge is such that the magnitude of its electrostatic field at the cathode is sufficient to overcome the speed of the emitted electrons and repel them back to the cathode.

If the temperature of the cathode is increased, the thermal agitation is increased and the electrons are emitted with a higher average speed. This speed is sufficient to overcome the repelling force of the electrostatic field of the previously existent space charge so that electrons join this space charge, and the space charge density consequently in-

For every Electron that goes to the Space Charge when Emission Saturation has been reached, an Electron is returned to the Cathode and the Space-charge density remains constant

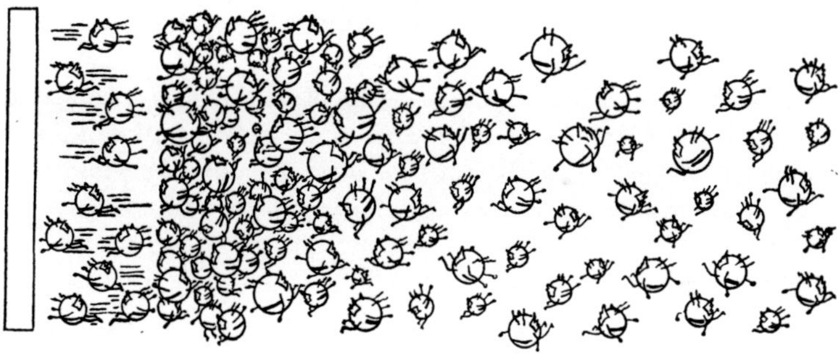

creases. When this increase in density nas again reached such a value that the repelling force of the electrostatic field of the space charge is sufficient to overcome the higher average velocity of the electrons emitted at the increased cathode temperature, and is again repelling the emitted electrons, a new state of equilibrium has again been established and *emission saturation* again exists.

In this way it can be said that the space-charge density automatically increases with the increase in emitter temperature. For every fixed temperature of the cathode, a state of equilibrium develops between the amount of emission and the space-charge, so that the space-charge density increases no further. Similarly, a reduction in temperature of the emitter causes a reduction in the density of the space charge by returning electrons to the cathode.

Current Is Space-Charge-Limited

So much for the subject of the formation of the space charge and the state of emission saturation. We shall have occasion in the next chapter to speak in more detail about its action. In the meantime, we can consider one additional significant detail. The space charge exerts a controlling influence upon the electrons emitted from the cathode in the tube, for unless something is done to prevent its action, the field of this space charge restricts the effectiveness of the emission of electrons from the emitter.

The function of the space charge is paramount, because in the operation of virtually all types of vacuum tubes, the emitted electrons travel across the inter-electrode space to other electrodes within the tube. Since the average cathode is capable of emitting more electrons per unit time than are usually called upon to move across the space between the elements, it is essential that some force exist to maintain control over the emitted electrons. This is done by the space charge and that is why the current flowing through a vacuum tube is often referred to as being *space-charge-limited.*

Taking all things into account, it can be said that the basic function of space charge, which is to be found in all tubes under all normal conditions, is to perform as a *limiting agency upon the movement of electrons from the cathode out into the inter-electrode space.*

Plate Current

Let us now put the emitted electrons to work. By this we mean that we want them to move through the space within the tube in a definitely directed manner. To do this we must exert some force upon these electrons. This force is obviously some form of attraction, for in their present condition as a space charge they are distributed throughout the space in the tube. We now desire to collect these electrons. To do this we must get the electrons to complete their journey from the cathode to another element which we shall place within the tube. This element, known as the *plate* or *anode,* is a metal surface which can be positively charged by connecting it to the *positive* terminal of a voltage source and *by connecting the negative terminal of the same voltage source to the cathode.*

With an arrangement such as this we can cause the flow of electrons between the cathode and the anode and this electron flow is called *plate current.* The flow of this plate current is a prime requisite in all vacuum tubes and it is by means of this current that the various uses of the vacuum tube can be achieved. This plate current is influenced by the space charge, or, to put it differently, the importance of the space charge can be appreciated by noting its effect upon the plate current. Having created that arrangement within our vacuum tube which enables investigation of the space charge—plate current relationships, let us now analyze what conditions exist inside the tube. At all times we shall try to correlate the action taking place in this tube with the subjects that have previously been discussed.

In the first place, we have arranged for the existence of certain electrostatic fields within the tube. By connecting the voltage source between the plate and the cathode—for the moment we are uninterested in either the shapes of these elements or their separation—we have duplicated the conditions described in Chapter 3 in consequence of which an electrostatic field exists between the plate and the cathode. The strength of this field is determined by the magnitude of the voltage difference between the plate and the cathode. If we now imagine the cathode to be hot and therefore emitting electrons, we have another electrostatic field present in the tube. namely, that due to

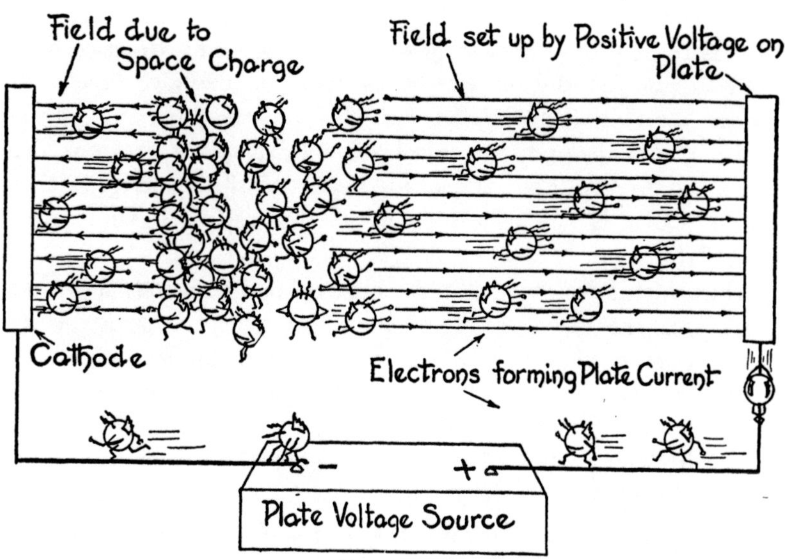

the space charge. This is in accordance with the previous discussion of the space charge. We then have present in this simple tube two electrostatic fields which require our attention. But before we discuss the actions of these two fields, let us see what conditions must be identified.

Continuing with our new tube system, we now have a completed vacuum-tube circuit, which is something we did not have in the previous example of the emitter alone located in the glass envelope. If a movement of electrons from the cathode through the inter-electrode space to the plate occurs, these electrons now can find their way back to the cathode, via the voltage source and so complete their path. You will recall that electric current was identified as a directed flow of electrons. Therefore the directed flow of electrons within our tube is the equivalent of electric current flow and, since the destination of these electrons within the tube is the plate, the current is identified as *plate current.* In the same way, the continued movement of electrons from the plate back to the cathode through the external system, is spoken of as *the flow of plate current in the external circuit.*

The voltage source which we show joined to the plate and the cathode is called the *plate-voltage source,* because it is applying the charging voltage to the plate, or, to say this differently, it is the voltage source which maintains the plate at a certain positive voltage with respect to the cathode. This is in contrast to other voltage sources which may be associated with a vacuum tube, as, for example, the voltage source which is responsible for heating the emitter system. Practice has resulted in the adoption of certain symbols to identify these voltage sources: the voltage associated with the emitter system is designated as the *"A"-voltage* and that associated with the plate circuit is identified as the *"B"-voltage.* We speak about this "A"-voltage, although it is not shown in our basic tube circuit illustration. It will, however, appear later.

A third factor deserving mention is associated with the space charge. Whereas, in the early part of this chapter, we stated that the space charge occupies the space within the tube, the insertion of the plate and the application of a voltage to this plate narrows down the limits of the space charge which is of interest to us. Now, we are solely concerned with that portion of the space charge which exists between the *cathode* and the *plate.* At first thought this statement may give the impression that the space charge is still spread throughout all the space within the tube. This is incorrect, for if we assume emission in the form of parallel beams from the entire surface of the cathode, and then further assume that the area of the plate is such as to cover the area of emission from the cathode, then we can imagine the space charge, with its original non-uniform distribution of electrons, to exist only between the cathode and the plate.

Fields Between Cathode and Plate

In line with this thought, there is another which we wish to advance in order to make comprehension of the subject easier. Technically, it is not absolutely correct but the error introduced is very small and, since you are familiar with the true status of conditions, no harm is done. What we have in mind is the following.

Since we shall be concerned with the actions of the various electrostatic fields within the tube, it may be difficult for you to visualize an electrostatic field due to a non-uniform space charge which reaches from a small distance from the cathode to the plate. No doubt you would be able to visualize this field better if you imagined the space charge to be concentrated in a small area between the cathode and the plate, much nearer to the cathode than to the plate, and then completely disregarded that portion which is represented by the low density portion, that is, the electrons near the plate. Since the actual density is very low in the area near the plate and extremely high near the cathode, the former contributes very little to actual space charge, whereas the latter is very much *the* space charge. So, taking these conditions into account, it is not a very great departure from the truth to visualize the space charge as being a concentrated layer of electrons, quite a few electrons thick, and located between the cathode and the plate very close to the cathode.

With this picture in mind, we now can investigate the action of the various fields within the tube. What fields do we have? First, we have the field due to the space charge. Being located between the cathode and the plate, the lines of force emanating from it spread in two directions. One direction is towards the cathode and the other direction is towards the plate, as shown in the last illustration. The lines which point in the direction of the cathode would tend to repel any electrons emitted from the cathode, including those which are in the space between the cathode and the space charge. The lines directed towards the plate would be in the direction from the space charge to the plate, thus indicating that electrons would move towards the plate.

Next, there is the electrostatic field due to the voltage applied between the plate and cathode. The lines of force of this field are directed from the cathode to the plate. The lines of force due to the positive charge upon the plate in effect neutralizes that of the space charge. These two forces due to the two fields tend to act in opposite directions on an electron at the cathode, and the resultant motion of the electron while under the influence of these two forces depends upon that force which predominates.

Summarizing what we have said, we find the following: both the field set up by the positively charged plate and that by the negative space charge are present in the tube. Further, the field set up by the positively charged plate acts upon the space charge. Now we are ready to note the effects of these fields.

Electrons To and From the Space Charge

Suppose that we establish the premise that we desire a movement of electrons from the cathode to the plate, that is, we desire the flow of plate current. If you recall that the existence of the space charge is dependent on the emitter temperature and that the space-charge density varies automatically with changes in the temperature of the emitter, and hence with the amount of emission, it seems that the space charge must always exist as long as emission takes place. Further, the density of this space charge remains constant for any particular value of emitter temperature or rate of emission. Under most operating conditions this is true, and departure from this condition will be dealt with separately.

If we disturb this state of equilibrium by removing electrons from one portion of the space charge, say that end which is nearer the plate, this will result in a reduction in the total magnitude of the field of the space charge. Its ability to repel the electrons emitted from the

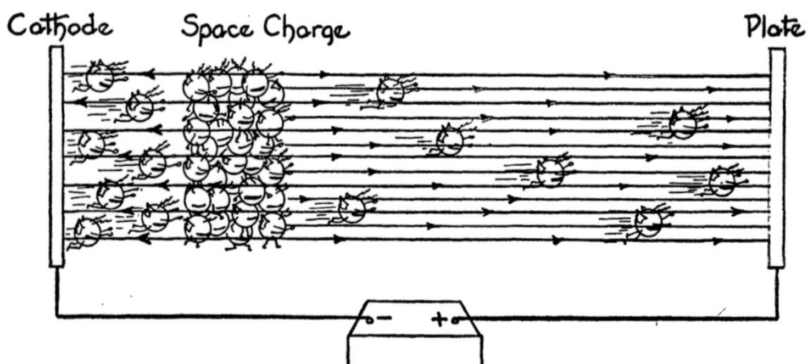

With a certain voltage impressed between the Plate and Cathode and with a certain Cathode temperature, as many Electrons enter the Space Charge from the Cathode as leave it for the Plate

cathode is therefore reduced, so that the number of electrons required to restore equilibrium will be added to the space charge in the zone near the cathode. More than the required number of electrons to offset those removed will not be added to the space charge, for the state of equilibrium previously mentioned is automatically maintained.

We visualize a condition wherein electrons are removed from one side of the space charge at a definite rate and electrons are admitted into the space charge at the same rate on the other side. If 1,000,000 electrons move from the space charge to the plate each second, another 1,000,000 electrons enter the space charge from the cathode each second. Thus there is a progressive movement of electrons from the one side of the space charge, through the space charge, and out of the space charge on the other side, *without altering the density of the space charge*. Such a directed movement of the electrons is equivalent to saying that the electrons emitted from the cathode travel over to the plate and comprise the plate current. But, it is also possible to say that the space charge acts as a reservoir of electrons, from which is drawn the number of electrons which is required to equal the instantaneous value of plate current, and that the cathode is the supply which replenishes the space charge. This is similar to a supply system, wherein the city draws water from a reservoir while a stream or lake automatically supplies the equivalent amount of water required by the reservoir to maintain a certain volume of water within it.

Let us now assume that the temperature of the cathode is such that it is emitting electrons profusely and at the same moment the plate has no voltage applied to it. Under these conditions there is no current flow through the tube, except for those *few* very high-speed electrons which do get to the plate and which we shall neglect for the present. We then apply a small positive voltage (20 volts) to the plate. Immediately an electrostatic field is set up between the plate and the space charge. We now note a small value of plate current, which indicates that there is a movement of electrons through the inter-electrode space. What is responsible for the movement of these electrons?

Two explanations are possible, although both conclude in the same way. Let us first give the simpler of these two. We stated that the electrostatic field due to the positive voltage upon the plate extended to the space charge. The strength of this field depends upon the plate voltage and consequently the magnitude of the force exerted upon the electrons in the space charge also depends upon the plate voltage. Therefore this field will act upon the electrons in the space charge so that a certain number of electrons will be drawn off from the space charge and go to the plate. Let us suppose that this is 10,000,000 electrons per second.

This action reduces the magnitude of the total space charge, and this in turn, reduces the intensity of the space charge field which is repelling the electrons that are emitted from the cathode. The result is that the number of electrons being repelled back into the cathode is reduced; however, the density of the space charge is fixed for any one cathode temperature. Consequently as many electrons are permitted to enter the space charge from the left as leave it from the right. This, in our example, is equal to 10,000,000 per second. As long as the plate voltage is maintained constant the movement of electrons from the space charge and the movement of electrons into the space charge continues. There is, therefore, a progressive movement of electrons through the space charge, namely, a steady flow of plate current.

The second view is slightly different, but ends up the same way. We again start with electrons being drawn off from the space charge, but we now speak about the positive field of the plate acting in opposition to the negative field of the space charge in the neighborhood of the cathode. The positive field of the plate acts in opposition to the negative field of the space charge at the cathode, and reduces the magnitude of the total repelling action sufficiently so as to permit as many electrons per unit time to enter the space charge as leave it to go to the plate.

Negative Voltage upon the Plate

When we started this discussion we stated that the condition required in order to attract electrons to the plate, is that a positive voltage be applied to the plate. What happens when a negative voltage is applied to the plate? If this is done, an electrostatic field is caused to exist between cathode and plate, and because this field is due to a negative plate voltage, it would tend to aid the space charge in repelling electrons, so that we can say that no electrons would flow.

It might also be well at this point to say that when no voltage is applied to the plate, and the plate is joined to the cathode thereby placing both elements at the same potential, there is nevertheless a small flow of electrons to the plate. This is due to the high-speed electrons which are emitted with sufficient velocity to reach the plate by penetrating the space charge. Although the value of such plate current is very small, it is nevertheless sufficiently great so that it must be taken into account in certain special applications of vacuum tubes.

You may realize that the two-element tube dealt with in this chapter is the conventional diode. This is right, but you should also understand that the general details relating to the behaviour of the space charge and the action of the electrostatic fields upon the plate current is not

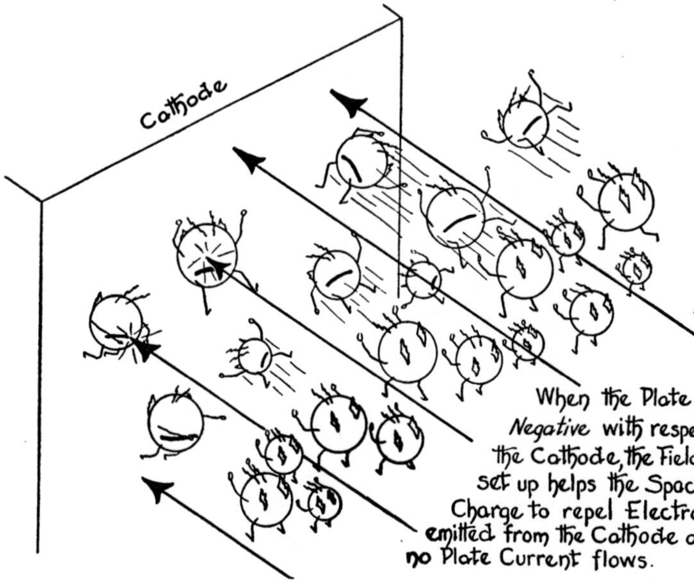

specifically limited to the diode. You will find these details also apply in a great measure to multi-element tubes, that is, tubes which employ more than two electrodes.

The introduction of elements cannot help but modify the conditions existing in the tube and these will be dealt with as they are reached. As to the diode tube that which has been said is preliminary detail. Specific details regarding constants and other operating data will be handled in the following chapters.

In connection with what is to follow, you will observe that a smaller number of animated illustrations appear than were used in the first four chapters. Some will naturally appear, but they will not be as abundant. The reason is that the nature of the subject does not lend itself to such illustrations in many cases, and we are forced, frankly against our will, to resort to the usual form of representation.

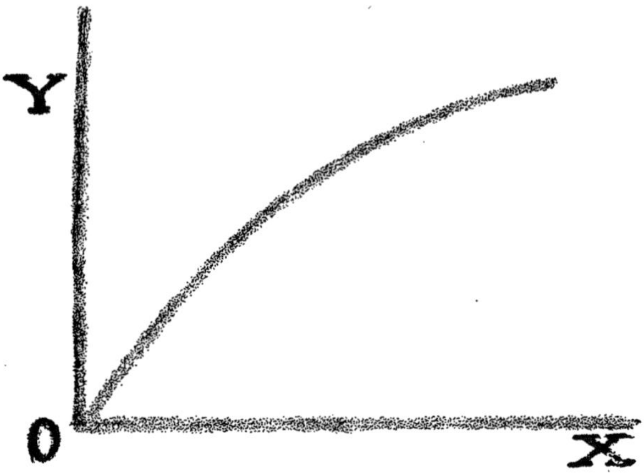

Chapter 5

FUNDAMENTALS OF TUBE CHARACTERISTICS

IT IS READILY APPARENT from all that has been said that the operation of the vacuum tube is based upon the directed motion of electrons between charged bodies. This has been referred to time and again, and we have identified this current as being the plate current. You will remember that although we have spoken quite frequently about tubes with but two elements within the envelope, we did say that the plate current as described was also to be found in vacuum tubes with more than two elements.

Also, we have shown that the charged surfaces between which these electrons move are produced by the application of fixed voltages, which we shall now identify by another name: *operating potentials.* Thus, the voltage which is applied between the cathode and the plate is known as the *plate voltage.* Since the force of attraction upon the electron originates at the positively charged plate, it stands to reason that there must be some relationship between the magnitude of the voltage applied to the plate and the number of electrons attracted to it in a unit time, say one second—in other words the value of the *plate current.*

Vacuum Tube Relationships

There is such a relationship, and it is called the plate voltage—plate current characteristic, being also referred to as the $E_b - I_b$ curve; however, that is only one of the relationships between the different factors associated with the vacuum tube. We have already described to you the conditions which determine electron emission from the heated cathode or filament in a vacuum tube, so that, considering the source of electrons by itself, we find another characteristic, namely, *cathode temperature versus cathode emission.* This cathode temperature may be called by two other names, depending upon the type of tube. For example, if the tube is of the filament type, the current through the filament determines its temperature, hence the characteristic associated with the filament would be *filament emission versus filament current.* If the tube were of the indirectly heated type, then the characteristic would be identified as *cathode emission versus heater current.*

For purposes of illustration it may be worthwhile to carry this a bit further by saying that the relationship between any two variable factors may be expressed as a *characteristic.* Thus, in the case of a filament or heater of a vacuum tube, the relationship between the current through the filament or heater and the voltage across it may be expressed as a characteristic,—that of voltage versus current. As it happens, the filament voltage—filament current characteristic is seldom used in engineering, for the design fixes the voltage which should be applied across a filament or a heater, and this determines the current flow through these elements which, in turn, determines the operating temperature.

Many characteristics are associated with a vacuum tube and it is important for you to understand the significance of these because they are the means whereby the manner of performance of the tube may be predicted in advance for various operating conditions. While it is true that engineering design fixes certain limits of operation for all of the vacuum tubes, and the exact conditions of operation fall within these limits, it is necessary to identify just what performance can be expected from the conditions which exist.

You may be aided in understanding the full significance of characteristic curves if you view them as being charts of *cause and effect.* This may be a little difficult to understand in the beginning, but after a while they become easy to read. To help you read the characteristic curves which appear later in this text, the following facts are given.

Characteristic Curves

In general, characteristic curves are associated with two reference lines, which in a way act as the boundaries of the chart: one of them horizontal and the other vertical. These two lines are joined in such a way, depending upon the form of characteristic curve being shown, that the bottom of the vertical line joins the horizontal line either at the left end of the horizontal line or at some point along the length of the horizontal line. Since we desire to show the simplest illustration, we shall assume that the bottom of the vertical line joins the left end of the horizontal line.

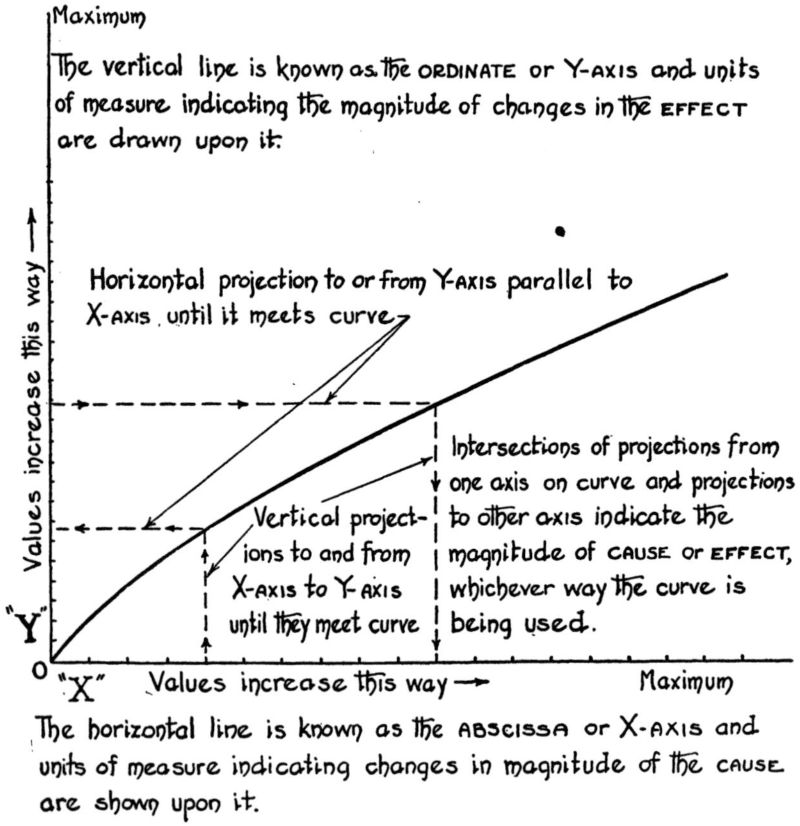

Certain names are associated with these vertical and horizontal reference lines. The former is known as the *vertical axis*, the *Y axis*, or *the ordinate*. The horizontal line on the other hand, is known as the *horizontal axis*, the *X axis*, or as the *abscissa*. Thus when we say that the vertical axis joins the left limit of the horizontal axis, we mean that the bottom of the ordinate joins the left end of the abscissa.

These two reference lines serve as scales upon which the units of measure (volts, ohms, amperes, etc.) representative of variations in the dependent variable (effect) and the units of measure (volts, ohms, amperes, etc.) representative of variations in the independent variable (cause), can be indicated. In other words, these reference lines as a rule are marked off in divisions, each division representing some value of the unit of measure employed. It may be anything from volts to pounds or gallons of flow of water, depending entirely upon the information shown upon the chart.

Generally changes in magnitude of the *cause* (independent variable) are shown upon the horizontal reference line or abscissa, and changes in magnitude of the *effect* (dependent variable) are shown upon the vertical reference line or ordinate.

In general, these scales of magnitude upon both the abscissa and ordinate start at zero (0), so that the junction of the two axes is the zero point upon both scales. This, however, is not necessarily true in all cases, because some charts require that the starting point of the units indicated upon either the abscissa or the ordinate, begin at some value other than zero. The units indicated along these axes are as you can readily appreciate, strictly dependent upon what information is to be conveyed to the reader. For the sake of simplicity we can assume that the scales on both axes start at zero.

The characteristic curve or curves, as the case may be, appears inside the space bounded by the two axes. Whether or not the curves will start at the common zero point and advance straight upwards, namely, advance diagonally away from the point of junction, or whether it is a curve, is of little consequence. Again, it is a matter of the relationship between cause and effect or the relationship between the units designated upon the two axes.

Accordingly, more than one characteristic curve may appear upon a single chart, as you will see in subsequent chapters, or different shapes or curves may appear upon the same chart. But no matter what the shape or the number of the curves, those which shall appear in this text are read in a certain way: by projections to and from the axes, with the junction of either projection upon the desired characteristic as the reference point from which the projection to the other axis is made.

Thus if we desire to note the effect of a change in the magnitude of the cause, we project upwards from the abscissa towards the characteristic. From the junction of this upward projection and the curve, we project a line which is parallel to the abscissa towards the ordinate. The junction of this horizontal projection with the ordinate indicates the magnitude of the effect in whatever units of measure are indicated upon the ordinate.

The reverse procedure can be used to employ the curve as the means of identifying how much change in cause is necessary to create a desired effect or change in effect.

Finally, simultaneous projections from both axes for various magnitudes of whatever is indicated upon the abscissa and the ordinate will intersect and establish points from which a characteristic curve may be drawn which will identify the relationship between the quantities indicated upon the two reference axes. Examples of these appear later on in this book.

Expression of Relationship of Characteristic Curves

For example, suppose you wanted to show by a curve how far an arrow would be shot by an archer for various degrees of pull on the string of his bow. Even if you have never used a bow and arrow, you

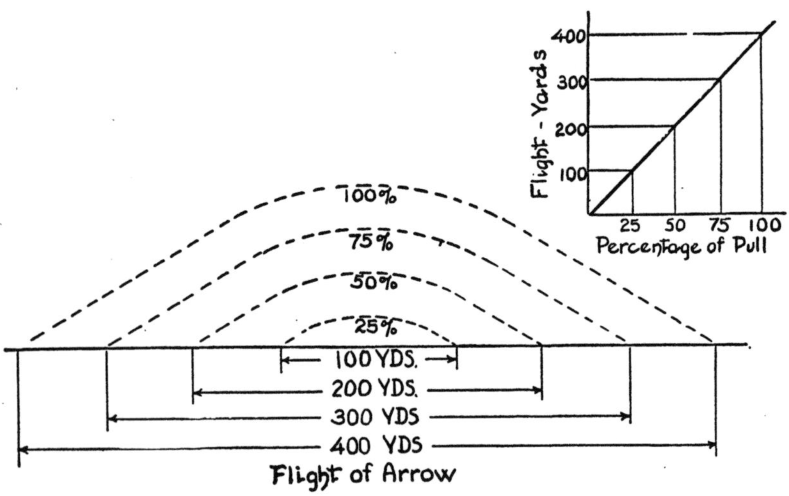

can easily realize that the further back from the bow the archer stretches the string on which the arrow is notched, the further will the arrow travel when it is released. For the sake if illustration, let us say that if he draws the string to ¼ of its full capacity, the arrow goes 100 yards when it is released. When he draws it half way, it travels 200 yards; 300 yards when he pulls it ¾ of the whole way, and 400 yards when he draws the bow to its full capacity. In other words, the greater the pull the archer exerts on the string of the bow, the further the arrow will travel.

The relationship between the pull of the archer and the length of the arrow's flight can be shown in the form of a curve, assuming the angle of the arrow with the ground is kept constant. One of the variables can be the percentage pull on the bow-string, with 100% being the full draw, and the other variable can be the distance in yards traveled by the arrow. Here then is a characteristic of cause and effect, where the pull is the cause and the distance of travel the effect.

The parallel between a vacuum-tube characteristic and a bow-string pull versus arrow flight chart is not very close, yet surprisingly enough, the two charts bear a close resemblance. Perhaps the exact shapes are not the same, but the ability to predict performance is existent in both —for that matter, it is existent in all characteristic curves. In fact, such prediction is a function of the majority of tube characteristics, for characteristic curves not only show what conditions exist in any one relationship of two variables, but they are also used to show what happens when the variable representative of the cause is changed by a known amount.

Characteristics Have Two Variables

As a general rule, such characteristic curves involve two variable factors. That which is the primary factor and which we have seen fit to consider as being the *cause,* is usually mentioned first. The *effect* is mentioned second in the name which identifies the characteristic. Thus, the plate voltage determines the amount of plate current, and consequently the characteristic curve which would picture this condition would be known as the *plate voltage—plate current characteristic curve.* If the curve shows the cathode temperature for varying heater current, the curve would be identified as heater current—cathode temperature curve, for it is the changes in heater current which vary the cathode temperature. As you may realize, these characteristic curves operate in both directions. If we consider the curve from the viewpoint of the cause and its resultant effect, we note the effect resulting from a change in the cause. But it is also possible to view this curve in another manner. We may desire to know what state must be created in the primary variable in order to result in a definite condition in the second variable. Thus we may examine a plate voltage—plate current curve in terms of the plate voltage that is required for a certain plate current, or the bow-string pull—arrow flight curve in terms of the percentage pull required for a certain distance of arrow travel, rather than the effect upon the plate current by a certain change in plate voltage, or the variation in distance of throw for different amounts of bow-string pull.

All of this can be described in another manner. The curves operate with equal facility in two ways, one manner being the resultant change in effect as the cause is varied, and the other being the required change in cause to produce a certain change in effect.

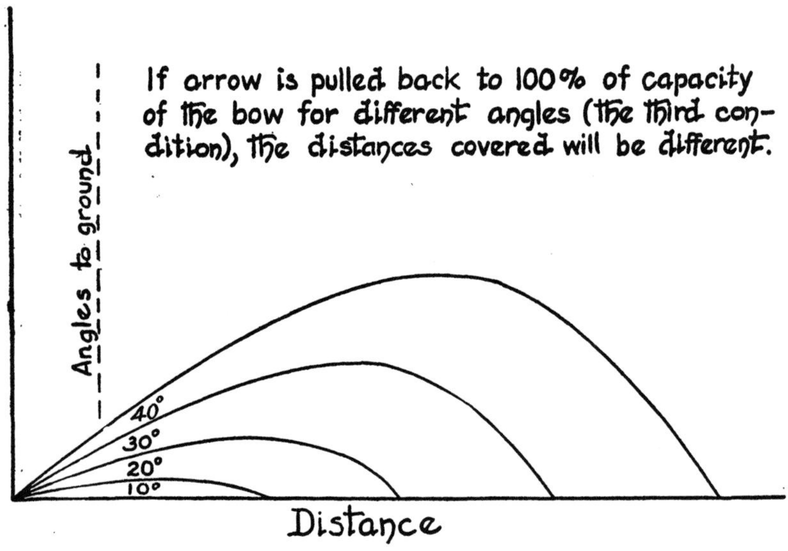

Although such characteristic curves illustrate the relationship between two variables, one being the result of the other, a third quantity is usually associated with the curves. This is a fixed quantity and is required in order to tell a more complete story. For example, getting back to our bow and arrow illustration, it is not sufficient to show bow-string pull against distance of arrow travel. It is important to identify the basic condition which exists and makes possible such a characteristic curve. In this case the fixed quantity is the initial angle with the ground at which the arrow is shot from the bow. Thus, if this angle is 20 degrees in one case and 40 degrees in another case, the distance the arrow will travel will be different.

This same applies to the plate voltage—plate current curve. It is necessary to state the temperature of the cathode, or the current through the filament or heater wires, or the voltage applied to the filament or heater wires. If the temperature, or current flow, or voltage associated with the emitter is not given, the plate voltage—plate current characteristic cannot be identified as being that for any one condition, for while the curve shown may be representative of behaviour, the exact values of plate current in milliamperes, microamperes, or amperes mean nothing. The exact conditions productive of the state of behaviour are not identified!

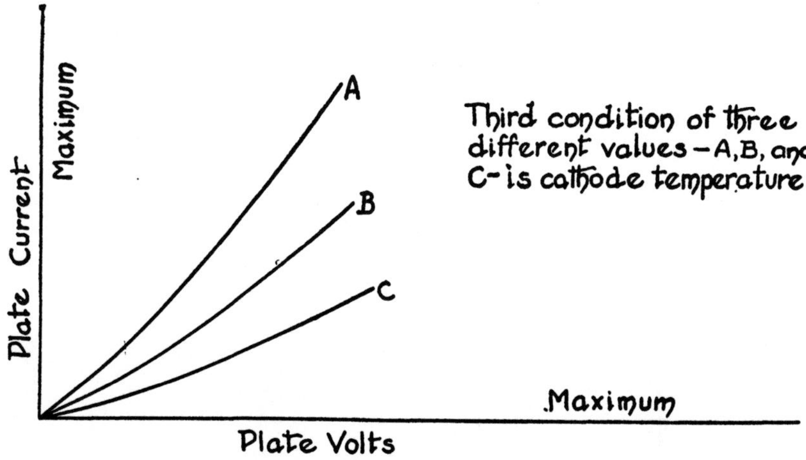

It is because of the above that you will usually see at least three quantities given in connection with any one vacuum-tube characteristic curve: the two variables illustrated in the curve and a third, each indicating the exact condition the other factor contributes to the behaviour illustrated in the characteristic curve.

Linear and Non-Linear Characteristics

When the relationship between the quantity being varied and the resultant changes in the second quantity appears as a straight line over the full range of variation, it is said that the characteristic being represented is *linear*. This can be described in a different way. If the resultant change in the second quantity is *directly proportional* to the variations in the first quantity, then the relationship between the two shown as a characteristic curve, i.e. a curve of behaviour, is linear. For example, if a d-c voltage is applied across a fixed resistor, and the current in the circuit is measured while the voltage is being varied, the relationship between the voltage and the current will be linear and, if this is shown as a characteristic, it will appear as a straight line over the full range of variation of the voltage and the resultant changes in the current. In this case, since the resistance remains fixed in value at all time, the changes in current flow are directly proportional to the changes in applied voltage.

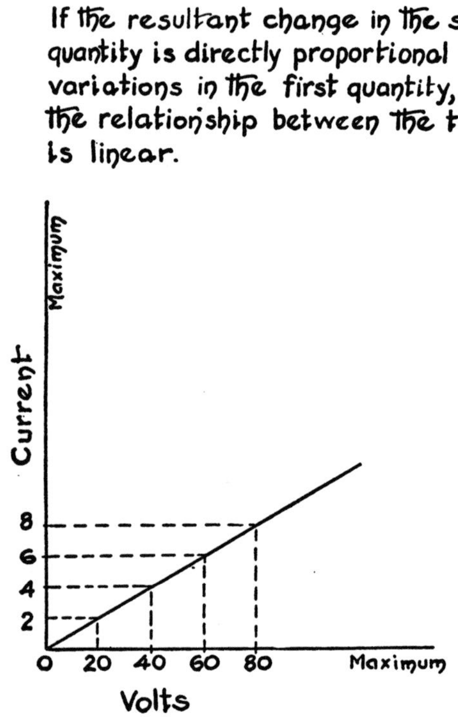

To state this quantitatively, if in a circuit the application of 20 volts causes the flow of 2 milliamperes, 40 volts causes the flow of 4 milliamperes, 60 volts causes the flow of 6 milliamperes, you can readily see that the change in current is proportional to the change in voltage —hence, the relationship between the two quantities is *linear*. Such linear relationships are not necessarily shown by a single line; instead, three, four, or more such linear characteristics may be contained upon the same *characteristic chart*.

Not many characteristics of vacuum tubes are linear throughout their length. They may be linear over a certain portion and this portion of the complete curve may be used. Those portions of the curve which are not straight, are known as *non-linear,* so that *tube characteristics may be linear over one portion and non-linear over another.*

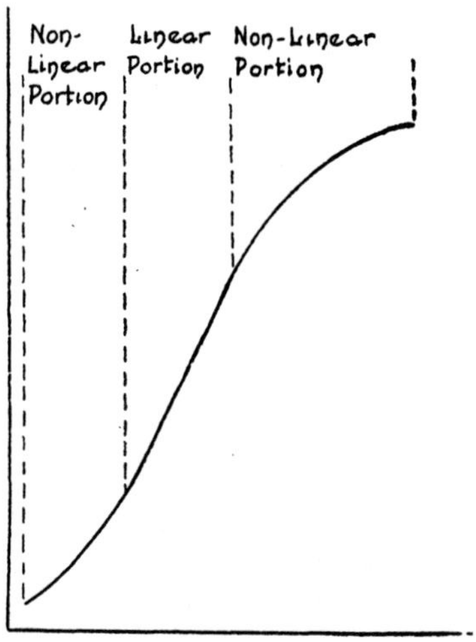

On the other hand, the entire characteristic may be non-linear, depending upon the conditions of operation. A quantitative example of a non-linear curve is given herein and it can be compared with the linear example also given. In contrast to the linear example, the conditions in the non-linear case are such that the change in effect is not directly proportional to the change in the cause. Stated differently, a uniform change in voltage does not produce a uniform change in current.

Static and Dynamic Characteristics

Linear as well as non-linear characteristic curves of vacuum tubes can be of two types. These are known as *static* and *dynamic* characteristics and differ in their shape as well as the actual numerical values which are shown associated with the variation in the quantities being represented. In other words, both static and dynamic characteristic curves exist for each vacuum tube. Thus, there is a static as well as dynamic characteristic curve illustrating the relationship between plate voltage and plate current, as well as similar curves for the other characteristics which are associated with vacuum tubes and which will make their appearance as we progress through this text.

Without recourse to specific illustrations, which would be premature if shown at this time, a simple explanation of the difference between the static and dynamic characteristics is that the static characteristic expresses relationships under conditions which are not usually typical of actual operation. The dynamic characteristic, on the other hand, expresses the relationships that exist when the tube is used in a manner which is equivalent to the actual operating conditions with respect to the associated electrical systems external to the tube.

This explanation may not be as clear as we should like it to be, but whatever confusion may exist in your mind will be cleared up later in this book when we discuss the different types of tubes. Unfortunately, we cannot, at this time, mention those conditions which are productive of the dynamic state in a vacuum tube without involving terms and

In this non-linear characteristic the change in effect is not proportional to the change in cause. Thus changing the plate voltage in equal steps of 20 volts does not result in equal changes in plate current.

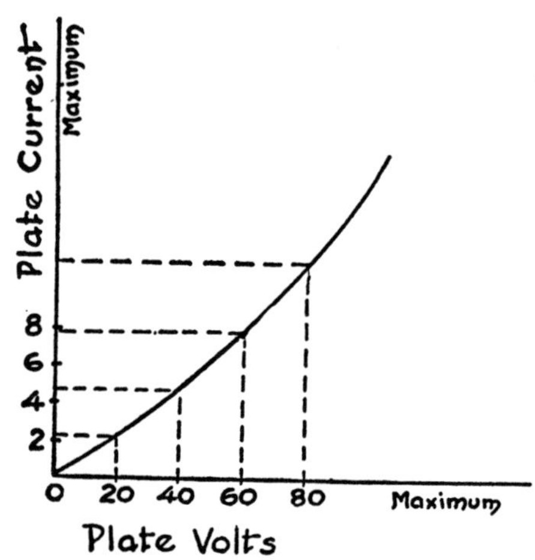

conditions which have not as yet been discussed. However, we felt that this was a good place to advise you of the fact that two general kinds of characteristic curves exist for relationships in vacuum tubes.

Number of Characteristics for Each Tube

The number of different types of characteristic curves which are established for any one tube depends upon the information desired, the more detailed and elaborate the required information, the greater the number of types of characteristic curves. The manufacturers of vacuum tubes in general have standardized upon a certain number, usually either one or two different relationships, and it is these which they furnish in the handbooks and reference guides. Since we have not discussed specific tube types as yet, it would be premature to speak about these specific kinds, but you shall become acquainted with them in due time.

Families of Curves

In many instances characteristic curves are shown as *families,* that is, a number of different curves which show the relationship between the same two quantities but under different conditions, are indicated upon the same chart. For example, again referring to our archer, we showed four curves illustrating the distance the arrow would fly when shot at a number of different angles with the ground. In like manner we showed different plate voltage—plate current curves for different values of cathode temperature. Families of characteristic curves are quite commonplace in tube literature and you will see many of them later in this book.

The most common and perhaps one of the most important vacuum-tube characteristics is that which illustrates the magnitude of the electron stream moving across the inter-electrode space within a vacuum tube for various values of positive voltage applied to the plate of the tube. This, as has already been mentioned, is the plate voltage—plate current characteristic.

That such should be the case is not difficult to understand if you bear the following in mind. It is in terms of the plate current that most applications of the vacuum tube are judged, for it is by means of the flow of this plate current through the various electrical devices associated with the vacuum tube and connected externally to it, that the various functions of the vacuum tube are put to use.

Whether the vacuum tube is a converter of electrical power of alternating character into electrical power of direct character, or is used inversely as an inverter, or if the device is one wherein electrical power of alternating character is utilized to control the release of energy from another electrical source and thus create a state tantamount to the process of amplification—in all instances, this is expressed in terms of the plate current which flows through the tube and through the devices externally connected to the tube. Hence you can expect to find very frequent reference to the plate current in vacuum-tube characteristic curves.

Resistance Within a Vacuum Tube

In a sense a vacuum tube is an unusual device for we find therein a condition which is difficult to duplicate elsewhere. This is the flow of electric current through an empty space, the inter-electrode space. A common statement is that electric current cannot flow unless a continuous path is provided for the moving electrons. This is true external to the vacuum tube, but inside of the tube we find free movement of electrons through a vacuum.

But when we say free movement of electrons we neglect one thing, namely, that a force is necessary to get these electrons to move. This is the force exerted by the lines of force of the electrostatic field due to the charged plate. But why this force? There must be something existent inside the vacuum tube and associated with the electrons which makes it necessary that a force be used to make the electrons move in the manner desired. There must exist within the vacuum tube some retarding influence which acts upon the electrons and which must be overcome in order to get the desired flow of plate current.

That which the lines of force of the electrostatic field within the vacuum tube must overcome to secure the definitely directed movement of a certain number of electrons represents the resistance to the directed movement or motion of the electrons within the vacuum tube. Thus, the factor of resistance is associated with the flow of plate current within the vacuum tube. In other words, the statement that electrons are capable of moving through a vacuum does not eliminate the presence of resistance. In fact, the resistance associated with the movement of the electrons within the vacuum tube plays a dominant role in the operation and application of the vacuum tube, as you will see later.

We have already associated the plate current flow with plate voltage and if we now associate the resistance with the place current, since the electrons moving through the space within the tube is the plate current, we find that in the vacuum tube, as in the external electrical system, we experience the three quantities of voltage, current, and resistance.

In this relationship between voltage, current, and resistance within the vacuum tube we find a very interesting condition, namely, a non-linear relationship between plate voltage and plate current. We shall not discuss this now, because we are not yet ready to do so, but we can state that the internal resistance of the vacuum tube does not behave like an external fixed resistance which remains constant in value for various values of applied voltage.

It is of interest to realize that the resistance to the flow of electrons within the vacuum tube, that is, the resistance to the flow of plate current, is to be found in all vacuum tubes regardless of the number of elements within the tube, and this factor is important irrespective of the function or application of the tube.

Power in Vacuum Tubes

Perhaps this is an ambiguous title for this portion of the text since there is more than one way of interpreting it, but proper identification in this text will narrow down its application. Since we have spoken about current, voltage, and resistance as applied to vacuum tubes, although strictly in a general way,—we might also mention power. Again you will find that this is closely affiliated with the plate current.

When electrons move across the inter-electrode space of a vacuum tube under the influence of the various electrostatic fields, they possess the *energy of motion,* otherwise known as *kinetic energy.* When they reach and strike the plate, they give up this energy and heat the plate. The amount of heat generated can be so great as to actually heat the plate to a white-hot temperature and eventually destroy it. Normally, the amount of heat generated in the plate under the impact of the arriving electrons does not result in any visible change in the color of the plate, except under certain conditions when the temperature of the plate may be raised so that it glows cherry red.

The heat must be dissipated, which means that the heat must be radiated from the plate as fast as it is generated. Now, since the heat is caused by the electron bombardment of the plate, it stands to reason that this heat depends on the plate current, that is, the total number of electrons striking the plate per second and the voltage applied to the plate, which in turn governs not only the number of electrons which strike the plate but also the speed with which these electrons move across the inter-electrode space, and hence the greater the kinetic energy possessed by the electrons which is converted into heat when they strike the plate, and then go out into the external plate circuit.

These electrons striking the plate do work in heating the plate. If we correlate the amount of work done with time, we arrive at the value of power consumed per unit time. For a static operating condition, the power dissipated at the plate of a vacuum tube is equal to the value of plate current multiplied by the plate voltage. Equilibrium is reached at the plate when the power radiated from the plate as heat is equal to the kinetic energy of the impinging electrons. This, too, is a characteristic of the vacuum tube, although it is of less importance in some applications than in others and hence is not always shown.

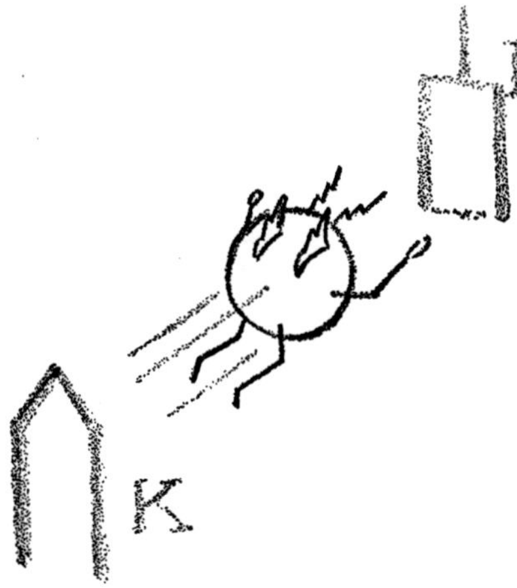

Chapter 6

THE DIODE

THE MATERIAL GIVEN thus far in this book has prepared us for what shall follow in this and subsequent chapters. We have, up to now, given certain facts relating to the operation of the vacuum tube: the manner in which and the reasons why, the electrons, so essential to the vacuum tube, are liberated; the basic reasons for the movement of electrons between charged surfaces, that is, the underlying basis for plate current within the vacuum tube. True, we have not as yet, mentioned *types* of vacuum tubes, but the specific type has very little to do with the principles which govern the motion of the electrons.

We are now ready to investigate what happens in the different types of vacuum tubes and to investigate the variation in the magnitude of the inter-electrode electron stream as the result of changes in the strength of the electrostatic fields which exist in the vacuum tube. In practical language, these electrostatic fields are referred to in terms of voltage,—*operating voltages* and *control voltages,*—and the changes in the magnitude of the electron stream within the vacuum tube are referred to as *changes in plate current.* Almost all of the essential conditions normally associated with the operation of the tube are considered in connection with the plate current. That which we call the operating characteristics, the behaviour of the tube, is usually interpreted in terms of the plate current.

This fact will become indelibly impressed upon you as you progress through this section. You will see how even the selection of the electrical components associated with the vacuum tube, but located outside it, are chosen because of the facts which are learned from a study of the behaviour of the plate current within the vacuum tube when the tube is operated in a certain manner. In fact, observation of the plate current by means of special electrical devices, gives an excellent insight into the actual performance of the vacuum tube.

The Diode

Of the many types of tubes which are in use, the simplest is the *diode*. This is a two-element tube, invented before 1900 by J. A. Fleming, and which consists of an *emitter of electrons* and a *collector of electrons*. The term diode, by the way, refers to *the number of elements* within the tube envelope rather than to the specific application of the tube. Many different names are applied to the diode which indicate the specific function of the tube in any particular electrical circuit. For example, the diode may be used in one circuit to perform the function of developing an automatic-volume-control voltage; hence it would be called the *avc tube* without necessarily stating that it is a diode. That it is a diode would become evident upon an examination of the wiring diagram of the system which would show the symbol for such a tube. There are many more such occasions wherein the diode is known as a *rectifier, demodulator, detector,* etc. So, when discussing the basic diode, we have no particular application in mind. These we will consider later.

The two elements which comprise the diode have more than one name. The electron collector is known as the *plate* or the *anode,* while the electron emitter, if of the indirectly heated type, is called the *cathode;* while if of the directly heated type, is generally spoken of as the *filament.* However, the electron emitter is called the *cathode* without any special consideration being given to whether the emitter is indirectly or directly heated. Today this usage seems quite natural in that by far the greater number of tubes in use are of the indirectly heated type. In fact, it is our intention, as we have already mentioned, to consider the electron emitter in the vacuum tube to be of the indirectly heated type, and we shall therefore, unless statement is made to the contrary, always have that in mind.

Cathode and Filament Structure

The types of cathode, heater, and filament structures that are used in the diodes we shall discuss are quite generally employed in most of the "small" tubes, so that it might be well to describe these constructional features in this chapter and thereby avoid duplication later. However, before discussing them, it might be well to mention that some special types of diodes do use cathode systems somewhat different in physical design from those to be illustrated.

Directly and Indirectly Heated Tubes

In the directly heated tubes, the filaments are of the general construction shown in Fig. 6-1. The filaments themselves are built in the

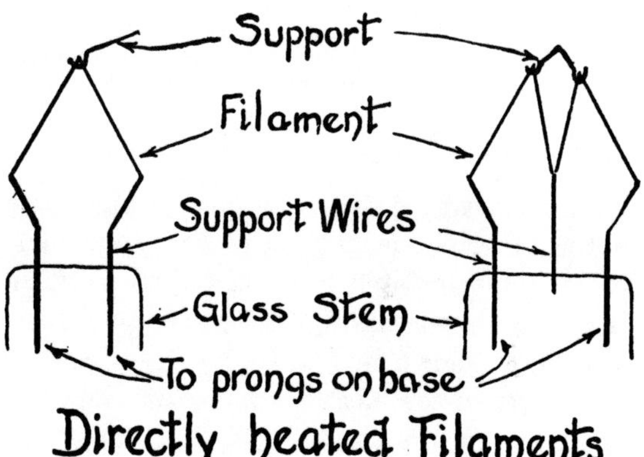

FIG. 6-1. Directly heated filaments are of two main types. That one shown at the left is an inverted V and the one at the right is an inverted W. The supports at the tops of the filament wires are erected from the tube base.

form of an inverted V or W, being mounted and held in place by suitable metal supports which rest in the glass stem of the tube. The filament voltage is applied across the prong terminals of the filament in the tube base and the current flowing in the filament heats it to the electron-emitting temperature.

In the indirectly-heated tubes, the cathode-heater arrangement may be one or the other as shown in Fig. 6-2. The cathode is a sleeve which is insulated from the heater wires but surrounds them. Upon this sleeve are baked a number of layers of an oxide coating that freely emits electrons. The heat from the heater raises the temperature of the cathode sleeve to the emitting value. The essential advantage of this type of emitter over the directly-heated type is that it permits heater operation on alternating-current power circuits. The cathode material cannot follow the rapid variations of the heater current, which therefore has very little effect upon the instantaneous temperature of the cathode. Hence a steady rate of emission can be maintained by the cathode with a uniform distribution of emission along its entire surface, despite the fact that the current through the heater varies periodically.

Furthermore, if the cathode is surrounded by a charged surface, there is a uniform distribution of the lines of force between the charged

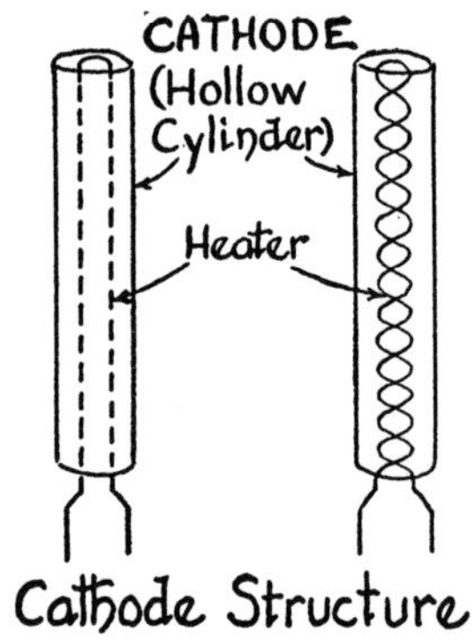

FIG. 6-2 (left). In an indirectly heated cathode, the heater wire is usually either U-shaped, as shown on the left, or it may be twisted throughout its length. The heater wire is enclosed within the electron emitting cylinder.

FIG. 6-3 (right). The symbol of the heater and cathode of a vacuum tube that is used in schematics is an inverted V under an inverted L.

body and the cathode along its entire length, because all of the cathode is at the same potential with respect to the surrounding charged surface. In some instances the heater wire within the cathode sleeve is an inverted U, whereas in other cases the heater is twisted throughout its length, as shown in Fig. 6-2.

When a cathode heater system is shown in an electrical wiring diagram, it appears as in Fig. 6-3, an inverted L over an inverted V. Since in the normal electrical system, the heater circuit is of little importance with respect to the other components associated with the performance of the circuit, it is often omitted from the schematic symbol, it being assumed that a simple reference upon the diagram is sufficient to indicate the method of connecting the tube heaters to the heater-current supply system. That shall generally be our practice, for once you understand the function of the heater and its behaviour under different conditions, there is little need for further discussion.

As to the plate or anode, its schematic representation is a short straight line, which may be located parallel to the vertical portion of the cathode, as shown in Fig. 6-4(A), or it may be parallel to the horizontal part of the cathode, as shown in Fig. 6-4(B). In the case of the filament-type tubes the plate is shown as in Fig. 6-4(C) and (D), being vertical or located above the filament. This is merely a matter of individual preference on the part of the person making the drawing.

A circle generally surrounds these elements to show that they are in an evacuated envelope. Generally the symbol for the heater is omitted in schematic drawings as this circuit is unessential to the functioning

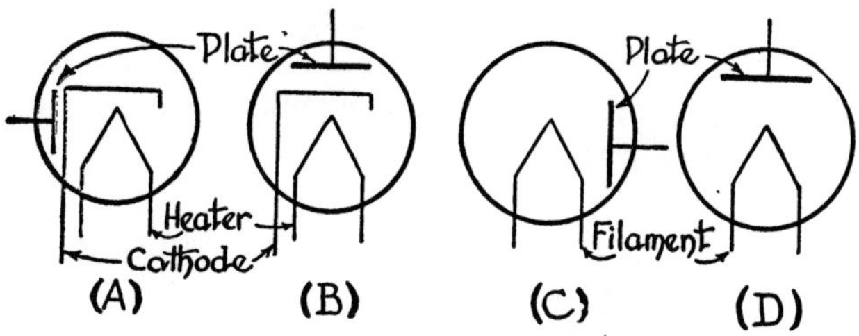

FIG. 6-4. The line indicating the plate may be drawn in any one of the four positions shown in the above schematic symbols of a diode.

of the other elements of the tube. Concerning the specific arrangement of emitter and plate within the actual tubes, the most general form of construction is to have the plate surround the entire cathode, as shown in Fig. 6-5.

FIG. 6-5. The cylindrical plate surrounds the cathode in the actual diode, as indicated on the right. These two electrodes are supported by wires that are embedded in the glass press at the bottom of the tube.

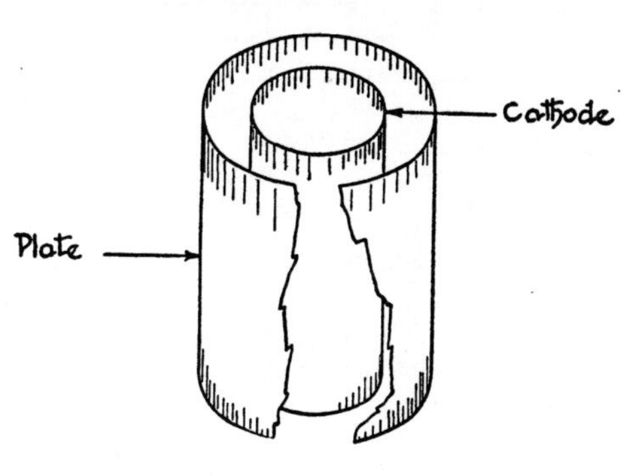

In connection with the actual construction of diodes, some employ two diodes located within a single envelope. In some, those of the indirectly heated variety, individual cathodes are available for each section. In other tubes, which use a filament as the emitter, two separate plates are available, and the two separate filaments are connected in series. Illustrations of these are given in Figs. 6-6(A) and (B) together with the schematic symbols. The tube which contains separate cathodes for each plate is known as a *duo-diode* (for example, the 6H6 G), whereas the tube which uses a common filament for the two plates is usually called a *full-wave rectifier* (for example, the 80).

The justification for such a distinction between the two tubes is due solely to construction and not utility, for the duo-diode can be arranged for use as a full-wave rectifier without any trouble. However, the fact that separate cathodes are available in the duo-diode gives it certain operating capabilities which are not possessed by the tube which has the two separate plates, but a single source of electrons. In the case of the duo-diode, there may be only a single heater winding serving the two cathodes.

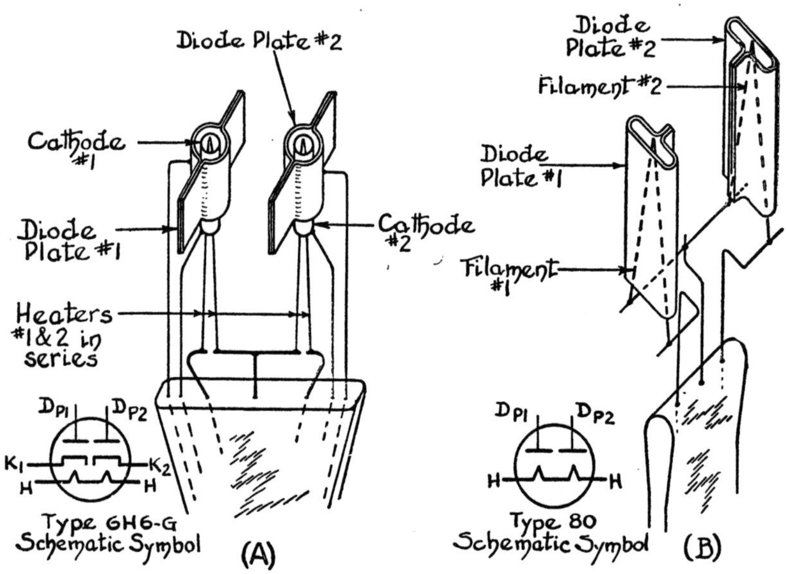

FIG. 6-6. The duo-diode tube shown in (A) has a cathode for each of its two plates and the 80 type rectifier shown in (B) has but a single continuous filament for both of its two plates.

The presence of more than one diode structure in an envelope does not in any way alter the considerations surrounding the basic two-element tube, and this applies equally well to the tube with the two plates but a single filament. Whatever is said about the single diode applies equally well to the other section. The same is true if the two plates of the two diodes are connected in parallel by some electrical connection and two cathodes are operated as one. The only difference resulting from such an arrangement is a change in the characteristics of the tube, but not in the basic principles of operation. Just what we mean by this in actual numerical values will be shown later in this chapter.

Function of the Diode

The main function of the two-element tube is as a *converter of power*. By this we mean that it is a device whereby electrical power in the form of alternating voltage and current can be converted into electrical power in the form of pulsating direct current, and subsequently to d-c voltage. This is the process of *rectification*. From the viewpoint of application, this description includes a very great number of different operations. In the field of communications alone, the various ways in which such power conversion is utilized are very numerous.

The principles governing the process of rectification in diodes have already been described, although it is true that when they were discussed we deliberately omitted mentioning the association between the phenomenon then considered and rectification.

You will recall mention in Chapter 4 that electrons moved from cathode to plate only when the plate was maintained positive with respect to the cathode. When the plate was made negative with respect to the cathode, current flow in the diode ceased.

Therein lies the process of rectification in the diode, for if we apply an alternating voltage between the plate and the cathode of a diode, current will flow through the tube, hence through any system connected to the tube—in a series of pulses corresponding to the periods when the plate is positive with respect to the cathode. This output current and corresponding voltage is unidirectional.

If we now view the vacuum tube as a conductor, it is evident that *the tube can conduct current in only one direction:* from the emitter to the plate. Therefore, we are justified in saying that the diode possesses the property of *unilateral conductivity*. If for the moment you were to visualize something which periodically makes the plate alternately positive and negative with respect to the emitter, and also remember the laws which govern the motion of electrons between charged bodies, you can then readily visualize the flow of plate current as a series of pulses—a pulse of current each time that the plate is made positive with respect to the cathode, and no current flow when the plate is made negative with respect to the cathode. In brief, this is the process of rectification of an alternating current into a pulsating direct current or the conversion of a-c power into d-c power. We shall discuss this in more detail later.

Electron Flow and Current Flow

Before we can continue any further discussion of current flow in the diode, one important point must be mentioned. It is important not only in connection with the diode, but with regard to all other tubes. We are referring to the modern conception of current flow in the vacuum tube based on the electronic viewpoint versus the old conception of electric current flow which is still prevalent in vacuum-tube discussions.

In view of all that we have said so far, particularly in Chapter 4, you naturally conclude that since the movement of electrons is from the emitter to the plate within the vacuum tube, and since the movement of electrons constitute electric current, the electric current flow within a vacuum tube (plate current) is from the cathode to the plate or from minus to plus. The movement of this current outside of the tube is from the plate to the cathode through whatever devices are connected between the plate and cathode.

That impression is entirely correct, yet we find it necessary to speak about another and contrary one. Frankly, we do not like to mention it, let alone discuss it, but it is about time that this old-fashioned and erroneous idea was completely cast out of text books. However, since so many authors still use it in their illustrations, we find it necessary to recognize that it existed at one time.

Back in the dim dawn of electrical knowledge, nothing was known of electrons nor of our present conception of the atomic make-up of matter. It was, however, necessary to specify the direction in which electricity flowed and so, quite arbitrarily, it was decided to say that electric current flowed from the positive pole of the battery to the negative pole of the battery through the external circuit. Today, however, we know that *electrons flow from the negative pole of the battery, through the external circuit, and back to the positive pole of the battery.*

As you can see, our modern idea is exactly opposite to the old one, but the old conception was universally adopted and has been in use for a long time. The result is that in many illustrations, in fact, most illustrations of current flow in vacuum-tube circuits, the current is shown as flowing from the plate to the emitter within the tube, and from the plus terminal of a battery or voltage source to the minus terminal of the voltage source through the external circuit. In order to be in line with modern thought, a schematic diagram founded upon the modern concept of electron flow should be shown, wherein the electrons flow from the emitter to the plate and the current flows from the negative pole of the battery to the positive pole of the battery through the external circuit.

We, too, are guilty of the practice of illustrating the old convention, but this is where we stop. We show but one schematic of the old convention and everything else is in line with the new. In other words, all of our drawings indicating movement of current in vacuum-tube circuits will conform with the modern concept of electron flow. Consequently, the direction of electric current flow in vacuum-tube circuits as shown in this book, may differ from illustrations shown in other vacuum-tube texts although the relative polarities will be the same in all illustrations. For you see that changing the direction of current flow from the old to the new concept does not change the polarity of points along the current-carrying circuits.

The reason why the relative polarities do not change when changing from the old to the new convention is because that point in the old convention which was declared as being positive, since the current left from that point, is now declared as being positive in the new convention because the current arrives there. Thus, while the direction of current flow is changed, *the polarities remain as before.*

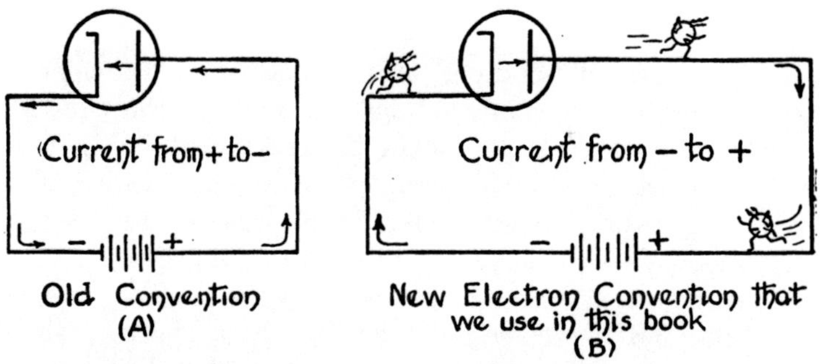

FIG. 6-7. The old convention of current flowing from + to — is illustrated in (A) and the modern convention of current flowing from — to + is shown in (B).

An example of the old and new conceptions of current flow are shown in Figs. 6-7(A) and 6-7(B). From now on we shall forget all about the old convention.

Plate Current in The Diode

Once again we shall discuss the flow of plate current in the diode tube, but this time we shall interest ourselves in the factors which determine the magnitude of this current, rather than in what makes it flow through the tube and through the external circuit. The basic reasons for the flow of current between the elements of the tube have already been discussed in Chapter 4, so that we need merely illustrate the similarity between the diode and the two-element structures previously illustrated. This is done in Figs. 6-8(A), (B), and (C). Since we shall speak about the flow of current through the circuit, a meter M is included in each illustration.

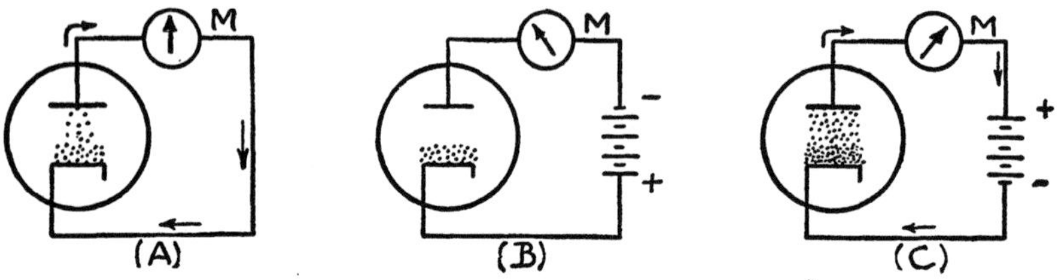

FIG. 6-8. High-speed electrons arriving at the plate form a small plate current, even though no potential is applied to the plate; this is shown in (A). No plate current results when the potential on the plate is negative, as shown in (B), and a current flows that is greater than that of (A) when a positive potential is applied to the plate; this is shown in (C).

In each of these illustrations, the cathode is the equivalent of the emitter discussed in Chapter 4, and the plate is the equivalent of the positively charged surface. The battery shown in Fig. 6-8(C) is the voltage source used in Chapter 4 which made the charged surface positive with respect to the plate. In the case of Fig. 6-8(B), the battery makes the plate negative with respect to the cathode.

The dots shown between the cathode and the plate in Figs. 6-8(A), 6-8(B), and 6-8(C), correspond to the space charge discussed in Chapter 4. In Fig. 6-8(C), the plate voltage is of such magnitude that a flow of plate current occurs. In Fig. 6-8(B), the plate having been made negative with respect to the cathode, the plate current is zero, and the space charge can be seen concentrated in the neighborhood of the cathode just as described in Chapter 4.

We now come to Fig. 6-8(A), which has not been mentioned. Basically, the conditions existing in this circuit correspond to what has been described in Chapter 4 but, as you can see, the battery has been omitted and the plate is joined to the cathode. What happens in this circuit? In the first place, since the plate and the cathode are joined and no voltage source is connected between the plate and cathode, presumably no difference of potential exists between these two elements. Since the two are joined, whatever the voltage may be at the cathode—incidentally, this is assumed to be zero—the same potential is supposed to exist at the plate; hence the plate is not exerting any attracting force upon the electrons in the space charge.

But the meter indicates the flow of a very small amount of current! This is not as mysterious as it may seem at first thought. In fact the answer has already been given, but since it may have been lost in the maze of other subject matter, repetition is justified. In fact, it is worthwhile to give this subject somewhat more thought inasmuch as it plays a very important role in the application of the diode tube in many systems.

Contact Potential

You may remember, while discussing the velocity with which the electron leaves the emitter, that we made the statement that these speeds varied; some electrons left the emitter at a very slow speed, whereas others left the emitter with a high velocity. In fact, we said that the speed of some electrons emitted during each unit of time was so great that they could penetrate right through the space charge and travel over to the plate without any attracting force existing at the plate, that is, without any positive voltage being applied to the plate, and hence with no electrostatic field set up by the plate.

It is these high-speed electrons which reach the plate and constitute a plate-current flow through the diode, and a corresponding movement of electrons through the external circuit. This current is indicated upon the meter.

This condition of plate-current flow through a diode without any electrostatic field being set up by the plate, is a very important one as far as practical operation is concerned. It means that a value of zero voltage upon the plate does not necessarily signify zero plate current. It is necessary to find a term which will express the *equivalent* voltage difference which can be imagined to exist between the plate and the cathode, and which is responsible for the flow of this amount of current. For, there is not supposed to be any voltage difference between the cathode and the plate since the cathode is a unipotential surface, and joining the plate and cathode in such a manner is supposed to place both at the same potential all along both surfaces.

Yet we know that current does flow in the circuit and, if we assume that the external circuit joining the cathode and the plate has resistance—which is a normal assumption—a voltage drop due to this flow of current can be expected to exist across this resistance. This voltage does exist and the higher the resistance of the external circuit which connects the plate to the cathode, the more easily it can be measured. As to just what you would call this voltage, the best that we can do is to refer to the name by which it is known, namely, *"contact potential."*

The use of this term to identify this voltage is not wholly correct, yet, for want of a better name, it is used in commercial tube literature, and we shall therefore use it on those grounds.

Determination of Diode Behaviour

All of this information about the flow of current in a diode, with a positive, a negative, and zero voltage at the plate, merely supplements the facts given in Chapter. 4. What we wish to determine is the behaviour of the diode when those conditions which cause the flow of plate current are varied. Only then can we establish facts relative to the manner in which the diode is used.

What are the conditions which determine the amount of plate-current flow in the diode? Since the magnitude of plate current is partly dependent upon the emission of electrons from the cathode or filament of the tube, the temperature of the emitter is a factor. In accordance with what was said in Chapter 4, the electrostatic field set up at the plate influences the effectiveness of the space charge, and hence affects the plate current. Consequently, we see that the value of the plate voltage has an effect upon the amount of plate current flowing through the tube.

We must therefore establish characteristic curves, or behaviour curves, showing the variation in plate current with emitter temperature and also characteristic curves showing the variation in plate current with plate voltage. These are the two basic characteristic curves of the diode, although, strange as it may seem, that which can be considered the more basic, namely, the emitter temperature—plate current characteristic curve, is infrequently used. In fact, tube manufacturers do not usually show such curves in their tube specifications, the reason being that tubes are designed to be used at a specific value of emitter temperature, as determined by the voltage applied across the filament or heater.

Emitter Temperature—Plate Current Characteristic

To establish the manner in which the plate current of the diode varies with the emitter temperature, it is necessary to set up a circuit using a diode wherein it is possible to create various emitter temperatures and to maintain the voltage applied to the plate at a constant value. The general shape of this curve can be predicted from what has already been said about the various conditions of emission.

First of all, we know that electron emission from metals, such as are used as emitters in vacuum tubes, requires a definite high temperature. In addition, we know that raising the temperature of the emitter will increase the amount of emission; however, because of the presence of the space charge within the tube, increasing the emission from the emitter does not necessarily mean that a greater number of electrons will travel over to the plate of the tube.

We have further learned that the flow of plate current (electrons) across the inter-electrode space, between the two surfaces of the diode, depends upon the relative magnitudes of the fields due to the space charge and the field set up by the plate when the plate is made positive with respect to the emitter. Under the circumstances, it would seem natural that a curve of plate current versus emitter temperature with a fixed plate voltage, would start at zero, rise, and then eventually flatten out at some value, beyond which there would be no further increase in plate current no matter how high the emitter temperature.

For example, with a fixed plate voltage of reasonable value, emission would start when an emitting temperature was reached. In this connection we are assuming that the temperature of the emitter is just high enough to cause sufficient emission, and a plate current which would be indicated upon a reasonably sensitive meter. At this point the space charge would be of negligible effect in offsetting the field due to the voltage upon the plate so that there is a progressive movement of the emitted electrons through the space charge to the plate.

As the temperature is raised, the emission would increase and, while the number of electrons located in the space between the emitter and the plate would also be increased, the field due to the plate voltage would still be capable of sufficiently nullifying the field due to the space charge, so that all of the electrons emitted would move over to the plate.

This condition would continue with further increases in emitter temperature and the plate current would also continue to increase. However, after a while when the emitter temperature has been raised to a sufficiently high level, the field due to the space charge would begin to offset the field which is due to the positive voltage upon the plate. Then the rise in plate current, for equal increases in emitter temperature, would no longer be as great as it was before, and the plate current curve would begin to level out.

The point where it would flatten would be that which is equivalent to the creation of a space charge which would have a field sufficiently

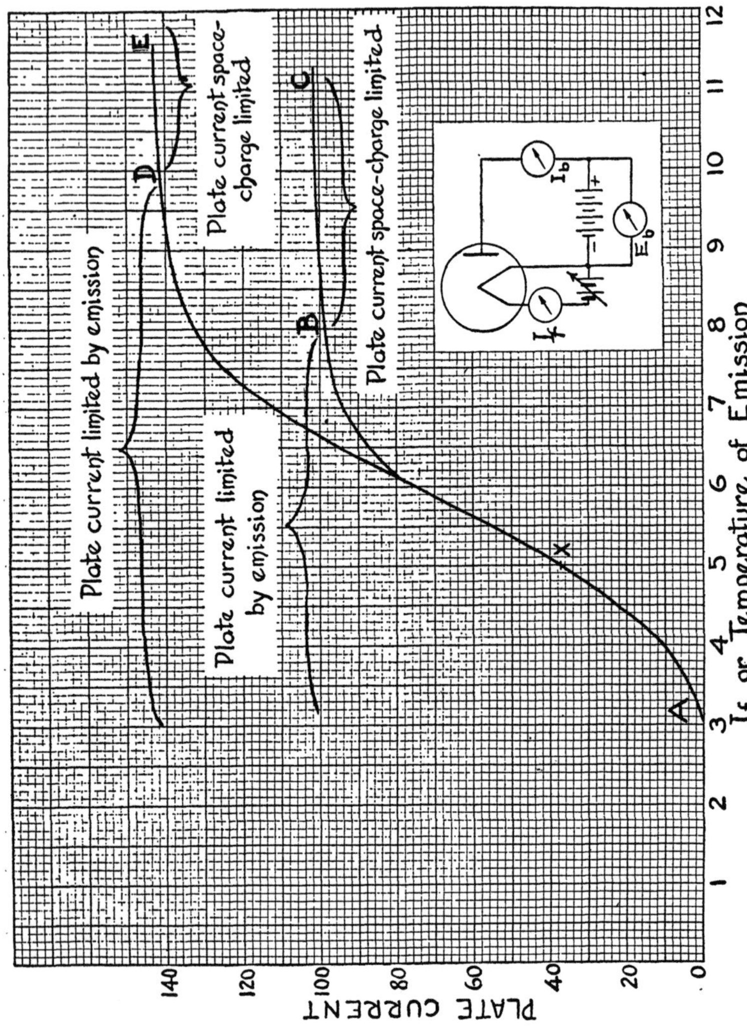

FIG. 6-9. Emitter temperature—plate current characteristic. These curves show that the plate current does not increase appreciably above a certain value even though the temperature of the emitter is increased.

great to offset at the emitter any attracting force which the field, due to the voltage applied to the plate, would have. From this point on, there would be a steady flow of plate current, the magnitude of which would be space-charge-limited and there would be no further increase in plate current for a further rise in emitter temperature.

Such a curve is *ABC* of Fig. 6-9, and which is made by using the accompanying electrical circuit. Although we have described this curve in terms of emitter temperature, it is more customary to plot such curves in terms of either filament or heater current indicated upon the meter "I_f". The horizontal axis of the graph can be designated in either filament current or in emitter temperature. The vertical axis is marked off in values of plate current. Inasmuch as nothing is gained by giving specific values of either filament or plate current, the values shown are arbitrary. The meter marked "E_b" is used to indicate the plate voltage, while the meter "I_b" is used to measure the plate current.

Referring again to this emitter temperature—plate current characteristic, the point *B* represents the saturation of the emitter for the existing plate voltage. It is evident from this characteristic curve that nothing is gained by increasing the temperature of the emitter or the filament (heater) current beyond that point along the horizontal axis which is equivalent to the point *B* on the characteristic, because there is no further increase in plate current. All emission greater than that productive of point *B* on the plate-current curve is excess. All the electrons emitted in excess of those which progressively advance to the plate are being repelled back into the emitter.

The means of utilizing higher emitter temperature than that corresponding to point *B* in the curve *ABC*, is self evident. Since the point *B* on curve *ABC* is due to the action of the space charge, the means of raising this point is by reducing the effectiveness of the space charge. To do this the field set up by the voltage upon the plate must be increased, which is done by the simple expedient of raising the voltage applied to the plate.

When this is done the rising portion of the plate-current curve is increased and a new value for temperature saturation or the start of the space-charge-limited plate current, is established. This is shown as curve *ADE*. Whereas with the lower value of plate voltage the amount of emission corresponding to point *B* at approximately 8 units on the temperature axis represented temperature saturation, with the higher value of plate voltage, more electrons were attracted to the plate and the space charge was not limiting the plate current. It was necessary to advance the emitter or temperature to 10 units before the space charge in the tube increased to the point where its field was capable of offsetting the effect of the field set up by the voltage applied to the plate, and a new and higher value of temperature saturation was attained.

Thus, we see from Fig. 6-9 that for every value of plate voltage there is a limit to the number of electrons which can be drawn over to the plate in unit time to form the plate current. There is consequently a limit to the value of filament or heater current which need be applied. Still another significant detail which appears in the curve, is that if the filament or heater current is not sufficiently great to create space-charge-limited current for any particular value of plate voltage, nothing is gained by increasing the plate voltage. For example, in Fig. 6-9 point X corresponds to 5 units of filament current. All the electrons emitted at this temperature are being attracted over to the plate when the lower value of plate voltage is being applied. Raising the plate voltage to the higher value does not produce any increase in plate current.

Plate Voltage—Plate Current Characteristic

This plate voltage—plate current characteristic curve is far more frequently employed than the one associated with the temperature of the emitter. Here again we find a very peculiar condition as it relates to the diode. Certain very significant theoretical facts can be gleaned from the characteristic curve, but unfortunately it does not have much practical significance because the special conditions created in making these curves are not usually experienced in practice.

Based upon the general discussion of the relationship between the space-charge field in a diode and the field due to the plate voltage, it would seem that a curve which would show the variation in plate current for variations in plate voltage would begin at zero for zero plate voltage. From the previous discussion associated with Fig. 6-8A, we know this not to be the case, for, due to the high-speed electrons which reach the plate without any attracting force being applied to the plate, a small amount of plate current flows with zero plate potential.

Granting that to be the case, it would seem that the plate-current curve would start rising from this small value of plate current for zero voltage, to higher values as the plate voltage was increased until a point is reached where all of the electrons being emitted would find their way to the plate. In other words, the value of plate voltage would then be sufficient to nullify completely the field due to the space charge, so that any further increases in plate voltage for any given value of emitter temperature would cause no further increase in the plate current. This condition is called *plate-voltage saturation.*

A plate voltage—plate current curve would appear like that shown as *ABC* in Fig. 6-10. The electrical system for producing it is shown in the accompanying schematic. E_b indicates the voltage being applied to the plate, and the meter I_b indicates the plate current. Since the emitter temperature is fixed, no meter to measure the filament or heater current is indicated.

Referring to the characteristic curve *ABC,* the point *B* corresponds to the point of plate-voltage saturation, which is reached for a voltage of approximately 300 volts. In contrast to the curve illustrating the relationship between emitter temperature and plate current, that portion of the plate voltage—plate current curve which is rising, represents the zone of space-charge-limiting action upon the plate current. Over all of this portion, as the plate voltage is being increased from zero to about 300 volts, the space charge is also controlling the current. At point *B* the field due to the plate voltage is sufficient to nullify the entire space-charge influence and the plate current becomes limited only by the available emission. From this point on, the plate takes all of the electrons emitted by the cathode.

As you can see from the curve, the point *B,* equivalent to the complete overwhelming of the space-charge control by a plate voltage of about 300 volts, is equal to the flow of about 67 milliamperes of plate current. Increasing the plate voltage beyond this point has been said to be productive of no further increase in plate current. Actually there is a slight increase, for there is always an increase in plate current as the plate voltage is raised. This increase is so small as the plate voltage is raised beyond the upper bend in the characteristic curve, as to justify the statement that beyond point *B,* which is the plate-voltage saturation point, there is no further increase in plate current for an increase in plate voltage.

To secure an increase in plate current for plate voltages in excess of 300 volts, it is necessary to procure a greater flow of electrons per unit time. This means that the temperature of the emitter must be raised. Assuming this to have been done, we can imagine a curve like that shown as *ADE* in Fig. 6-10, Point *D,* corresponding to a plate-current flow of about 85 milliamperes and a plate voltage of about 375 volts, is the point of plate-voltage saturation.

Increasing the emitter temperature has again restored the action of the space charge and it is once more exerting control over the plate

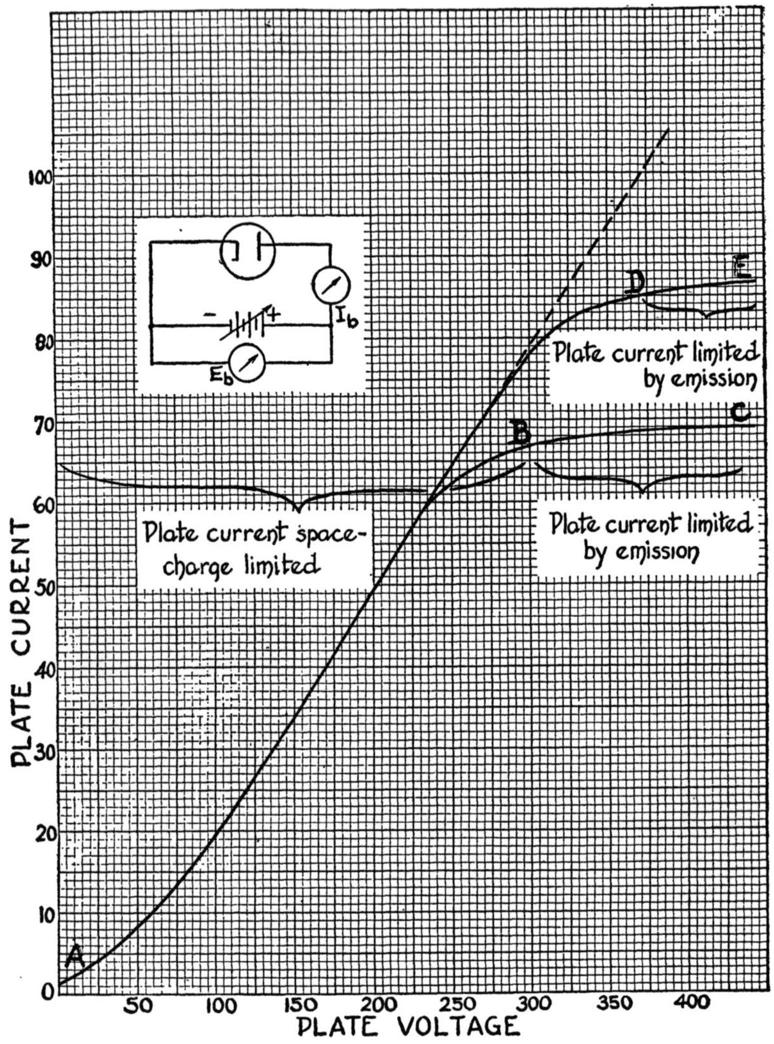

FIG. 6-10. Plate voltage—plate current characteristic. These curves show that the plate current does not increase appreciably above certain values no matter how much the plate voltage is raised, as only a limited number of electrons are emitted.

current against the field due to the plate voltage, until that voltage reached a value of approximately 375 volts. At that point the plate current represented the total number of electrons emitted per unit time by the emitter. Further raising of the plate voltage beyond 375 volts is of little aid in increasing the plate current, because all of the electrons liberated are already traveling to the plate.

From the theoretical viewpoint, the plate voltage—plate current characteristic teaches us that for every diode which is operated at a certain emitter temperature (filament or heater current), there is a certain value of plate voltage beyond which it is useless to increase, for at this value all of the liberated electrons are moving over to the plate.

From the practical angle, however, a different condition occurs. The tube used to develop the plate voltage—plate current characteristic of Fig. 6-10 had a tungsten filament. Were this type of emitter replaced by one with an oxide coating or a thoriated-tungsten type of emitter, it would have been very difficult to find values of plate voltage which would draw off all of the electrons being emitted, that is to reach a point of plate-voltage saturation, without first ruining the tube and thus nullifying all of the work. In other words, modern tubes, most of which have oxide-coated filaments or cathodes, do *not* display a plate-voltage saturation characteristic, but rather show continually rising plate-current curves as the plate voltage is increased, as indicated by the dotted line in Fig. 6-10, until the plate current is so great as to damage the tube.

Under the circumstances, it is needless to elaborate upon the fact that in practice the values of plate voltage applied to two-element tubes are far removed from anything which would even approach the condition of plate-voltage saturation. This, however, plus the condition previously mentioned, does not defeat the value of such a plate voltage—plate current characteristic curve. At least the theoretical data are given and you have become acquainted with the manner in which such curves are used to interpret conditions existing within the tube and the action of the various agencies functioning within the diode. From this point on we shall concern ourselves with the more practical aspects of tubes, and the values of current and voltage which are actually experienced in the daily use of diodes.

Resistance of the Diode

Since the application of a positive voltage to the plate of a diode tube results in the flow of plate current and this current is not infinite but is held within bounds and varies with the plate voltage, it stands to reason that the tube must have some value of resistance. This it does, and the diode, like all other vacuum tubes, possesses two kinds of opposition to the flow of current: *d-c plate resistance and a-c plate resistance.* Of these two classifications, the latter is by far the most important.

D-C Plate Resistance of the Diode

As the name indicates, the d-c plate resistance of the diode is that opposition to the flow of current which is offered by the tube when a d-c voltage is applied to the plate. It is that quantity which can be determined by the application of Ohm's law for resistance, employing the applied d-c plate voltage and the measured value of d-c plate current as the known quantities in the equation.

For example, Fig. 6-11 illustrates a typical plate voltage—plate current characteristic for a typical 6H6 duo-diode tube, using just one pair of elements. By one pair of elements is meant one cathode and its associated plate. The plate voltage applied to the tube is shown upon the horizontal axis of the graph and the resultant plate current is shown upon the vertical axis. While this is the same as is found in the plate voltage—plate current characteristic curve illustrated in Fig. 6-10, the general shape of the two characteristic curves is quite different. This comes about as the result of the special conditions mentioned in connection with Fig. 6-10. The curve shown in Fig. 6-11 is a closer approach to the actual curve with which you will work.

You will remember that we qualified the conditions indicated in the curve of Fig. 6-10 to the extent that the plate-voltage saturation points were not to be expected when using oxide-coated filament or cathode type tubes. The 6H6 is an oxide-coated tube and it is evident in Fig. 6-11 that plate-voltage saturation is a long way off. In fact, judging by the values of plate current which prevail for the comparatively low values of plate voltage, it is doubtful if the plate-voltage saturation point could ever be reached. More than likely the tube would be damaged long before that value could be reached by the application of sufficient plate voltage.

With a comparatively low value of plate voltage, say 28 volts, applied to the plate, the formidable value of about 66.4 milliamperes of current already flows in the tube. And judging by the slope of the plate-current curve, the steep portion is yet to come, that is, if we increased the plate voltage above 32 volts. There is little doubt that were this plate voltage increased to say 50 or 60 volts, the plate current would rise to such a value as to destroy the tube. This does not mean that diodes are not used at plate-voltage values greater than 32 volts. Much higher values of plate voltage are employed, but when

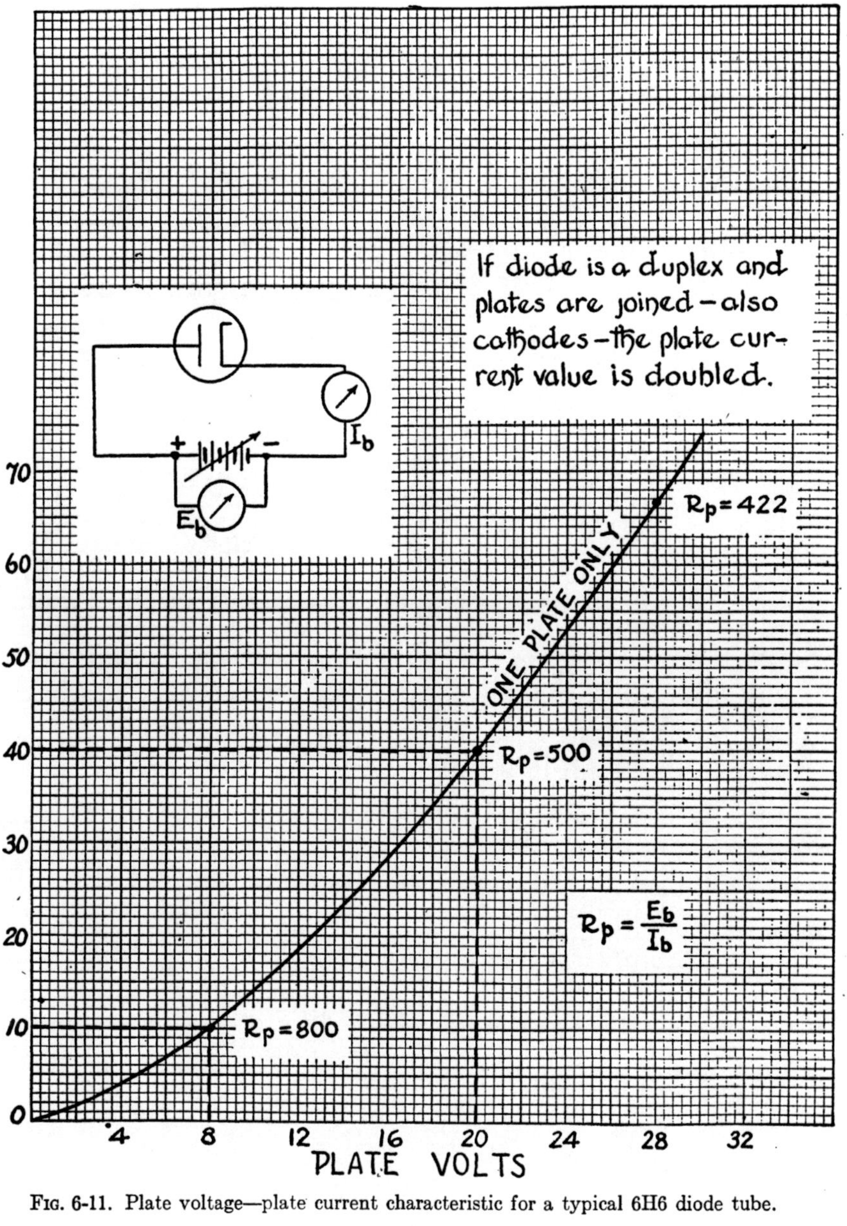

FIG. 6-11. Plate voltage—plate current characteristic for a typical 6H6 diode tube.

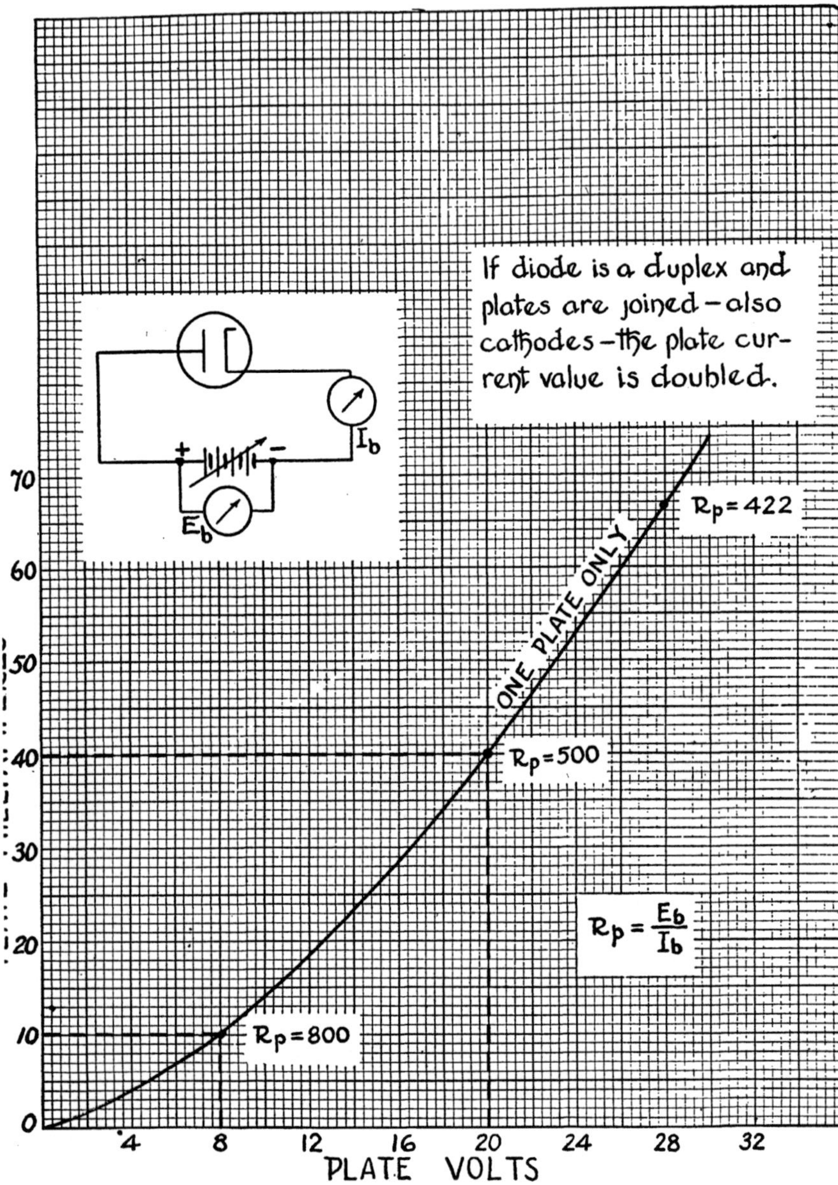

FIG. 6-11. Plate voltage—plate current characteristic for a typical 6H6 diode tube.

that is done other elements which tend to keep the plate current to a safe value are also used in the circuit. Such additions will be discussed later.

As we discussed earlier in this text, the plate-current curve of a diode, like that of other vacuum tubes, is of great importance. In this particular instance it serves to furnish information relative to the opposition offered by the tube to the flow of plate current; in other words, it furnishes information about the d-c resistance of the tube. For example, with 8 volts d-c E_b applied to the plate, the plate current I_b, as shown upon the vertical axis, is 10 milliamperes or 0.01 ampere. According to the Ohm's law, these values of plate voltage and plate current represent a d-c plate resistance of

$$R_p = \frac{E_b}{I_b} = \frac{8}{.01} = 800 \text{ ohms}$$

With 20 volts d-c applied to the plate of the tube, we note from the curve that the plate current is 40 milliamperes or 0.04 ampere. This corresponds to a d-c resistance equal to $20/.04 = 500$ ohms. With 28 volts d-c applied to the plate, the plate current according to the curve is 66.4 milliamperes or 0.064 ampere and the equivalent d-c resistance is $28/.0664 = 422$ ohms.

Examining these three values of d-c plate resistance for three values of applied d-c plate voltage, we note a peculiar condition. The opposition which *the diode offers to the flow of the plate current is not constant,* as we are ordinarily accustomed to experiencing in conventional d-c systems. Said differently, *the resistance offered by the diode to the flow of the plate current is not linear.* Judging by the relationship between the plate voltage, plate current, and the equivalent d-c resistance, it appears that the resistance of the diode *decreases* as the plate voltage is increased, and *increases* as the plate voltage is decreased. Were the resistance of the diode linear over the entire range of plate voltage, the plate-current curve would be a straight line instead of the curved line appearing in Fig. 6-11. Such a straight line would indicate that the d-c plate resistance remained constant over whatever range of plate voltage and plate current is embraced by the upper and lower limits of the straight line.

Of further interest in connection with the plate voltage—plate current characteristic given in Fig. 6-11, is the condition created when the two plates of this duo-diode tube are joined in parallel and the two cathodes are connected in parallel. The plate current for such parallel connection is equal to twice the amount that would occur if the same value of plate voltage were applied to a single pair of elements. Thus, if the plate current for the single diode is 10 milliamperes with 8 volts d-c applied to the plate, the parallel connection of the two plates and the two cathodes results in 20 milliamperes of current, and a d-c resistance of 400 ohms. In the same way, the values of plate current corresponding to any point along the plate current curve of Fig. 6-11 for any value of plate voltage, are doubled when both sections of the duo-diode are connected in parallel and the plate resistance is half of that existing when just one pair of the two diode sections is employed.

You may wonder why the plate current corresponding to zero plate voltage is zero in Fig. 6-11, whereas in Figs. 6-8(A) and 6-10 we mentioned the flow of a small amount of current. There is no conflict between these two conditions. Inasmuch as the value of plate current is so very low when the plate is zero, we felt that no misunderstanding would result by showing the plate current at zero, since it would in no way interfere with the interpretation of the facts shown in the graph.

A-C Plate Resistance of the Diode

When we first mentioned the opposition which the diode tube offered to the flow of plate current we said that two types of resistance existed. The d-c plate resistance has been discussed; now we shall look into the a-c plate resistance. Referring again to the basic function of the diode as a converter of a-c electrical power into d-c electrical power, it would appear reasonable if we said that in most instances the voltage which is applied to the diode tube plate is of alternating character rather than the d-c voltage which we discussed in connection with Fig. 6-11. Hence it is logical that any quantity which is associated with the manner in which the tube is used, is more important than some other which is only infrequently involved in the various technical considerations of the tube. The a-c plate resistance is, as a rule, the only term associated with the resistance of the diode when the term "resistance" is employed without qualification. The same is true also for other types of vacuum tubes; that is, when the term "resistance" is employed, it practically always refers to the a-c resistance.

Before we start discussing the a-c plate resistance of the diode tube, we might as well say a few words about a matter of notation which is used in connection with such a-c quantities. When investigating the action of alternating potentials or currents in vacuum tubes, the simplest means of simulating the application of an alternating potential is to vary a d-c voltage about a fixed value. The maximum variations in voltage on both sides of some average value, would represent the peak a-c voltages. The magnitude of these variations is, of course, a matter of individual preference in any experimental work; as a rule it is kept quite low.

When referring to such small changes in applied voltage it is customary to denote the change by the small letter "d," which signifies a "small change of." Thus if we are speaking about a small change in voltage E, we may write "dE" to symbolize the entire expression "a small change in the voltage." Similarly, if I represents current, the symbol "dI" would signify "a small change in current." We can of course extend the idea to whatever quantities are desired. Thus if we wish to represent the resistance represented by "a small change in voltage" divided by "a small change in current," the expression would appear as

$$r = \frac{dE}{dI}$$

Because of what is done in ordinary algebraic calculations where a similar letter appears in both the numerator and the denominator, that is, the cancellation of the same letter from both the numerator and the denominator, as for example

$$\frac{3a}{4a} \text{ can be rewritten as } \frac{3}{4}\cdot$$

we have to mention that such cannot be done with the letter "d" for it is not an algebraic unknown in the usual sense, but rather as we have pointed out, signifies an operation which is expressed by the phrase, "a small change of."

In accordance with the above, we show a typical plate voltage—plate current characteristic curve in Fig. 6-12 for one section of the 6H6 diode, which, as you can see, is exactly the same curve as was used in Fig. 6-11. To establish the a-c resistance at any point along this curve, we vary the plate voltage on both sides of some definite value. Thus we know from Fig. 6-11 that the d-c plate resistance of this diode is 500 ohms when the plate voltage is held constant at 20 volts d-c. What is the a-c resistance when the mean value of this assumed a-c voltage is 20 volts and the swing in voltage is 1.6 volts on each side of the mean of 20 volts? In other words what is the a-c plate resistance when de_p is equal to 3.2 volts?

To get the answer we check the plate current for a d-c plate voltage of 21.6 volts and find that it is 45 milliamperes or .045 ampere. Then the plate voltage is reduced to 18.4 volts and we find that the plate

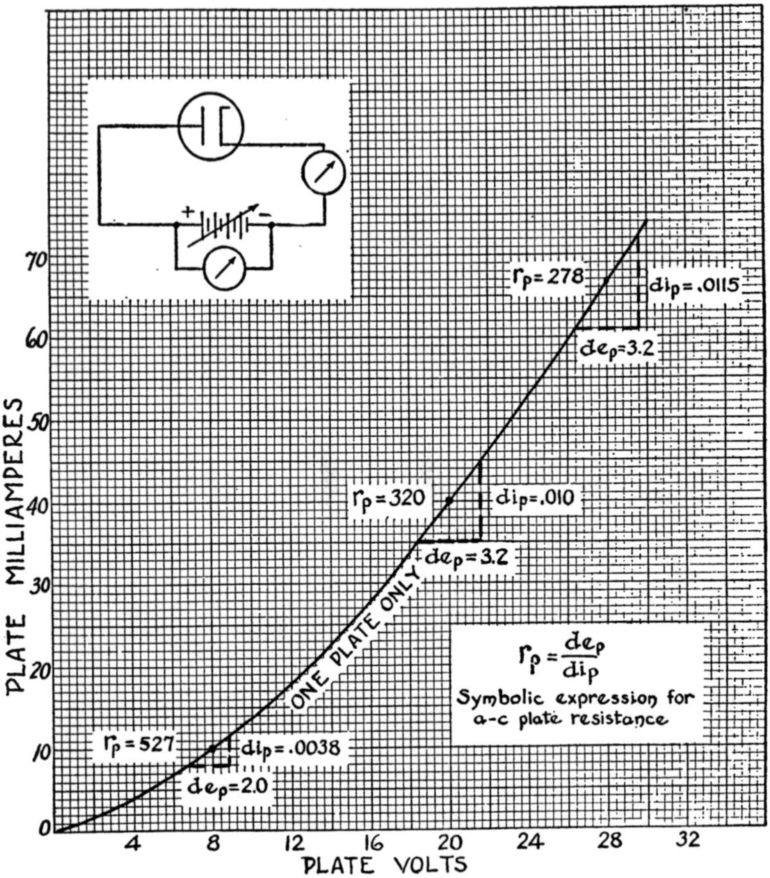

FIG. 6-12. The a-c plate resistance of a diode can be determined from the plate voltage—plate current characteristic as explained in the accompanying text.

current is 35 milliamperes or .035 ampere. We now have the two limits of voltage and the two limits of current. Then the symbol "de_p" represents the *change in plate voltage* or the difference between 21.6 and 18.4 or 3.2 volts. And the symbol "di_p" represents the corresponding *change in plate current* or the difference between .045 and .035 or .010 ampere. The a-c plate resistance r_p then can be expressed as

$$r_p = \frac{de_p}{di_p} = \frac{21.6 - 18.4}{0.045 - 0.035} = \frac{3.2}{0.010} = 320 \text{ ohms}$$

This value of 320 ohms is the a-c plate resistance for a value of plate voltage which fluctuates between 18.4 and 21.6 volts, whereas the same point is equal to a d-c plate resistance of 500 ohms when the plate voltage is held constant at 20 volts.

Employing the same method we secure a value of 278 ohms a-c plate resistance at the 28 volts plate-voltage point for a change in plate voltage of 1.6 volts each side of the mean value of 28 volts. This compares with a d-c resistance of 422 ohms when the plate voltage is maintained constant at 28 volts.

You may have noted the difference in symbols used to denote d-c and a-c plate resistance. The former is shown with a capital "*R*" and a small sub-letter "*p*", whereas the a-c plate resistance is designated by a small "*r*" and the sub-letter "*p*". This is commonplace in vacuum-tube literature. In fact capital letters are usually used to identify *d-c fixed quantities,* whereas small letters indicate *varying* or *a-c quantities.* (See page 407.)

Thus we find that the d-c plate resistance and the a-c plate resistance differ appreciably. *The same is true for all types of vacuum tubes,* not only diodes, and this is well to bear in mind. As a rough approximation, the a-c plate resistance is equal to about one-half of the d-c plate resistance. Furthermore, as you can see from the illustrations given in Fig. 6-12, the a-c resistance is also related to the plate voltage, decreasing as the plate voltage is increased, and increasing as the plate voltage is decreased. The exact value of a-c resistance depends upon the point of operation selected on the plate-current curve. Thus, if the point of operation is at a plate voltage of 20 volts, the a-c plate resistance is greater than when the operating point chosen corresponds to a mean plate voltage of 28 volts.

Static and Dynamic Diode Characteristics

The various schematic diagrams of diode systems given in this chapter were typical of circuits employed in the development of the plate voltage—plate current characteristic curves and those required for a simple discussion of the plate current. They did not, however, represent circuits as used in practice. In order that a diode be capable of performing its normal function as a converter of electrical power of a-c character into power of d-c character, its external circuit must contain a *load.* It is through this load that the diode tube current flows outside of the tube, and the voltage drop developed across this load is then representative of the so-called "output" of the tube.

When such a load resistor is added to a diode circuit, as R_b in Fig. 6-13, the operating characteristic of the tube undergoes a major change, that is, the shape of the plate voltage—plate current curve is materially altered. It now becomes the *dynamic* characteristic rather than the *static* characteristic which exists when there is no load.

The reason for this change in characteristic is simple to understand. When there is no load in the circuit or $R_b = 0$, the current in the tube circuit is determined entirely by the resistance of the tube itself. If an external resistance, which we identify as the *load resistance,* is added to the circuit, as in Fig. 6-13, the total opposition to the flow

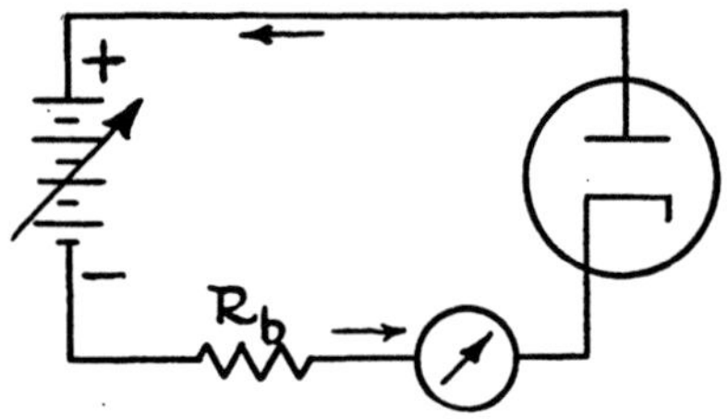

FIG. 6-13. The load resistance, R_b, and the tube resistance are in series and they form the total opposition to the flow of the plate current. The internal resistance of the battery is neglected.

of current is no longer the resistance presented by the tube itself, for it now includes the load resistance as well as the plate resistance. If we select a value for this load resistance which is many times the internal resistance of the tube, the effect of the tube resistance upon the amount of plate current flowing in the entire system is made negligible. Accordingly, if the external load resistance is of such a character that it maintains its resistance regardless of the amount of current flowing through it, that is, irrespective of the voltage applied across its terminals, the plate voltage—plate current characteristic of the system is changed from a curved line to a substantially straight line.

We may show these various conditions upon a separate graph, as illustrated in Fig. 6-14. The dashed line represents the static characteristic of plate voltage and plate current, that is, when the load resistance is zero. Solid line (1) indicates the voltage—current relationship when the load resistance is 1,000 ohms. This line still has a considerable curvature, but much less than that possessed by the static line. When the load resistance is 10,000 ohms, we obtain solid line (2). This line is fairly straight, its main curvature appearing at the region of low plate voltage where the internal resistance of the tube is the highest. The line representative of the 100,000-ohm load resistance is so straight throughout its entire length and lies so close to the horizontal axis that it would have been impractical to draw it.

Concerning the straightening of the plate voltage—plate current characteristic of the diode by the addition of the load so as to produce the required linear dynamic characteristic, the higher the value of load

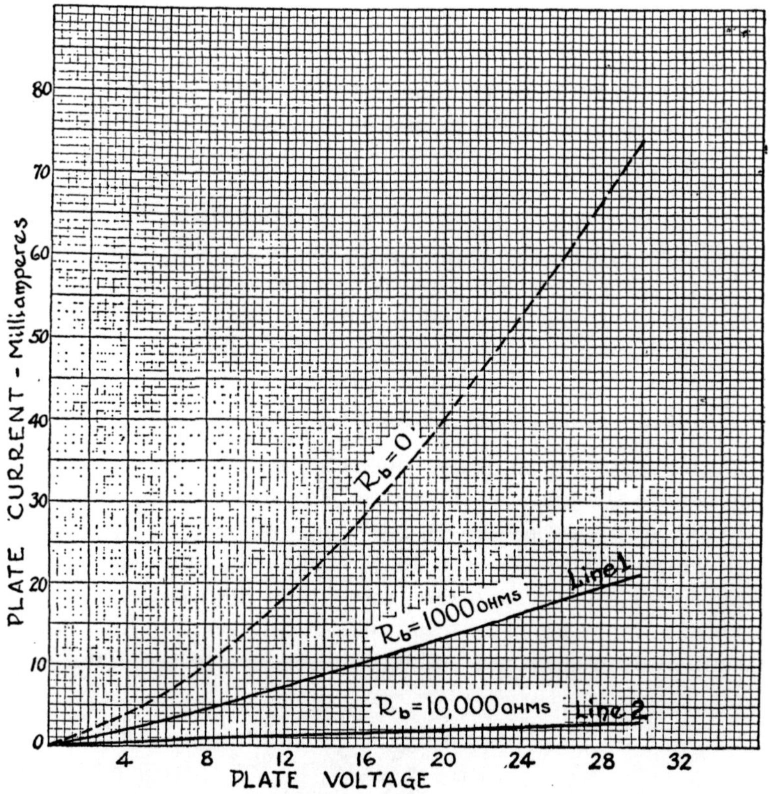

FIG. 6-14. Lines 1 and 2 are the plate voltage—plate current characteristics of a diode having a load resistance of 1000 and 10,000 ohms respectively.

resistance, the straighter is this dynamic curve. At the same time, however, the lower is the amount of plate current which flows in the circuit. This is not harmful, provided that the value of the load resistance is kept within reasonable limits.

Having read all of this information about the difference between the static and the dynamic plate voltage—plate current characteristics, and the effort to make this curve straight, you may wonder why this is done. The primary reason is that when the diode characteristic is straight, its action in certain portions of radio communication systems is substantially free from the production of distortion and this is greatly to be desired. There are, however, various applications of the diode, in which the exact nature of this characteristic is not of particular significance, although it should be understood that some straightening action takes place concurrently with the use of a load, and some sort of a load is required in order to make the diode of practical value.

It is also of importance to realize that the higher the load resistance used in a diode circuit, the greater is the permissible voltage which may be applied to the plate without fear of causing the flow of such high values of current that will damage the tube. You can see from the plate current values shown in Figs. 6-11 and 6-12 that the application of several hundred volts to the plate will cause the flow of such very high values of plate current as to damage the tube, unless high values of load resistance are used, which reduce the plate current to permissible values.

A-C Applied to the Diode Plate

It is necessary to speak briefly about the application of an alternating voltage to the plate of the diode. The reason is that everything we have said so far has shown d-c voltage applied to the plate, although we mentioned that the basic function of the diode was as a converter of a-c power into d-c power. The characteristics we have mentioned and developed by means of d-c voltages upon the plate are those which exist when a-c voltages are applied to the plate, for, in

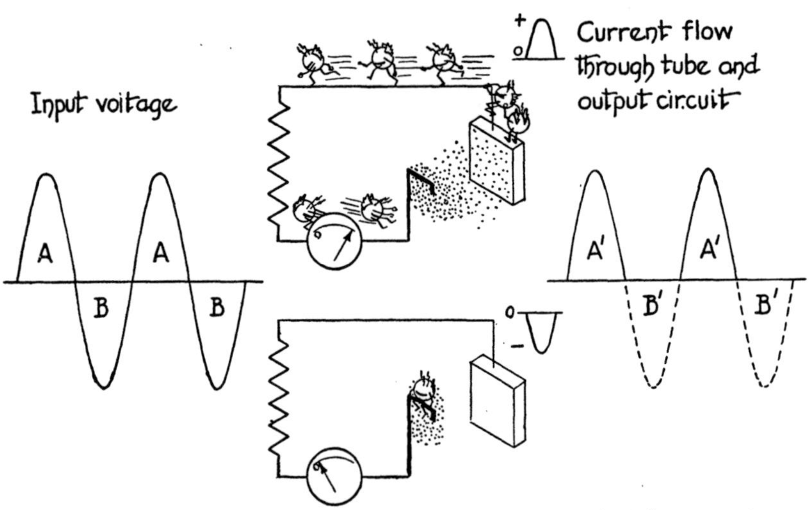

Plate current flows during positive alternation and plate is positive with respect to cathode. Plate current does not flow when plate is negative with respect to the cathode. Thus output current and voltage is unidirectional although input is alternating.

the final analysis, the tube operates only on one half of the a-c voltage wave, that is, during that time when the plate is positive with respect to the cathode. Therefore, the conditions which exist in the tube during that positive half cycle can be simulated by the application of d-c voltages.

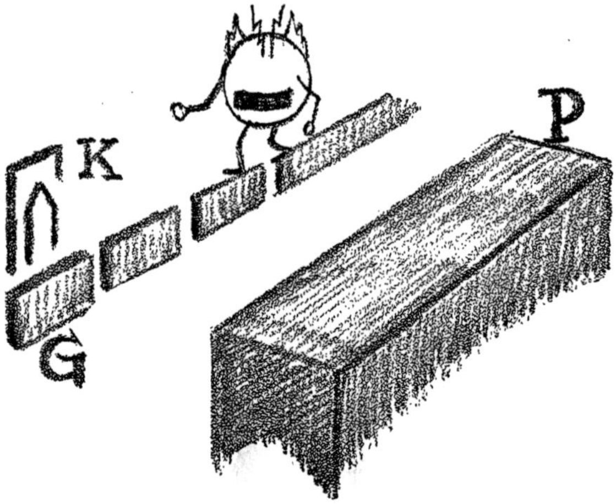

Chapter 7

THE TRIODE

THE TRIODE, or three-element tube, is an elaboration of the diode which we discussed in Chapter 6. For a number of years prior to the addition of the third element to the diode by De Forest, the two-element tube was the only thermionic device employed in connection with radio communication. But when in 1907 De Forest added his *control grid,* he provided that missing link to the art of both radio and wired communication which has made the present-day development of electronics possible. By adding this third electrode to the diode, which had limited applications, he evolved a device which has had the most far-reaching effects and was destined to influence the daily life of every human being.

We have seen that the diode consists of two electrodes: an emitter of electrons and a collector of electrons. These two electrodes also appear in the triode, in fact, they constitute the basis of the triode. It is the addition of the third electrode in the space between the emitter and the plate, and the active functioning of this third electrode which makes up the triode.

It may strike you as peculiar that we should mention such a condition as "active functioning" in connection with this third element, yet such is the case. If this electrode is not permitted to perform its proper function as the control element, as would be the case if it were joined to either the plate or the cathode outside of the tube, the tube with three elements would act as a simple diode. To realize the properties of the triode the control grid must be in active service and performing its allotted task.

Grid Structure

This control grid is a metal structure which may take any one of a number of different shapes, depending upon the design of the tube. For example, it may be a helix, it may be a ladder-shaped structure, or it may consist of a lattice network. As has already been mentioned, its location within the tube is between the emitter and the plate, and it is usually closer to the cathode than to the plate. When indicated upon a wiring diagram, the triode is represented by any one of four

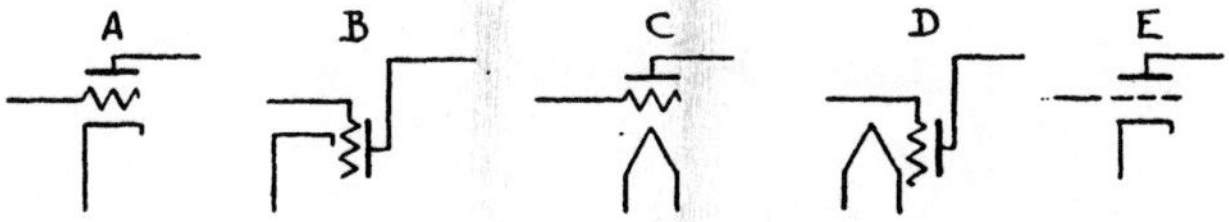

FIG. 7-1. The old convention for the triode symbol was the use of a zigzag line for the control grid, but the grid is now represented by a dashed line as shown in (E).

different symbols, as shown in Fig. 7-1, with a zig-zag or dashed line representing the control grid. Whichever type of symbol is used depends upon the individual preference of the illustrator, since there has been no definite and rigid rule. The control grid however, is so frequently indicated by a dashed line in the latest standards instead of the zig-zag form, that it will be so shown in the illustrations in this book.

As to the specific relationship between the control grid and the other electrodes in the triode, there is no hard and fast rule, for this is a matter of tube design. In most cases the grid structure completely surrounds the emitter, as shown in Fig. 7-2. There are also other arrangements, such as a grid structure which is divided into two halves, one half being located on one side of the emitter and the other half being located on the other side of the emitter. This last arrangement was very common in the early days of radio before the development of the modern type of tube. In those days most of the tubes were of the directly heated filament type.

In addition to a variety of grid shapes, other considerations are associated with the grid. These relate to such items as the diameter of the wire used for the spirals, helix-shaped, and ladder-type structures, the spacing between these wires, and the size of the holes in the screens and lattices. All of these items are related to design to obtain the desired tube performance, and are taken care of by the manufacturer. These factors are beyond the control of the user of the tube, except as it reflects upon the final tube performance and characteristics, which subjects will be dealt with later in this book.

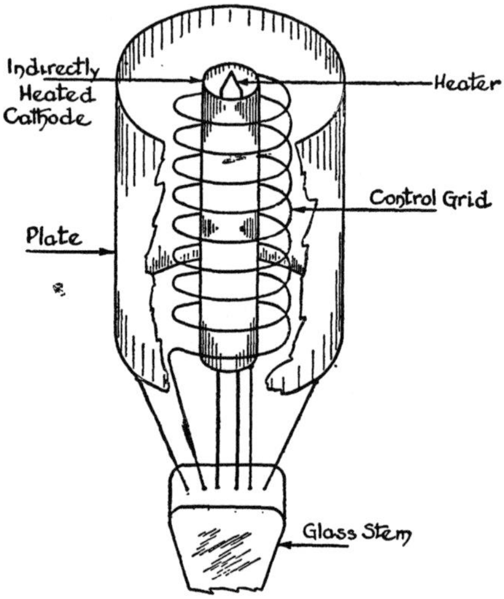

FIG. 7-2. In general, the control grid is a helix that surrounds the cylindrical cathode and in turn is surrounded by the cylindrical plate. These electrodes are supported by wires embedded in the glass stem of the tube base.

The Function of the Triode

As in the case of the diode, the possible applications of the triode are very numerous, but all of these applications are founded upon one broad basic function. In the diode, the fundamental function is as a converter of a-c power into d-c power. In the case of the triode, the basic function is that of an amplifier of current, or voltage, or electrical power. Many instances of its application are not specifically of amplifying character, yet its basic action as well as the final result utilizes its amplifying capabilities. This can be expressed in another way; for example, it is a device which is capable of releasing a much greater amount of electrical power in its output system than is applied to its input system. We appreciate that these terms *input* and *output* are new to you, and hence may not yet be clear, but, if you will bear with us for just another paragraph or two, both will be discussed and you will then have a better conception of the basic definition.

In this process of amplification the control grid plays a most important role. Due to its location, it is capable of exerting a tremendous influence upon the movement of electrons in the space between the emitter and the plate, and consequently upon the tube plate current. It does this by acting upon the space charge by either working against the attracting force of the field set up by the plate voltage or else by aiding the attracting force of the voltage upon the plate. When the field of the grid aids the space charge, it decreases the movement of electrons to the plate and therefore tends to reduce the plate current. On the other hand, when it is acting against the space charge, it accelerates the movement of electrons towards the plate, and hence tends to increase the plate current.

But most important of all is its ability to exert this great controlling action upon the plate current *with comparatively low voltages applied to the control grid.* In other words, a comparatively small negative voltage applied to the control grid is capable of completely nullifying the attractive force of a twenty-times greater or even hundred-times greater positive voltage applied to the plate, thus causing complete cessation of electron movement to the plate. In the same way, the application of a small positive voltage to the grid can cause such a large increase in the movement of electrons from the space charge to the plate, as would be produced by a twenty-times or even a hundred-times as much increase in the voltage applied to the plate.

What we have said is but a brief and general description of the action of the control grid and by no means an explanation of just why these conditions occur. That subject is left for later in this chapter, after we have introduced a few other items relating to the triode.

Input and Output Circuits of the Triode

The triode, as well as all other types of vacuum tubes, can be divided into two major parts. One part is that representative of the *input* system of the tube and the other part is that representative of the *output* system of the tube. Speaking in generalities and regardless of the exact application of the triode, the input system is that portion of the tube circuit into which the electrical voltage, current, or power is fed into the tube. Generally this is the control grid—cathode circuit of the triode, as shown in Fig. 7-3.

On the other hand, the output system of the tube is that portion from which the electrical power, current, or voltage, is taken out of the tube. This is usually the plate—cathode circuit of the triode, as indicated in Fig. 7-3. If this triode were of the directly heated type so that the cathode was replaced by a filament, then the word "filament" would appear wherever the term "cathode" was originally used. As you can see, the emitter circuit is common to both the input and output circuits and this is a point which will be referred to again later in this chapter.

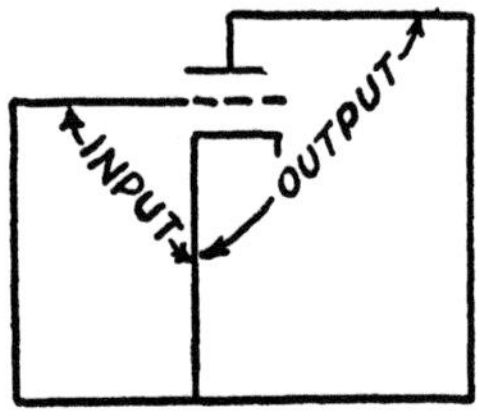

FIG. 7-3 (left). The input circuit of a triode comprises the grid and cathode circuits; the output circuit includes the plate and cathode circuits. Note that the cathode circuit is common to both input and output circuits.

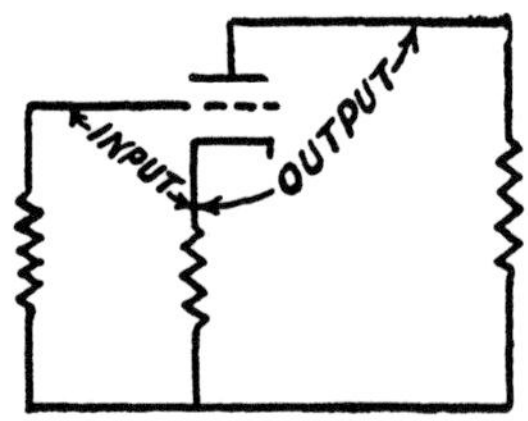

FIG. 7-4 (right). The resistor at the left is said to be in the input or grid circuit and the one on the right is said to be in the output or plate circuit. The resistor in series with the cathode is common to both the input and output circuits, and when necessary can be used as the output circuit. (See Chapter 13.)

Supplementing the previous broad division of input and output circuits of the triode, it is possible to be more specific and to narrow down the input as being the grid circuit and the output as being the plate circuit respectively. Thus, if some device, say a resistor, is spoken of as being located in the input circuit of a triode, it would be located in the grid circuit as shown in Fig. 7-4. And if another resistor is said to be located in the output circuit of the triode, its location would be in the plate circuit, as shown in Fig. 7-4. The resistor in the cathode leg is common to both the input and output circuits.

Now, our reason for first giving the broad division of input and output circuits and then narrowing down this division is well founded. It is due to the fact that not every application employs the grid circuit and the plate circuit as the input and output systems respectively. Instead, the cathode circuit may be used as the input system in one case, while in another it may be used as the output system. At the moment this may seem confusing, but it will be cleared up in a later chapter. In the meantime, let us continue with other details.

The tube circuits shown in Figs. 7-3 and 7-4 are, of course, inoperative as illustrated, due to the absence of the operating potentials.

The Triode Operating Potentials

Proper operation of the triode requires the application of three operating potentials or voltages. The need for two of these voltages is evident from the opening statement used in this chapter: the triode is an elaboration of the diode. Since this is the case, it stands to reason that a voltage is needed to drive the heating current through the filament or the heater wires in order to raise the emitter temperature to the correct value. This heater or filament voltage, depending upon the type of tube that is used, is usually referred to as the "A" voltage, as indicated in Fig. 7-5.

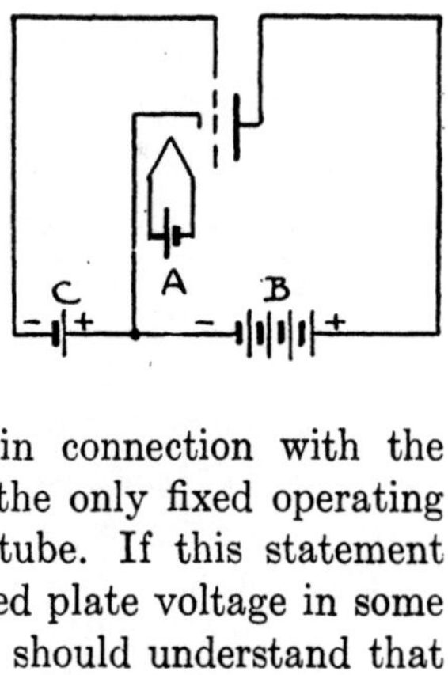

Fig. 7-5. The operating voltage sources of a triode are designated as follows: the "A" battery supplies the heater or filament current; the "B" battery supplies the positive potential on the plate; and the "C" battery supplies the grid bias.

Ordinarily, this terminology is not used in connection with the diode, perhaps because in most cases that is the only fixed operating potential which is normally used with that tube. If this statement surprises you, in view of the presence of a fixed plate voltage in some of the diode circuits shown in Chapter 6, you should understand that those fixed voltages were used only in order to develop the basic tube characteristics. During normal use, the voltage which is applied between the emitter and the plate of the diode is an a-c voltage representative of the a-c electrical power being fed into the tube for conversion into d-c electrical power.

In the triode, the situation is different. Here, three elements are given fixed operating potentials, and it is a matter of convenience to identify them when referring to them. But even in the case of the triode it has become the custom to refer to the "A" voltage only when using a d-c source of supply for filament type tubes, for when referring to an a-c supply to heat the heater wires, the voltage is usually spoken of as the *"heater voltage."*

Since we require a flow of electrons to the plate of the triode and these electrons must be attracted to the plate, a voltage must be applied to the plate and this voltage must be positive with respect to the emitter. This voltage, shown in Fig. 7-5, is usually referred to as the "B" voltage or *"plate supply voltage"* and this name applies without regard to the type of voltage supply source used. Incidentally, when the tube employs more than three electrodes, so that more than one electrode is positively charged by being connected to a voltage-supply source, it is customary to identify the voltage applied to the tube electrode in terms of the electrode, as for example "plate voltage" when it is applied to the plate. The name "B" voltage would then be applied to the complete source of the various positive voltages being distributed to the various electrodes.

An operating voltage is also associated with the control grid. Up to this time we have said nothing which would seem to indicate that such a fixed voltage is required upon the control grid. The reasons for this requirement will unfold themselves as we delve into the actual conditions existing within the tube. In connection with this fixed and steady grid voltage, it should not be confused with the previous reference to the negative and positive voltage which was respectively applied to the control grid and its action upon the plate current. That description, meagre though it was, explained the control action of the grid, but did not include the reason for the steady operating voltage.

This steady voltage which is applied to the control grid is spoken of as the "C" voltage or *"grid bias"* or *"C-bias,"* as shown in Fig. 7-5. The illustration indicates the use of a battery form of voltage source, but this is not a rigid requirement, for other means of securing this fixed voltage are also employed. The polarity indicated, that is, with the grid kept negative with respect to the emitter, is a requirement in the majority of triode applications. The reason for this as well as the arrangements which depart from this condition, will be explained later in this chapter.

To attempt to quote specific values for these operating voltages is impractical because they vary in accordance with tube types and the operating conditions desired. You should always keep in mind however, that improper operating voltages on any one of the three elements will impair the performance of the tube, no matter what its purpose or application.

Electrostatic Fields in the Triode

We have now reached that point in this discussion of the triode where we can explain the manner in which the control grid achieves its remarkable results. In order to describe this action we shall refer you back to the diode. We said in the opening sentence of this chapter that the triode was an elaboration of the diode. Now we can advance a step beyond that and say that a triode is essentially a diode with a third element added to the tube and located between the emitter and the plate.

The thought behind this statement is to bring to your attention two very significant facts. The part played by the emitter is the same in the triode as it is in the diode and also, the relationship between the plate and the voltage upon it is the same in the triode as it is in the diode. Therefore, it follows that the basic electrostatic fields which exist in the diode are likewise to be found in the triode. These are two in number as you will recall: one due to the space charge and the other due to the voltage upon the plate. It is true, as you shall soon see, that these two electrostatic fields are not the only ones existing in the triode, but that does not alter the fact that these two basic fields which are present in the diode, are also present in the triode.

You will remember that during the description of the diode tube characteristics, we brought certain facts to light. For example, we showed how the plate resistance, both a-c and d-c, decreased as the plate voltage was increased and the plate current increased. This same condition exists in the triode, and while the control grid makes certain contributions to this effect, it still does not alter the basic conditions associated with variations in plate voltage.

Furthermore, we showed how the plate current in the normal diode was always space-charge-limited. This too, for the majority of applications, is a basic condition for the triode. Although the control grid may determine the exact value of plate current by aiding or bucking the action of the voltage upon the plate, nevertheless the fact remains that the plate current in the triode is space-charge-limited.

In the same way, if we imagine a triode which is being operated in such a manner that all of the electrons emitted find their way to the plate, the value of plate current would then be controlled by the electron emission, or, to say this differently, the plate current is limited by the emitter temperature. Once more, the grid, by virtue of the voltage that is applied to it, may help the plate in creating this state, but it would not alter the fundamental condition as described in connection with the diode.

From this, you can see that there is much similarity between the diode and the triode, although not necessarily in actual performance, but rather in the basic conditions existing within the tube. There are, of course, certain conditions to be found in the triode which do not have their counterpart in the diode, but this is due to the presence of the control grid and its action. You should remember that the control grid in the triode is not an electrode which alters the basic relationships existing between the action of the space charge upon the emitted electrons and the action of the electrostatic field due to the plate voltage on the space charge, but rather, that it is an electrode which is capable of controlling the extent to which the electrostatic field due to the voltage upon the plate can act upon the space charge.

An important thing to understand about the control grid in the triode is that it is another charged surface, with an electrostatic field of its own which is interposed between the plate and the space charge. The electrostatic lines of force due to the voltage upon the plate, thread through the lines of force of the electrostatic field which are due to the voltage upon the control grid. Both of these fields terminate upon the space charge, and the effect upon the space charge with respect to electron movement from the space charge to the plate *depends upon the relative intensities of the grid and plate fields at the space charge.* Let us see just what this means when the grid is negative with respect to the cathode, when it is positive with respect to the cathode and when it is at zero potential or at the same potential as the cathode.

Negative Voltage on the Grid

Suppose that you examine Fig. 7-6. In this drawing we show the grid in the form of a series of large dots with spaces between them, the circles indicating the solid portions of the grid and the space between indicating the space between these solid portions. Further, a battery marked "C" applies a small voltage to the control grid with such polarity as to make the grid negative with respect to the cathode. Incidentally, in all of this discussion, we can forget about the field which exists between the space charge and the cathode or electron emitter, for as long as a space charge exists the action taking place between the emitter and the space charge, as far as electrons being repelled back to the emitter is concerned, depends upon what is happening in the zone between the space charge and the plate, which naturally includes the control grid. If conditions are such that electrons move from the space charge to the plate, then electrons emitted from the cathode will be admitted into the space charge.

Getting back to Fig. 7-6, and for the moment neglecting the field due to the voltage on the plate, let us analyze the action of the grid. With even a very small voltage applied to the grid, that surface becomes a charged body and has an electrostatic field of its own. Hence, if we show the direction in which the lines of force of the electrostatic field set up by the voltage on the control grid will act upon any loose

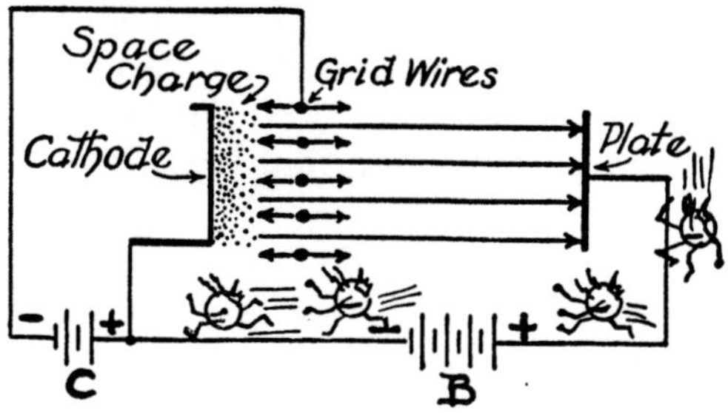

FIG. 7-6. The arrows pointing to the left from the grid wires represent the field that tends to repel electrons back into the space charge.

electrons, it must be in a direction which is *away* from the control grid. Thus the arrows which originate at each of the grid wires point away from the grid.

For the present we are not interested in those lines from the grid which point towards the plate, for once the electrons get past the grid they are under the control of the field set up by the plate voltage, as well as being repelled towards the plate by the force of the grid field in that direction. We are, however, definitely interested in the lines which point towards the space charge. Since the arrows indicate the direction in which a negative charge would move under the influence of the field in question, it is evident that the field set up by the negative voltage on the control grid would tend to prevent electrons in the space charge from approaching the grid and in that way prevent their moving towards the plate. In fact, we can go further and say that the field set up by the negative voltage applied to the control grid would tend to repel the entire space charge,—actually move it closer to the emitter,—at least repel those space-charge electrons nearest the control grid deeper into the space charge, thus making the latter more concentrated.

Let us now consider the field due to the voltage on the plate. This field in effect terminates upon the space charge, at the same location where the field due to the negative voltage upon the control grid terminates. Thus we have an attracting field and a repelling field opposing each other at the space charge. Which of these is going to predominate over the space-charge electrons depends, of course, upon the relative magnitudes of the two fields. In this consideration, we are not much concerned with the electrostatic lines of force from the plate which are blocked or screened by the solid grid wires, for there is plenty of room between these wires for the plate field to pass through and reach the space charge.

The field existing inside the triode is shown as an anaglyph on Plate 3. It will be observed that the field between the cathode and plate is shown as straight lines midway between the two grid wires. The field is bent away from these grid wires when it is near them because of the negative charge on the grid. An examination of the field is instructive since the arrows on the lines indicate the direction in which an electron would move.

Whether the field due to the voltage on the plate will be able to attract electrons from the space charge now depends upon the extent to which the field due to the negative voltage upon the control grid offsets it. In this connection we must again refer to the discussion in Chapter 3 relative to the force exerted by an electrostatic field existing between two parallel plates upon a test unit charge. The two parallel plates in this case are first, the plate and the cathode, and second the control grid and the cathode, the test charge being an electron which is between the grid and cathode.

If you recall, the force acting upon the test charge varies inversely as the distance between the two charged surfaces. Thus, for like values of voltage applied to the plate and to the control grid of Fig. 7-6, the plate voltage being plus and the grid voltage being minus, the strength of the field due to the grid voltage will be tremendously greater at the location of the test charge in the space charge than the strength of the field due to the voltage upon the plate. This is because the grid, as a charged surface, is so much closer to the cathode.

Purely as the result of the separation between the control grid and the plate with respect to a test electron in the space charge, a value of negative voltage exists which can be applied to the grid so that the resultant field exerts a force upon the test charge which will just offset the attracting force of the plate field, although the grid voltage is *very much less* than the voltage applied to the plate. The relative values of the grid and plate voltages depend upon the separation between the grid and the plate, and this is a variable factor which depends upon the physical design of the tube. In one tube the negative grid voltage required to offset the attracting force of the plate completely at the space charge may be $\frac{1}{10}$ of the plate voltage, while in another tube it may only be $\frac{1}{100}$ of the plate voltage. This relationship is an

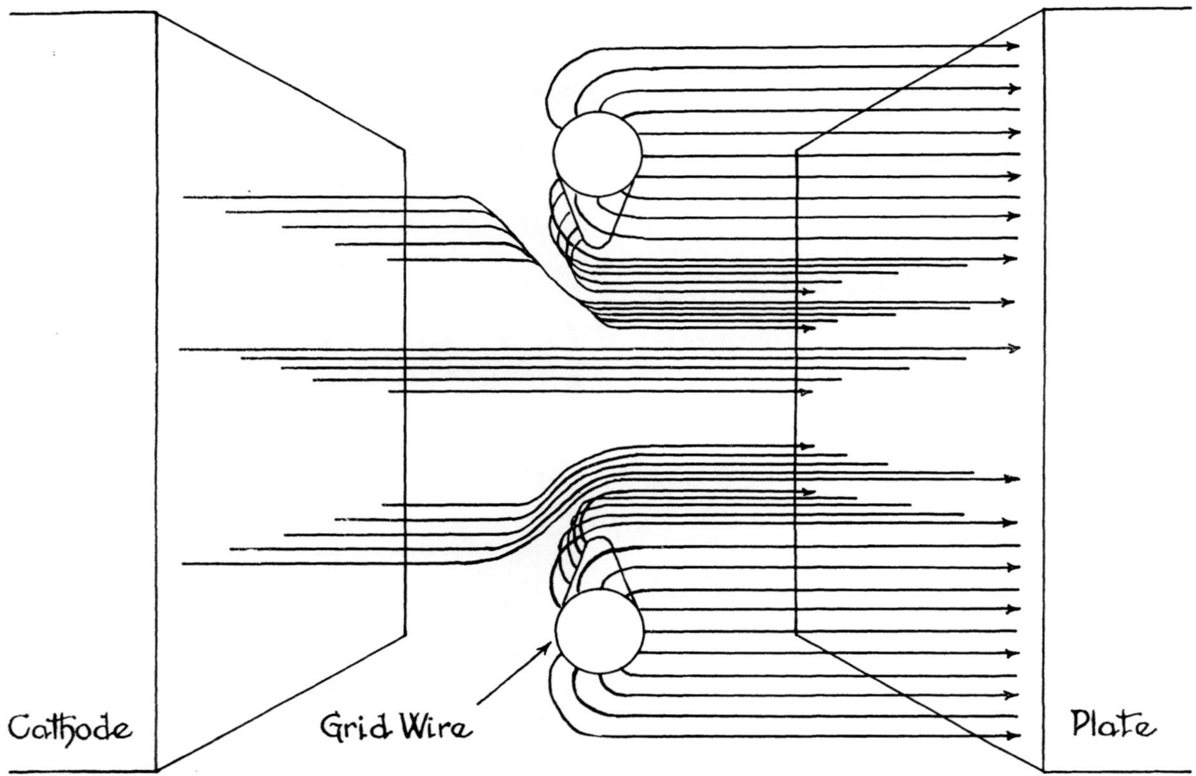

PLATE 3. The arrangement of the lines of force in a triode, in which the grid is negative with respect to the cathode.

extremely important matter, for it illustrates the control of the grid voltage upon the plate current.

In the discussion regarding Fig. 7-6 we concerned ourselves with a value of grid voltage which would offset the plate voltage and thus stop the movement of electrons from the space charge to the plate, which means the stoppage of plate current flow. But suppose that the value of the grid voltage, still negative of course, is not so high as to offset completely the attracting force of the plate voltage. What then? The answer is that no matter how low a negative voltage is applied to the control grid, provided that it is more than zero, that electrode has an electrostatic field which is working against the field due to the voltage on the plate. It may not offset the entire field due to the plate voltage, but it will offset some of the force of the plate field and in that way reduce the number of electrons attracted to the plate from the space charge in a unit time.

This item of unit time as it relates to the flow of electrons is important, for the action of the grid voltage, if it is not great enough to prevent completely the movement of electrons to the plate, is enough to reduce the number passing through the grid. Since the value of current is determined by the *number* of electrons which move past a certain point in one second, this reduction in the number of the electrons is equivalent to reducing the current flow. Thus any value of negative voltage applied to the control grid has the effect of reducing the plate current flow below that value which would exist if the grid were absent or if the voltage upon the grid was zero or positive. A sufficiently high negative voltage applied to the grid will reduce the electronic flow to the plate to such a negligible quantity that for all practical purposes it may be said that the plate current flow has completely stopped.

Another detail to be seen in Fig. 7-6 is that which relates to the grid circuit, that circuit which contains the grid voltage source. In all of the discussion so far we have spoken about plate current flow in the tube. What about current flow in the grid circuit? Recognition of the fact that when the grid is negatively charged its electrostatic field tends to repel electrons, is in itself evidence of the fact that there can be no flow of current in this circuit, for there is no movement of electrons from the space charge to the wires of the grid. There may be, as has been explained, movement of electrons from the space charge *through* the spaces of the grid to the plate, but as long as the grid is negatively charged there should be no flow of electrons to the grid itself, and consequently no flow of electrons through the grid circuit and C-battery back to the cathode.

As you see, we say that "there should not be a flow of electrons to the grid" rather than the positive statement that "there is no movement of electrons to the negatively charged grid," because in certain tubes there may be a movement of a comparatively few electrons to the grid, and hence the flow of a minute value of *grid current,* as shown in Fig. 7-7, even when the grid is negatively charged. This movement of

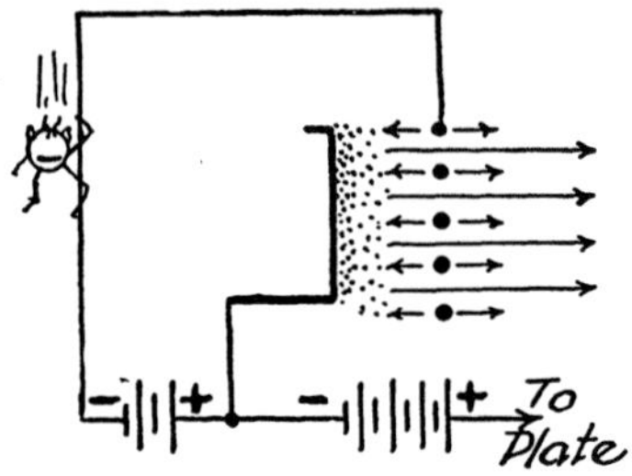

FIG. 7-7. In some instances, electrons do flow in the grid circuit when the grid is negatively charged because the grid bias is insufficient to repel the electrons that are attracted to the plate because of the high plate voltage.

electrons to the grid may occur for a number of reasons, one of which is that a small amount of negative grid bias may not be sufficient to repel the high-speed electrons which are being attracted to the plate due to a high plate voltage. We shall discuss this in more detail later.

Positive Voltage on Grid

What happens if the grid voltage is changed from negative to positive? While it is true that this is not the customary method of operation, the fact remains that it is occasionally done, and furthermore, the conditions which shall be described can be correlated into a corresponding action accomplished in another way, so that the subject does deserve attention.

Changing the grid voltage from negative to positive does not alter the existence of either the electrostatic field due to the plate voltage, or the field due to the space charge. And since the application of a positive voltage to the grid results in the setting up of an electrostatic field at the grid, the original fields exist in the tube. However, the direction of the grid field is now reversed from what it was before. This is shown in Fig. 7-8.

If you compare Figs. 7-7 and 7-8, you will note a significant difference in the direction of the grid fields. In Fig. 7-7, the arrows which indicate the grid field point away from the grid, both between the space charge and the control grid, as well as between the control grid and the plate. In Fig. 7-8 however, the arrows indicating the grid field between the control grid and the space charge point towards the control grid, or in the same direction as the arrows designating the plate field.

Since we have already devoted considerable space to the discussion of the action of electrostatic fields it is not necessary to describe in detail the fundamental behaviour of these various fields. It should suffice, therefore, to say that the grid is now positive with respect to the cathode and would therefore attract electrons towards the grid. The field set up at the control grid is capable of acting upon electrons which are not directly in line with these grid wires, as well as upon those electrons directly in line with these grid wires. At the same time the electrostatic field due to the voltage upon the plate has penetrated to the space charge through the spaces in the grid. The electrons in the space charge are therefore acted upon by two attracting fields, one due to the field of the positive grid and the other due to the field of the positive plate.

The net result of this combined action is two-fold. Electrons are attracted to the grid and reach the grid wires, while at the same time most of the electrons which have been attracted towards the grid, pass

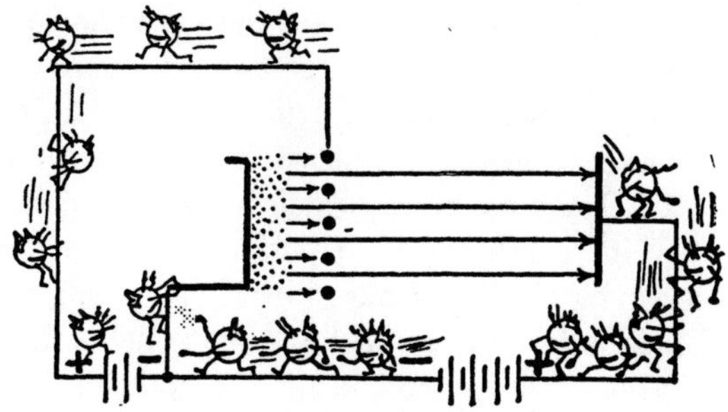

FIG. 7-8. When the grid bias is made positive with respect to the cathode, then electrons are attracted to the grid and current flows in the input circuit. Notice that the direction of the field from the grid is reversed from what it was in Fig. 7-6.

through the spaces in between the grid wires and move towards the plate. These electrons, as you can see, are under the influence of both attracting fields. But this is not the complete story. Bearing in mind the facts we learned in Chapter 3 concerning the relationship between force and distance as far as the action of an electrostatic field upon a test unit charge is concerned, the attracting force of the grid greatly increases the number of electrons which move towards the plate. The reason is due to the difference in separation between the grid and plate with respect to the space charge. The grid is closer to the space charge and can therefore exert a much greater attracting force upon a test unit charge than can the plate. In fact, a positive voltage upon the grid can increase the movement of electrons from the space charge towards the plate to an extent equal to that of a plate voltage many times the value upon the control grid. The relationship is the same as existed when the grid voltage was negative, except that instead of opposing the plate voltage with the negative grid, a positive voltage upon the grid aids the plate voltage.

There is also another condition created when the grid voltage is made positive. If the grid is made positive with respect to the cathode, the only difference between it and the plate is that the grid does not present a solid surface. But the grid does have solid surfaces in the form of the grid wires, and, as you would expect, each of these is a miniature plate. When electrons are attracted to the grid and they strike its wires, it is the equivalent to a flow of current in the grid circuit, in other words *grid current*, and there is consequently a flow of electrons around the grid circuit back to the cathode. This is shown in Fig. 7-8.

Now, it would seem that increasing the value of the positive voltage applied to the grid would result in a steady increase in plate current, for the greater the positive grid voltage, the greater is the number of electrons leaving the space charge for the plate, and this is generally true. If you examine Fig. 7-8, you will note that the electrons which have passed into the space between the control grid and the plate now are under the influence of two opposing electrostatic fields. Since the field extends from the grid wires and this field tends to attract electrons, the grid field between the control grid and the plate tends to attract the electrons which have moved into that zone.

With a low value of positive voltage applied to the grid this effect is negligible, for a high initial speed is given the electron by the combined action of the grid and plate fields in the zone between the control grid and the space charge. In addition, the attracting force upon the moving electron is increasing as the electron approaches the plate. The result is that over a range of positive grid voltages, starting at some low value and gradually increasing, there will be evident a continual increase in plate current, but little by little, as the values of positive voltage upon the grid increase, the plate current increases less and less rapidly.

With high plate voltages such flattening off of the plate-current curve is difficult to reach and is rarely used in practice. Even with low plate voltages, and particularly if the grid voltage is more positive than the plate voltage, the attracting force of the grid in the zone between the plate and grid will not only be sufficient to pull some of the electrons moving towards the plate back to the grid, thus reducing the plate current, but in addition any electrons which are emitted from the plate as the result of the impact of electrons striking it, will be attracted back to the grid. This is a *secondary emission effect.*

The preceding paragraph requires some qualification, for the secondary emission effect just described usually only occurs in certain special tubes. The more conventional tubes that are ordinarily used, are subjected to a special treatment which tends to prevent this secondary emission. In consequence, although the plate-current curve exhibits a tendency to flatten off, the effect is not very definite. For, before it is possible to obtain any appreciable flattening off of plate current, a large current flows for a moment, the tube becomes ionized, and the useful life of the tube is terminated.

Accompanying the state of affairs described as existing between the grid and the plate is, of course, the increase in the number of electrons which lodge upon the grid wires. As the positive grid voltage increases, the grid current increases. This condition is important for a number of reasons which shall be described later in this chapter.

Zero Voltage on the Grid

Having described the action taking place with a negative voltage on the grid and the conditions created when the grid is made positive, that which is left is the condition arising from zero voltage on the grid.

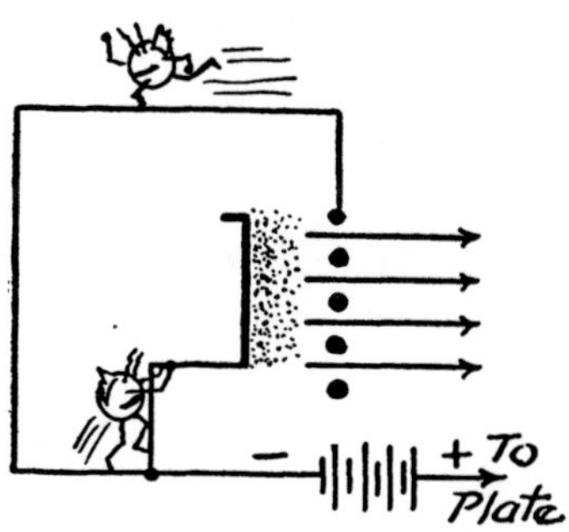

FIG. 7-9. Even though no bias is applied to the grid, a few of the electrons on their way to the plate may strike the grid wires and go into the input circuit.

It stands to reason that when the grid voltage is zero, there can be no electrostatic field set up at the grid, so that the grid cannot have any effect upon the space charge or upon the plate current. Essentially that is true, but a qualification must be made. This is made necessary by the fact that the grid structure does not permit the wholly unimpeded movement of electrons. Due to the fact that the grid has solid wires or surfaces, some of the electrons which are attracted over to the plate by the plate field, shown in Fig. 7-9, lodge upon the grid wires.

Normally, the value of grid current caused to flow under such conditions is not very great, being in the order of a few microamperes, but such a definite stipulation of grid current value cannot be taken too rigidly, for the actual value depends upon a number of factors, such as the type of tube, the abundance of emission from the cathode, and the plate voltage. With respect to the latter, it has been found that the lower the plate voltage, the less the grid current. This appears reasonable on the grounds that the lower the plate voltage, the fewer the electrons which advance towards the grid under the influence of the attracting action of the plate field.

"Free" Grid

While we are speaking about zero grid voltages, there exists the possibility of confusing that term with another condition. We are referring to the absence of an operating voltage upon the grid because the circuit is open, that is, the grid is "free" as shown in Fig. **7-10.**

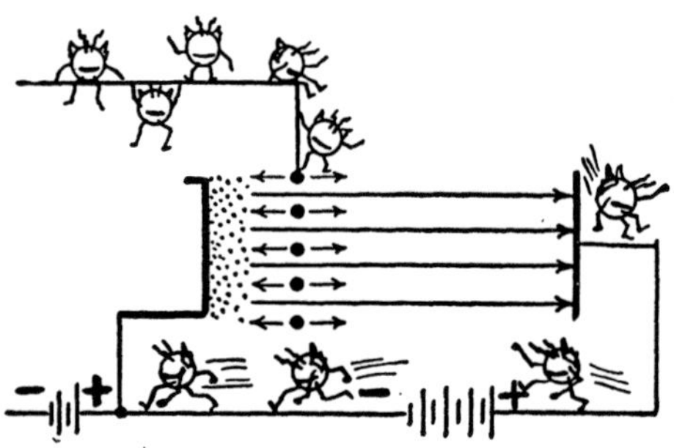

FIG. 7-10. When the grid circuit is open, the grid is said to be "floating." Electrons on their way to the plate, strike the grid wires and remain there. As there is no circuit through which they can flow, they accumulate on the grid and build up a negative charge.

Assuming that the proper emission takes place and the proper plate voltage is applied to such a tube, the presence of the free grid creates a peculiar condition which must be distinguished from that existing when the grid circuit is complete but there is no voltage applied to the grid.

An open grid, which is nowhere connected in the circuit, has a very material effect. As we have mentioned a number of times, the solid portions of the control grid intercept the movement of electrons from the space charge to the plate. With a reasonable negative voltage upon the grid, the electrons which would normally strike the wires of the grid are repelled from the wires, so that practically none lodge upon them. But with either zero or positive voltages upon the grid, these electrons reach the wires and grid current flows.

When the grid is free or open, electrons which strike the metal portions of the grid come to *rest* upon it, for they have no path to travel. The grid will, accordingly assume a negative charge. This negative charge on the grid charge will depend upon the magnitude of the space charge, the intensity of stray fields in the neighborhood of the tube, the amount of leakage of electrons from the grid inside and outside the tube, etc.

What happens to the plate current when the grid circuit is opened and the grid is left "floating"? The plate current may rise or it may decrease, depending upon the magnitude of the charge upon the grid prior to the opening of the grid circuit. If, before the grid circuit was opened, a positive voltage, a zero voltage, or a sufficiently low negative voltage was applied to the grid so that a relatively large plate current flowed, then upon opening the grid circuit the plate current would decrease. This is due to the fact that some of the electrons, in their passage from the space charge to the plate, would strike the solid grid wires and, having nowhere else to go, would remain trapped upon the grid. A negative charge would consequently build up on the grid, and this charge being more negative than that possessed by the grid before it was left to float, would act in opposition to the attracting field of the plate, so that the plate current would decrease. For very low values of plate voltage or low values of heater or filament voltage, the accumulation of electrons on the grid would cause the grid to become sufficiently negative so that the plate current would be reduced to a negligible quantity.

On the other hand, if the grid was sufficiently negative so that only a small plate current flowed prior to the opening of the grid circuit, then when the grid is disconnected the plate current would increase. For in order to acquire a potential determined by its surrounding conditions, the excess of electrons are knocked off the grid by the electrons which strike the grid wires, until the grid now possesses only that number of electrons so that the charge on the grid is of such magnitude that the grid now is at its environmental potential. At this stage the floating grid is less negative than it was before it was disconnected from the circuit and, since its field does not oppose the field of the plate to the same degree that it did when it was connected in the circuit, the plate current will increase.

A vacuum tube circuit in which the grid is free or floating is rarely employed because of its erratic behaviour. Changes in the stray fields in the vicinity of the tube, a change in the amount of light shining upon the tube, a variation of humidity of the air surrounding the tube, as well as other factors, all result in a different value of plate current.

It is for this reason that some conducting path is almost always found between the grid and cathode.

Summary of Voltages Applied to the Grid

Summarizing the conditions associated with voltages applied to the control grid, it might be well to point out the pertinent details already mentioned and to introduce a few which could have been spoken about during the discussion, but were kept for this part of the chapter in order to avoid confusion.

Naturally, the *control action* of the control grid upon the plate current is the paramount item. But even then, the most important thing to understand is the fact that whatever control action is accomplished by the control grid, whether it is to increase or to decrease the plate current, this action is equivalent to a much greater change in plate voltage to cause the same change in plate current. It is this condition which gives rise to the property of amplification possessed by the triode and other multi-element tubes. The reason is found in the separation between the grid and the plate with respect to the point where the grid field and the plate field act upon the electron. Generally speaking, the greater this separation between the grid and plate the greater is this form of effectiveness of the grid; or to put it differently, the higher the amplification available within the tube.

In connection with the voltage applied to the grid, you must, of course, appreciate that while we showed the control action of the control grid by securing the voltage from batteries, that is not a necessary condition. In other words, exactly the same type of action takes place when an a-c voltage is applied to the grid during the alternately positive and negative cycles of that voltage. In this case, however, the magnitude of the voltage active upon the grid at any one instant varies between the positive and negative peak values of the a-c voltage.

You have no doubt gathered from all that has been said that the plate current flow within the vacuum tube is not very steady, and that a variation of the space charge occurs, depending upon the polarity of the grid. Also, that the motion of the electron from the cathode, through the space charge, and then to the plate, while under the various influences, is a sort of progressive movement. That is correct and it results in what we class as normal operation over a very great range of frequencies, from 0 up to perhaps about 200 megacycles. Beyond that frequency, the time required for the electron to move through the tube and to pass through its various motions becomes a limiting factor.

This subject receives further discussion later in this book in Chapter 15, which discusses special purpose tubes and applications.

Why the Grid Bias

Earlier in this chapter, in connection with Fig. 7-6 to be exact, we referred to a voltage applied to the control grid, as a *grid bias.* At the same time we also spoke of the action of a negative voltage shown in Fig. 7-6 upon the electrostatic fields existing in the triode. Later on, in connection with Fig. 7-8, we changed the polarity of the battery, which was the source of the voltage fed to the grid, and again discussed the action of a voltage upon the electrostatic fields within the tube.

Thus, it would seem as if the grid bias could be of two kinds: one which would make the grid negative with respect to the cathode or emitter, and another which would make the grid positive with respect to the cathode or emitter. At the same time it would seem as if the grid bias, and what we later referred to as the *voltage applied to the grid,* were one and the same. Suppose that we clear this up first and the rest will then be easier to understand.

In view of the fact that a chapter covering the operation of vacuum tubes follows later in this book, we want to comment that what is being discussed at this time is not intended as application data, but rather the clarification of a possibly confusing point. In connection with the operation of a vacuum tube of the triode or multi-element type, one primary condition exists in virtually all cases when the tube is used as an amplifier unless special conditions, which need not be dealt with here, exist. The basic application of the triode, as well as the other tubes, is as a device which does not consume power in its input circuit, although power may be available from the output circuit. This condition is not offered as an example of securing something from nothing. But it so happens that in the case of vacuum tubes used as conventional amplifiers the action of the grid is such that it is permissible to say that under normal conditions the input circuit is purely voltage operated; there is a negligible current flow, hence negligible power is consumed in the grid circuit.

In order that such a condition be obtained, it is necessary, as you have seen in Fig. 7-6, that the control grid be kept at a negative potential with respect to the cathode or emitter. When the grid is negative to a reasonable extent with respect to the emitter, there is a movement of electrons *through* the spaces of the grid electrode as shown in Fig. 7-11, but there is substantially no movement of electrons to the control grid itself. Thus it is possible under such conditions—and for the present forgetting about the minute value of grid current which we have said exists with very low values of negative grid voltage—to say that the grid circuit is possessed of very high resistance. You can compare this to the ordinary circuit containing a battery and a extremely high value of d-c resistance wherein there is virtually no current flow, as shown in Fig. 7-12.

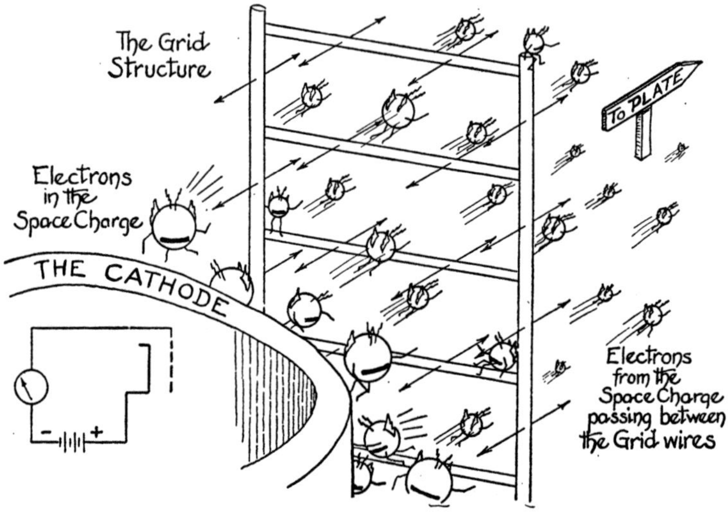

FIG. 7-11 (above). When the grid bias is negative, electrons are repelled by the grid wires and continue on to the plate.

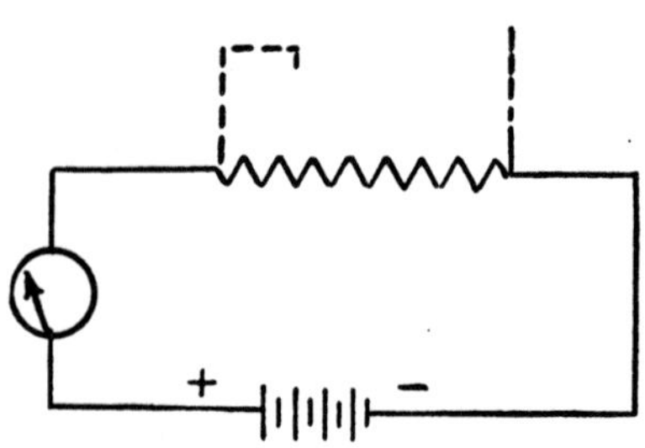

FIG. 7-12 (left). A negative grid bias is the equivalent of a very high resistance between the cathode and grid that prevents any current flow.

However, if the grid is permitted to become positive, the conditions described in connection with Fig. 7-8 develop, and there then exists a definite flow of electrons to the grid wires and a movement of electrons through the grid circuit external of the tube, back to the cathode as shown in Fig. 7-13. This we identified as grid current. Now, if such a flow of current takes place in the grid circuit, it stands to reason that the grid circuit is no longer a high-resistance circuit, but rather is now a comparatively low-resistance circuit, as shown in Fig. 7-14. The grid-circuit resistance then equals the voltage upon the grid divided by the value of current flow, as determined by Ohm's law.

Whereas the circuit of Fig. 7-11 involves voltage, but no power, for there is no current flow, the circuit of Fig. 7-13 involves both voltage and current, and hence the consumption of a certain amount of power occurs. Now, in normal operation of the triode as an ampli-

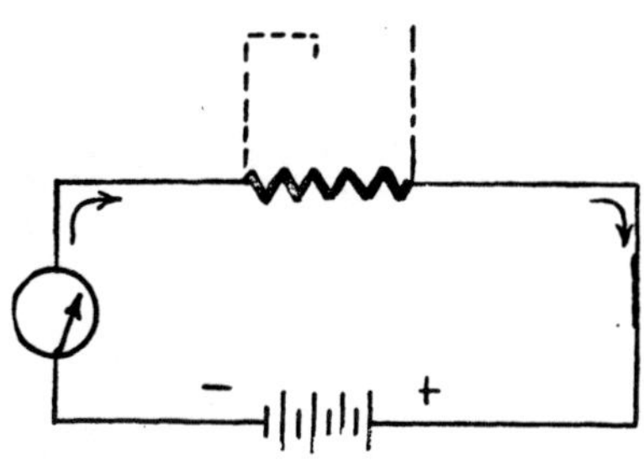

FIG. 7-13 (above). When the grid bias places a positive charge on the grid, electrons are attracted to the grid wires and flow through the input circuit back to the cathode.

FIG. 7-14 (right). A positive bias is the equivalent of a low value of resistance between the cathode and grid through which electrons can easily flow.

fier, the device or system which is the source of the signal voltage that is applied to the grid generally is not capable of furnishing this power without undergoing a change in its operating behaviour. This is brought about by the fact that such devices are usually designed to operate from a very high-resistance circuit, so that practically no current flows through the device. But, when the conditions in the grid circuit are such that grid current flows, this grid current would flow through the device across which the signal voltage exists. This is highly undesirable for reasons to be discussed later. Furthermore, when grid current flows, the resistance represented by the emitter-grid circuit is comparatively low, and if that device which is the source of the signal voltage is intended to operate into a high resistance, operation is impaired.

Supplementary to the above is also the following which must be considered. The signal voltage normally applied to the grid of the tube is of alternating character, that is, it is a voltage which varies between plus and minus with respect to the cathode. During the time that it is minus with respect to the cathode, the correct resistance conditions prevail in the grid circuit, but when it is plus, current flows, that is, if no means are employed to keep the grid always negative, or at least never positive with respect to the emitter. To set up such a condition is the function of the grid bias.

The grid bias sets up the initial voltage relationship between the control grid and the emitter of the tube so that, if a signal voltage which varies between negative and positive, as shown in Fig. 7-15, is

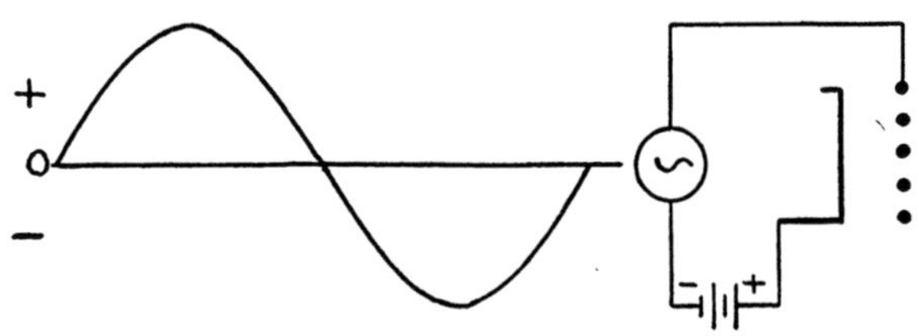

FIG. 7-15 (left). Even though an alternating voltage, varying between plus and minus, is impressed on the grid of a triode, as indicated in Fig. 7-16 (right), the grid is maintained negative with respect to the cathode by the grid bias, as explained in the accompanying text.

applied to the tube, the grid is never permitted to go positive. In other words, the grid bias is a fixed d-c voltage which is in effect in series with the input signal voltage, as in Fig. 7-16; consequently, during the negative half cycle of the signal voltage, the grid is negative with respect to the emitter by an amount equal to the grid bias plus the instantaneous value of the signal voltage, and during the positive half cycle the grid is negative with respect to the emitter by an amount equal to the difference between the grid bias and the instantaneous value of the signal voltage. During the instant when the signal voltage is zero, the grid is negative by an amount equal to the grid bias.

The grid, therefore, remains negative at all times, and the value of the grid bias is always set, at least that is the attempted condition, so that it is greater than the peak value of the input a-c signal. There is, of course, another definition for the grid bias, namely that it establishes the point of operation of the tube. Just what is meant by this definition will be discussed in connection with the various operating characteristics of the triode tube. If, for the moment, we again return to the subject of electrostatic fields, we can say that the grid bias applied to the tube sets up the initial electrostatic field at the grid. Then this field is either made stronger or weaker by the input signal voltage applied to the control grid. So much for the general subject of electrostatic fields in triodes and the facts related to the details associated with such illustrations. From this point on we shall speak about the actual characteristics of the triode, in other words, the "how" rather than the "why" of the triode tube.

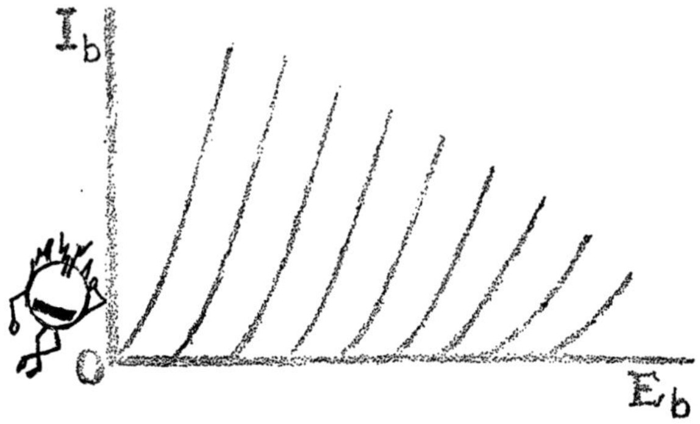

Chapter 8

STATIC CHARACTERISTICS OF TRIODES

LET US CONSIDER the action of the electrostatic fields within the triode from another viewpoint. We know that the application of a voltage of varying polarity and magnitude to the control grid of a tube results in the appearance of corresponding varying electrostatic lines of force at the control grid. The greater the value of this grid voltage, the greater the number of electrostatic lines of force associated with the grid field. The same is true of the lines of force set up at the plate of the tube by the application of the plate voltage.

Unfortunately, we cannot assign convenient numerical values to designate the quantity of these lines of force. Neither can we assign convenient numerical values to the resultant force that is due to the influence of the respective fields upon the electrons within the space charge. Yet we know that these two fields do act upon the electrons within the space charge. Therefore we must evolve some means whereby we can assign numerical values to represent the actions which take place within such a triode tube, and in that way establish the manner in which the voltage applied to the grid influences the plate current, the relative importance of different values of grid and plate voltage with respect to their combined action upon the plate current. In other words, we shall illustrate the manner in which a variation in voltage upon the grid is transferred to the plate circuit. To say this differently, we desire some means whereby we can show in graphical fashion the manner in which the plate current is affected by variations in the grid voltage.

We do this by means of what are known as *grid voltage—plate current* characteristic curves. These curves show the manner in which the application of certain voltages to the control grid affect the plate current. These data are quantitative and therefore permit of a definite determination of performance. At first thought it may seem to you that such curves are repetitions of the electrostatic field actions described earlier in the preceding chapter. This is not the case, for the discussion of the electrostatic fields is the "why" of the tube performance, whereas the grid voltage—plate current characteristic curve is the "how" of tube behaviour. The proper interpretation of these curves makes the prediction of tube performance possible and enables us to make a selection of the necessary external associated components so as to produce predetermined results.

Triode Circuit Element Notations

Before actually starting our discussion of the characteristic curves of a triode, it would be well for us to explain something about the notation employed when describing the action of vacuum tubes. It is customary in vacuum-tube literature to use certain forms of identification of the various circuit elements during the analysis of tubes. While each of these elements may be referred to by its full name and by means of an identifying description, definite conveniences are attained by utilizing certain notations to indicate the elements as well as their operating conditions. This is quite important, as you will see, for during the analysis of vacuum tubes, we work with steady as well as instantaneous values. For example, in Fig. 8-1 a simple triode circuit is shown. You will observe that we have designated the plate supply voltage supply as E_b, the plate voltage that is effective between the plate and the cathode as e_b, the grid bias supply voltage as E_c, and the grid voltage that is effective between the control grid and the cathode as e_c. Since in the case shown, E_b and e_b are equal, and the E_c and e_c are equal, you might ask why we show such apparent duplications of symbols for voltages that are identical.

There is, however, a reason. You shall find, as we continue this investigation of the operation of the triode, that it is often necessary to distinguish between different voltages which exist in the same circuit, and while in Fig. 8-1 E_b and e_b are the same, and E_c and e_c are the

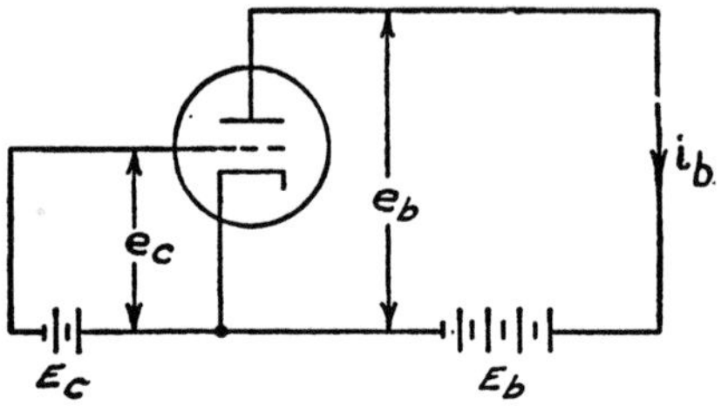

FIG. 8-1. The voltages and current in this triode circuit are designated as follows: E_b and E_c are the plate supply and grid bias voltages respectively; e_b, the voltage effective at the plate; e_c, the voltage effective at the grid, and i_b, the plate current.

same, there will be times when they are different. In fact, you will find cases where more than just two such voltages exist in the complete plate circuit or in the complete grid circuit.

We are, for the most part, utilizing the symbols which have been standardized by the Institute of Radio Engineers. This practice has several advantages. In the first place, these symbols are rapidly replacing those which were more loosely used in the past. If, therefore, you should desire to investigate more extensively any topic covered in this book, you would be able to do so without undergoing the mental gymnastics that would be made necessary by a change in symbols. While it is true that this is an elementary book, the use of the proper symbols serves to cultivate your acquaintance with a letter notation which will be of much use if you desire to advance your knowledge of vacuum tubes by referring to more advanced texts. Furthermore, the use of symbols will enable you to understand better the exact functions of the circuits and circuit elements.

However, we do want to say that while we shall generally keep within the boundaries of the IRE symbols, there are times when, for the sake of clarity, it will be necessary to deviate slightly from this practice, and we shall then employ the letter symbols developed by Dr. H. J. Reich.

Returning to the symbols or designations shown in Fig. 8-1, although some of them may not become entirely clear to you until further discussion, we define them as follows:

E_c = control-grid supply voltage (grid-bias voltage)
e_c = instantaneous total grid voltage, sometimes called instantaneous effective grid voltage
E_b = plate-supply voltage, or the B-voltage

e_b = instantaneous total plate-voltage, sometimes called instantaneous-effective plate-voltage

i_b = instantaneous total plate-current.

These are by no means all of the letter symbols employed in conjunction with the triode and others will be defined as they are introduced in subsequent illustrations. For your convenience as a reference, all of the symbols used in this book are shown in the Appendix in the rear of this volume.

Grid Voltage—Plate Current Characteristic Curves

Considering the nature of the triode, we know that three basic factors control the plate current: the emitter temperature, the control-grid voltage, and the plate voltage. The first of these we can promptly forget, for it is a general procedure in all vacuum tube operations, unless used under special conditions in which case it would be so stated, to consider the emitter temperature fixed and to be of such a value as to provide all of the necessary emission for all normal conditions. Accordingly, it is merely necessary to stipulate the emitter temperature in terms of either heater or filament voltage, or heater or filament current, upon whatever tube characteristic that is being illustrated.

This then leaves the control-grid voltage and the plate voltage as the two variable factors which deserve recognition in developing the grid voltage—plate current curve.

But it would not make sense, in the effort to establish the numerical relationship existing between the effects of the control-grid voltage and the plate voltage upon the plate current, to make both voltages simultaneously variable. Since we desire to know the effect of the grid voltage upon the plate current, the grid voltage is made variable over a range while the plate voltage is held fixed at a predetermined value. The resultant curve (or curves) is then identified by the factors that are variable, and is known as the grid voltage—plate current curve. If we wanted to know the effect of the plate voltage upon the plate current, then the plate voltage would be made variable and the grid voltage would be held fixed. In this case, the resulting curve would again be identified by the variable voltage, and would now be known as the plate voltage—plate current curve.

How the Curve is Developed

The development of a grid voltage—plate current characteristic curve is a relatively simple procedure. All that is necessary is to arrange for the application of known values of grid voltage, negative and positive, a fixed value of plate voltage, and a fixed value of heater or filament voltage. We then measure the value of plate current which flows in the plate circuit for various values of positive and negative grid voltages. The circuit used with a typical cathode-type triode is shown in the enclosed area in Fig. 8-2. The voltmeter V connected across E_c indicates the applied grid voltage. You will note that the grid battery is in two parts and so arranged that the grid can be made either negative or positive with respect to the cathode to an extent limited by the battery voltages.

The voltmeter V across E_b indicates the voltage applied to the plate, and the current meter I indicates the magnitude of plate current flow. The heater and its associated meter are not shown, for it will be assumed that the heater is operated at the proper emitter temperature.

The process of making the curve does not call for much skill. The simplest method is first to make written notes of the plate current for various values of negative grid-voltage, taking measurements of the plate current for each increase in grid voltage. After these have been noted, the performance is repeated for various values of positive grid-voltage, *always keeping the plate voltage fixed at the original value.*

The figures or observations are then recorded upon the chart. In arranging the chart, the plate current, in milliamperes or microamperes, usually the former, is located upon the vertical axis in equal divisions, as shown upon Fig. 8-2. The grid voltage is located upon the horizontal axis, with the zero point (0) selected so that it is about midway along the horizontal axis. This also is shown in Fig. 8-2. This is followed by actually marking the measured values upon the chart, the plate-current-value points being determined by the points along the chart where projections of the values of grid voltage and plate current meet. Thus, if the plate current for zero grid-voltage is 36.5 milliamperes, the point on the plate current axis corresponding to 36.5 milliamperes is projected to the right until it crosses the upward projection of the zero-(0)-grid-voltage line. This is point B in Fig. 8-2. Then suppose that the next measurement was made with 0.5 volt negative grid-voltage and the observed value of plate current was 32.5 milliamperes. Projecting the point equal to 32.5 milliamperes of the plate current axis until it meets the upward projection of the —0.5 volt grid-voltage line identifies this point for the plate current curve. This is point C in Fig. 8-2.

In this way the various points corresponding to the observed values of plate current for increasingly negative values of voltage are indi-

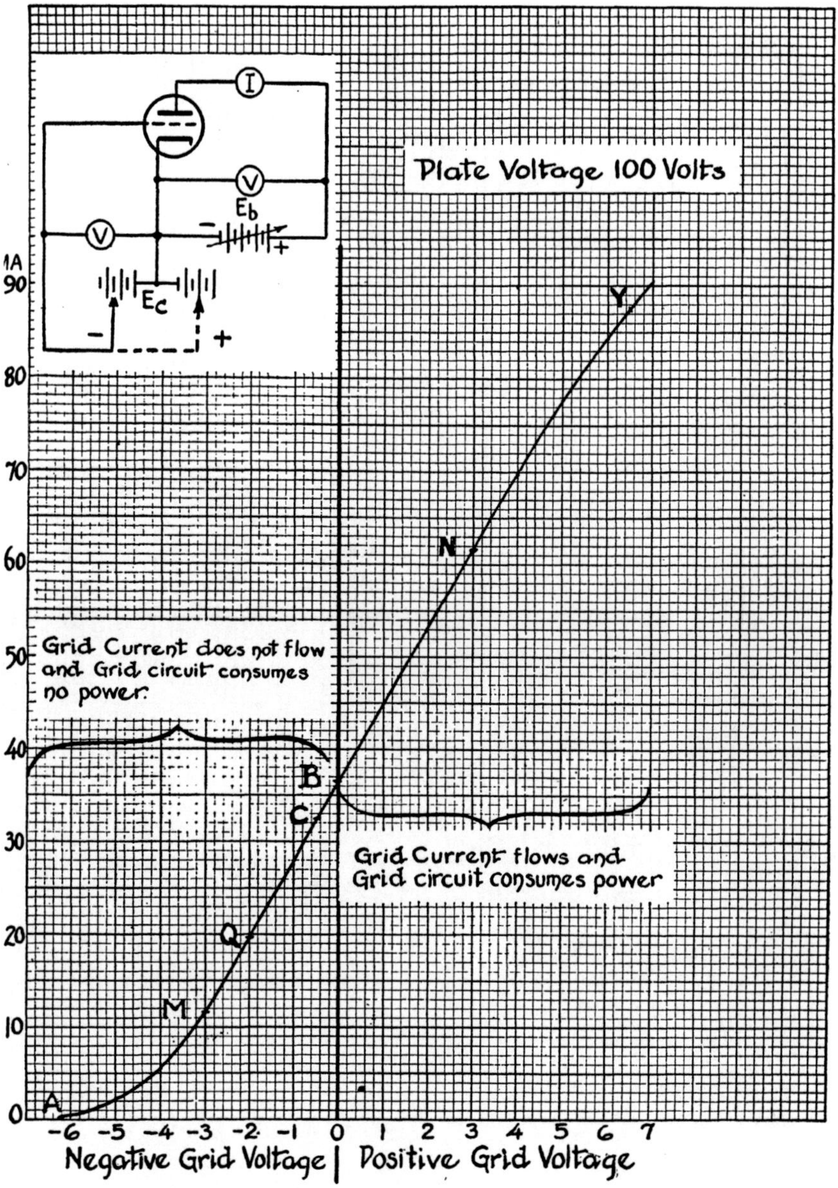

:. 8-2. Grid voltage—plate current characteristic of a triode. The schematic shows the loca-ns of the various meters in the circuit.

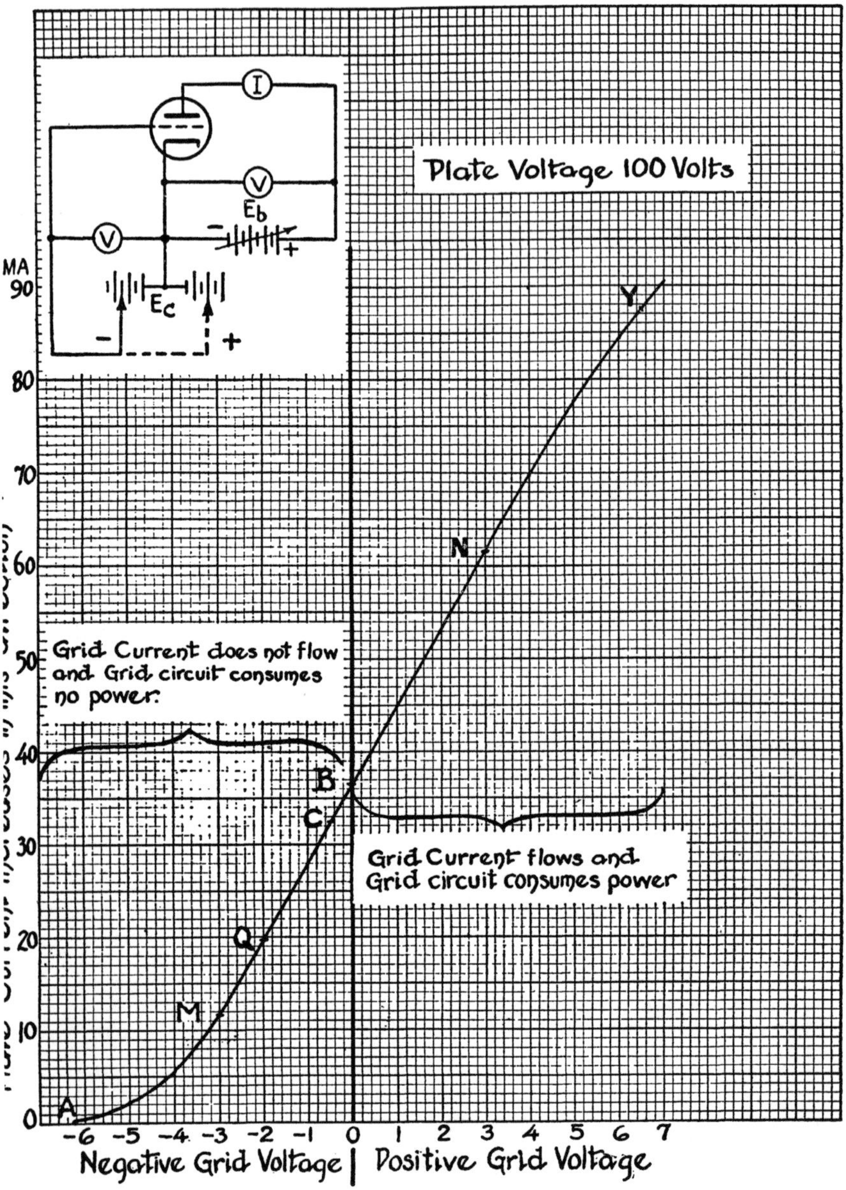

'IG. 8-2. Grid voltage—plate current characteristic of a triode. The schematic shows the loca- .ons of the various meters in the circuit.

cated upon the chart. The maximum limit of negative grid voltage is, of course, that value which results in plate current *cutoff*, which in this illustration corresponds to slightly more than —6.0 volts. Joining all these points together, produces the grid voltage—plate current characteristic for this tube over the zone of negative grid voltage between zero and cutoff value, for the fixed value of plate voltage used, which we shall, for the sake of simplicity, assume to be 100 volts. This portion of the total curve we identify as *ACB*.

Now we are ready to spot the points of the grid voltage—plate current curve for positive voltages upon the grid. The procedure is the same as before, but now the points corresponding to the values of plate current are located to the right of the zero-(0)-grid-voltage line or in the positive-grid-voltage zone and we develop the curve *B-Y*, so that the complete grid voltage—plate current characteristic for the tube used with 100 volts on the plate then becomes the curve *ACBY*.

This curve is a diagrammatic and quantitative representation of the action of the electrostatic fields present in the tube being discussed. This we have indentified as being an oxide-coated emitter type of triode. Whereas in the discussion of the electrostatic fields we spoke in terms of general behaviour, this graph takes this behaviour and interprets it into actual numerical values.

Concerning these numerical values, there are certain details which you must understand. While it is true that the curve *ACBY* is a typical curve for an oxide-coated emitter type of triode, the actual numerical values are not necessarily typical. By this we mean that it would have been possible to select some other tube of this general type and employ 40 volts on the plate instead of 100 and develop a curve which would resemble *ACBY* shown in Fig. 8-2, but involve other values of plate current for like values of grid voltage. Thus the point of plate-current cutoff might in the case of this other tube operated at 40 volts on the plate, take place at say —3.5 volts instead of slightly more than —6.0 volts on the grid; and also zero grid voltage might cause the flow of only 21 milliamperes of plate current while the point *Y* might be at 70 milliamperes of plate current.

How the Curve is Used

Now that we have the curve of Fig. 8-2 what can we do with it? What information does it furnish? How can this curve be used? Looking at the shape of the plate-current curve, we note certain peculiarities. Over a certain portion of its total length, it is straight, whereas over other parts it bends. In other words, the plate-current curve is not linear. If we correlate the straight and bent portions of the curve with the plate current and grid voltage, we find that over the straight portions uniform changes in grid voltage cause uniform changes in plate current. For example, the length of the plate-current curve between zero grid voltage and −3.0 volts on the grid is the same as the length of the curve between zero grid voltage and +3.0 volts on the grid. In other words the grid voltage—plate current curve is linear over this range of grid voltage, as shown by the line between points *M* and *N*.

If you will remember, we stated earlier in this discussion that the grid-voltage source need not necessarily be from batteries. Although it is true that what is here shown is d-c grid voltage, they can just as readily be a-c voltages with peak values equal to the d-c values shown. In other words, this curve shows that for this specific tube, uniform variations in plate current would occur in the plate circuit when an a-c voltage with a peak value of 3.0 volts was applied to the grid and the grid-bias voltage was zero. This can be stated differently as follows: if the grid swing was 3 volts peak with zero grid bias, a similar increase and decrease in plate current would be obtained.

In connection with this kind of information being obtained from the grid voltage—plate current characteristic curve, it should be understood that we do not intend to convey the impression that it represents the conditions of operation as an amplifier. All that we are now doing is to secure as much data as possible from the curve as shown and under the conditions stipulated as existing when the curve was drawn. For the present, we are neglecting the comments previously made about the detrimental effects of operating in the zone of positive grid voltage, that is, using the tube with grid current flowing. Some of the details relating to such conditions have been mentioned and more are to follow, but we desire to stress the point that the curve of Fig. 8-2 does show the existence of a useful portion of this plate-current curve in the zone of positive grid voltage.

Continuing with the comments that can be made about this grid voltage—plate current curve, we see from Fig. 8-2 that there is a distinct limit to the magnitude of negative grid voltage which can be applied to the tube for any one value of plate voltage. In this case it is slightly more than −6.0 volts at which point plate-current cutoff takes place. If we compare the overall change in plate current for the full range of −6.5 volts and for the full range of +6.5 volts upon the grid, it is clearly evident that the change in plate current in the upwards direction (in the positive zone) from that current existing with zero grid voltage, far exceeds the decrease in plate current (in the negative zone) from that current existing for zero grid voltage.

We also note in connection with the curve of Fig. 8-2 that the contour of the plate-current curve below the straight portion is not such as to cause uniform changes in plate current for uniform changes in negative grid bias. For example, the change in plate current between −3.5 volts and −4.0 volts is much greater than between −4.0 volts and −4.5 volts, although the difference is 0.5 volt in each case. This is an extremely important condition with respect to the operation of vacuum tubes. To illustrate why this is so, consider the following.

The Point of Operation

Let us assume that we have a circuit like that shown in Fig. 8-3. This is similar to that of Fig. 8-2, except that we have added a source

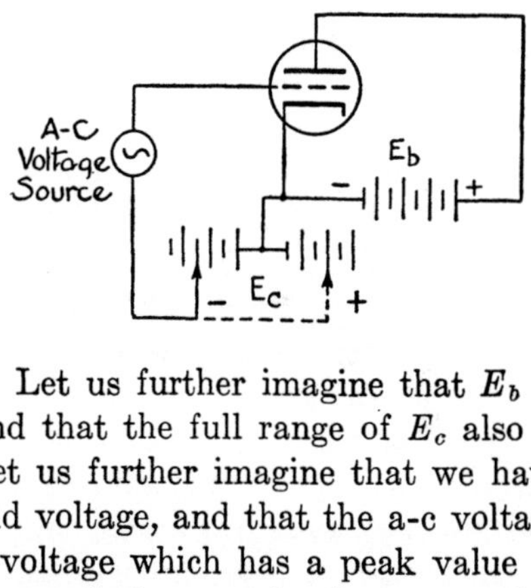

FIG. 8-3. An a-c voltage source has been added to the circuit of Fig. 8-1; this enables a peak value of plus or minus 3 volts to be impressed on the grid.

of a-c voltage in the grid circuit. Let us further imagine that E_b in Figs. 8-2 and 8-3 are the same and that the full range of E_c also is the same in both illustrations. Let us further imagine that we have adjusted E_c in Fig. 8-3 to zero grid voltage, and that the a-c voltage source in Fig. 8-3 is producing a voltage which has a peak value of 3.0 volts in each alternation. Suppose that we apply these operating conditions to our plate-current curve in Fig. 8-2. Since the steady grid-bias voltage is zero, the point of operation upon the plate-current curve is indicated as B. By *point of operation* is meant that value of plate current which is determined by the grid bias with no a-c signal voltage applied to the control grid. These stipulations establish the operating point as corresponding, in this case, to a plate current of 36.5 milliamperes. When the a-c voltage is applied, the peak negative voltage reduces the plate current to 11.5 milliamperes and the peak positive voltage increases the plate current to 61.5 milliamperes, an increase and decrease of 25 milliamperes.

Let us now visualize another condition. Suppose that instead of setting the grid bias at zero voltage, it is made 2.0 volts negative.

Then, according to the definition of the operating point, our curve shows that the plate current for no a-c voltage input would be equal to 19.5 milliamperes, as indicated by the point Q. The application of the grid bias established a new operating point. If now, we considered the application of an a-c voltage of 3 volts peak, it would mean that during the negative alternation, the maximum value of negative voltage present upon the grid would be 5 volts, for as we have said, (See Fig. 7-16) the grid bias and input signal voltages are additive. On the positive swing of the signal voltage, the grid would change in voltage from -2.0 volts to $+1.0$ volt, for this gives a total change of 3.0 volts.

Now, whether or not such a state of affairs would be satisfactory or unsatisfactory, when compared with the previous example is beside the point at the moment. That which is of special interest is the fact that the grid voltage—plate current curve is capable of supplying information concerning the ability of the tube to operate over certain ranges of grid-voltage variation when the plate voltage and grid bias are of certain values, and also the effect of the bias in establishing the operating point with respect to the signal voltage which may be applied to the grid. For example, we see from the curve that if the grid bias is set at -6.5 volts, the application of an a-c signal to the control grid will result in no current flow during the negative alternation of the input signal, whereas current will flow in the plate circuit during the positive half of the a-c voltage applied to the grid.

Again we desire to repeat that these references to the action of the grid bias and a-c voltage applied to the grid apply to the operation of the tube only as shown in Fig. 8-3. Whatever modifications are introduced by the addition of other components in order to use the triode in a receiver or amplifier is not being discussed at this time. What you should bear in mind are the general kinds of data that can be gathered from a grid voltage—plate current characteristic. When the time comes for a correlation between these basic facts and the actual applications of the various tubes, we shall remind you of what has been said in this chapter and show you how these basic details are subject to but minor changes.

To impart just one more thought concerning the operating point on such a grid voltage—plate current characteristic, it is evident that by proper selection of the grid bias, it is possible to move the operating point up and down along the curve between whatever limits are selected. Hence, it stands to reason that the grid voltage—plate current characteristic is capable of furnishing the information which determines the operating capabilities of the tube. This will become more evident later, when we shall talk about *grid families* of curves.

Continuing with the information that is obtainable from the grid voltage—plate current characteristic curve, we observe certain facts. For example, we note that the plate voltage is given as 100 volts. Also, that with zero grid bias, the plate current is 36.5 milliamperes. But why specifically 36.5 milliamperes? Because as a consequence of the structure of the tube, the design of the elements, the placement of the elements, and the various operating potentials, there comes into existence *within the tube* a definite value of resistance which tends to limit the magnitude of the plate current. In this case, its value, which is determined by the conditions already mentioned, for this particular tube is such as to cause the flow of 36.5 milliamperes of plate current when the plate voltage is 100 volts and the grid bias is zero.

We further note from Fig. 8-2 that without changing the plate voltage, but employing a grid bias of −2.0 volts, the steady plate current is reduced to slightly less than 20 milliamperes. Bearing in mind that no change has been made in the structure of the tube or the plate voltage, it becomes apparent that the grid bias must have an effect upon the internal plate resistance. Making the grid negative seems to increase the d-c plate resistance and making the grid positive seems to decrease the d-c plate resistance, as indicated by the corresponding decrease in plate current, although the plate voltage is maintained constant. That the grid bias has a material effect upon the plate resistance of a tube is entirely correct, as will be seen later in this chapter, when we discuss this subject at greater length.

Although there are numerous details which we can determine from a single such grid voltage—plate current characteristic curve, there is one important detail which is associated with such a curve, but not when there is only one such curve. We are referring to the determination of that change in grid voltage which corresponds to a change in plate voltage in order to produce the same result upon the plate current, in other words, the amount of amplification which is available from a triode. These data cannot be secured from a single grid voltage—plate current characteristic curve, for this requires more than one such curve. We shall speak about the development of this shortly, but for the moment let us bring your attention to another type of tube and its grid voltage—plate current curve.

E_C-I_P Curve for Tungsten-Filament Tube

When we were discussing the diode plate voltage—plate current

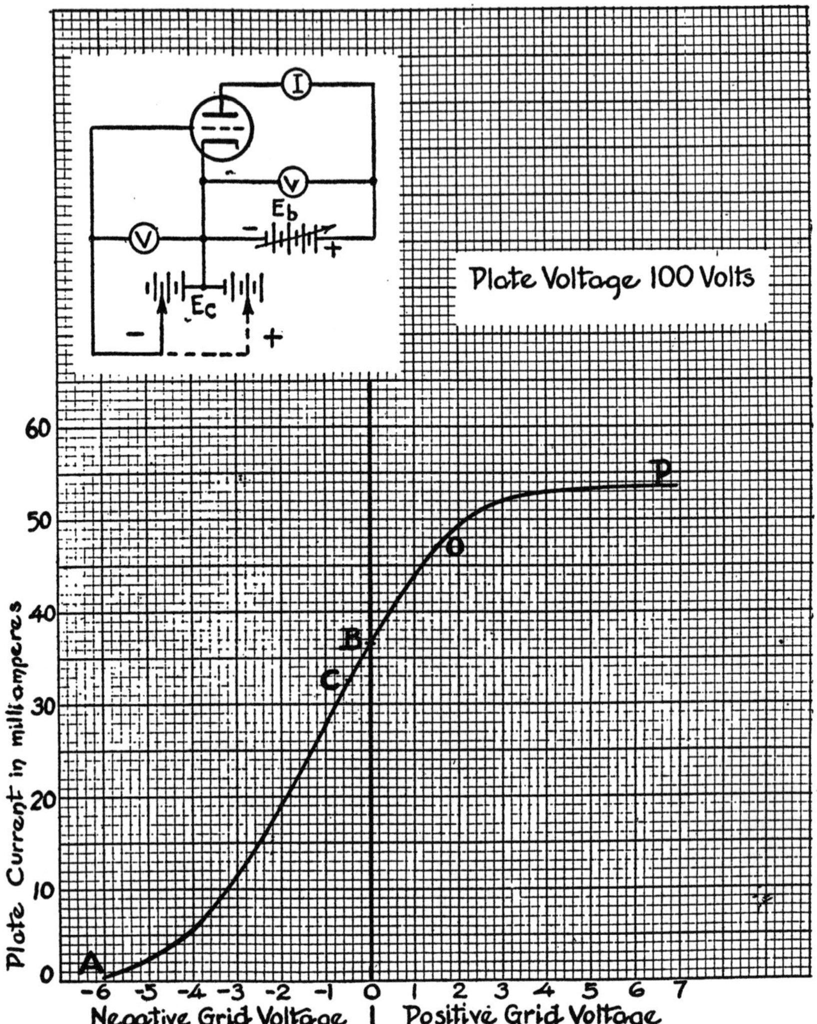

FIG. 8-4. A grid voltage—plate current characteristic curve for a typical tungsten filament type of triode.

characteristic curve associated with Fig. 6-10, we mentioned one very significant fact that was associated with the type of emitter used in the tube. We saw that the tungsten-filament type of emitter is subject to the condition that when a certain plate voltage has been applied to the tube, there is no further increase in plate current for still higher values of plate voltage, due to the fact that all of the emitted electrons are being attracted to the plate. In the case of the oxide-coated emitter, such a condition does not exist because the emission is sufficiently high to accommodate whatever plate voltage is applied, and the tube usually burns out before plate-voltage saturation is reached.

We have a somewhat similar condition in the case of the triode, but in this instance it is very important to bring up the subject of the tungsten-filament type of triode because it seems to have become common practice to illustrate the basic grid voltage—plate current characteristic while using such a triode. The net result is that the type of curve which has been used to represent the basic grid voltage—plate current characteristic differs from Fig. 8-2, which employs an oxide-coated emitter.

What we have in mind is shown in Fig. 8-4 as curve *ACBOP*. This is the grid voltage—plate current characteristic for a typical tungsten-filament type of triode which corresponds to the cathode type shown in Fig. 8-2 when both are operated at a plate voltage of 100 volts. Since the two tubes correspond to each other, you might expect that the values of plate current for similar values of grid voltage would be the same. This is the case over a certain portion of the characteristic curve, starting at the point of plate-current cutoff and continuing all through the negative zone of grid voltage and partly into the positive zone of grid voltage.

It is in the positive zone of grid voltage, at the point *O* along the plate current curve of Fig. 8-4 that the difference between the two tubes arises. We note that in the case of the tungsten-filament emitter there is a distinct limitation in the magnitude of the plate current. Up to about 1 volt of positive grid voltage, the increase in plate current corresponds to the action of the oxide-coated emitter type of triode. But beyond this point the increase in plate current of the tungsten-filament tube for increased values of positive voltage applied to the control grid, becomes less and less until at about 4 volts positive upon the control grid, there is no further material increase in plate current. In other words, further increases in positive voltage applied to the control grid are not capable of accelerating any additional electrons to the plate.

Strangely enough, and although the curve of Fig. 8-4 does not show it, the leveling action upon the plate current does not continue for an indefinite period with constantly increased values of positive grid voltage. After a certain value is reached, which value is greater than the maximum positive grid voltage indicated in Fig. 8-4, the plate current actually decreases due to the fact that with sufficient voltage upon the grid; the positive field of the positively charged control grid exceeds the strength of the field of the positively charged plate so that electrons which would ordinarily tend to move through the grid spaces and move toward the plate, are attracted back to the grid. Further, the secondary-emission electrons which are liberated from the plate in consequence of the high-speed electrons which strike it, are attracted to the grid, and hence move in the opposite direction to the normal plate current; therefore the effective value of plate current is reduced.

In contrast with the tungsten-filament type of tube, we shall note the behaviour of the plate current in the coated-oxide emitter type of tube. In this latter tube it seems virtually impossible to find the point of plate voltage saturation, for the tube usually burns out before that point is reached. As can be seen in Fig. 8-2, the upper portion of the curve *BY* bends slightly, but the tube burns out before an upper bend, which even remotely approaches that of Fig. 8-4, can be reached. The reason for the continual increase in plate current for increased values of positive grid voltage is the tremendous emission obtained from the oxide-coated emitter. This emission is so great that even for high values of positive grid voltage the plate current remains space-charge limited so that the plate current continues to increase as the grid is made more and more positive, until the tube burns out without reaching the limit in plate current. As has been stated, the curve *BY* in Fig. 8-2 is the one associated with modern oxide-coated emitter types of tubes rather than the flattened curve of *BOP* of Fig. 8-4 which occurs for tubes no longer used in great quantities in radio receivers or transmitters.

It is possible to find a number of interesting comparisons between the two types of tubes shown respectively in Figs. 8-2 and 8-4. In the first place, assuming equal emission capability over the normal operating range, there is no essential difference between the two tubes when operated in the zone of negative grid voltage, so we can promptly dismiss that portion of the curve. As to the behaviour of the plate current in the zone of positive grid voltage, it is easy to see that the tungsten-filament type of tube is limited in its operating capabilities. It is true that we have not spoken at any length about just what this operation may be, but whatever it is, we can see without very much trouble that if a voltage applied to the grid is supposed to cause a change in plate current, which is a basic operating principle of the tube, it does so over a much greater range in the case of the oxide-coated emitter tube than in the tungsten filament tube. This much is very evident in the two curves of Figs. 8-2 and 8-4.

The other general information which could be gathered from a grid voltage—plate current characteristic curve, and which was described as applying to the tube shown in Fig. 8-2, is also true in the case of the tube shown in Fig. 8-4. You probably realized this, but we mention it lest it escaped your attention.

In concluding our discussion of the grid voltage—plate current characteristic, one significant detail must be stressed. Inasmuch as the plate voltage used in developing each individual curve is of a fixed value regardless of the magnitude of the plate current, the curves can be said to be of *static* or fixed character. Furthermore, as you can readily see, the only components associated with the vacuum tube are those within the envelope of the tube and those associated with the operating voltage sources. In other words, these curves were not developed under the dynamic conditions that represent an actual operating state which would require components external to the tube, that is, under conditions of load.

For all types of tubes, grid voltage—plate current characteristic curves exist for both static and dynamic conditions. So far we have dealt with the static conditions and the dynamic will follow. As far as the static grid voltage—plate current characteristic is concerned, we have thus far shown but a single curve, a curve which was representative of but one value of plate voltage. Since the plate voltage is an important factor in determining the plate current, we should have several such curves for different values of plate voltage in order to establish the complete static behaviour of the tube. Therefore we have to advance from the single grid voltage—plate current characteristic curve to a family of such curves.

Grid Family of Characteristic Curves

A number of such static grid voltage—plate current characteristic curves, each of which is obtained for a different value of plate voltage by the use of the circuit shown in Fig. 8-2, is usually spoken of as the *grid family* or *static transfer characteristics.* Essentially, each of these curves individually is like that of Fig. 8-2 and as a source of information is equivalent in value, as well as limitations, to what has been said about the curve in Fig. 8-2. But when a number of these curves are correlated, the family as a whole is productive of far more information than we are able to secure from any one individual curve. Such a grid family is shown in Fig. 8-5, and the curves are those for the 6J5 tube. This tube was chosen because it is typical of many that are used.

Looking at Fig. 8-5 and comparing it with either Figs. 8-2 or 8-4, certain differences are quite apparent. As is readily evident, the curves are not carried into the zone of positive grid voltage, the reason being that since normal operation of the triode is predicated upon the absence of grid current, it is not necessary to include the positive grid voltage zone. Therefore this positive zone is omitted in this and subsequent curves. Only in special cases will it be included.

Another peculiarity which you may note is that the values of the plate current are considerably less than those shown in Figs. 8-2 and 8.4. This discrepancy is of very little importance, for we already said that specific values of plate current depend upon a number of different conditions. In this case the difference is nothing more than the arbitrary selection of plate-current values of Figs. 8-2 and 8-4, whereas the curves in Fig. 8-5 indicate actual values for a specific tube. In still other curves of actual tubes in use you may find plate currents which do not exceed 100 microamperes.

So much for the differences between Figs. 8-5 and 8-2. What is really of greater importance, is the information which can be secured from the grid family of curves and the correlation of this family with the plate family. Most textbooks start their discussion of families of tube characteristics by speaking about the grid family, whereas in actual practice, the plate family of curves is of far greater importance and therefore appears much more frequently, in fact is more easily available. We shall do the same in this book, although with a slight departure from the conventional. We shall present the grid family and plate family side by side and show you how the various static characteristics are secured from each of these families. It is hoped that in that way you will be able to assimilate the details more readily.

Suppose then that we examine Fig. 8-5. The fact that a group of curves is shown representing plate-current flow for different values of fixed plate voltage and different values of grid voltage is self evident and requires no discussion, but there are a few significant details which have future application. For example, we see that various combinations of plate voltage and grid voltage result in the same value of plate current; for example, 8.5 milliamperes of plate current is obtained with 350, 300, 250, 200, 150 and 100 volts on the plate, and —12.8, —10.6, —8.2, —5.6, —3.2 and —0.8 volts respectively upon the grid. This can be interpreted as the increase in grid voltage that is required to keep the plate current constant when the plate voltage is increased, or is the required decrease in grid voltage to keep the plate current constant when the plate voltage is decreased.

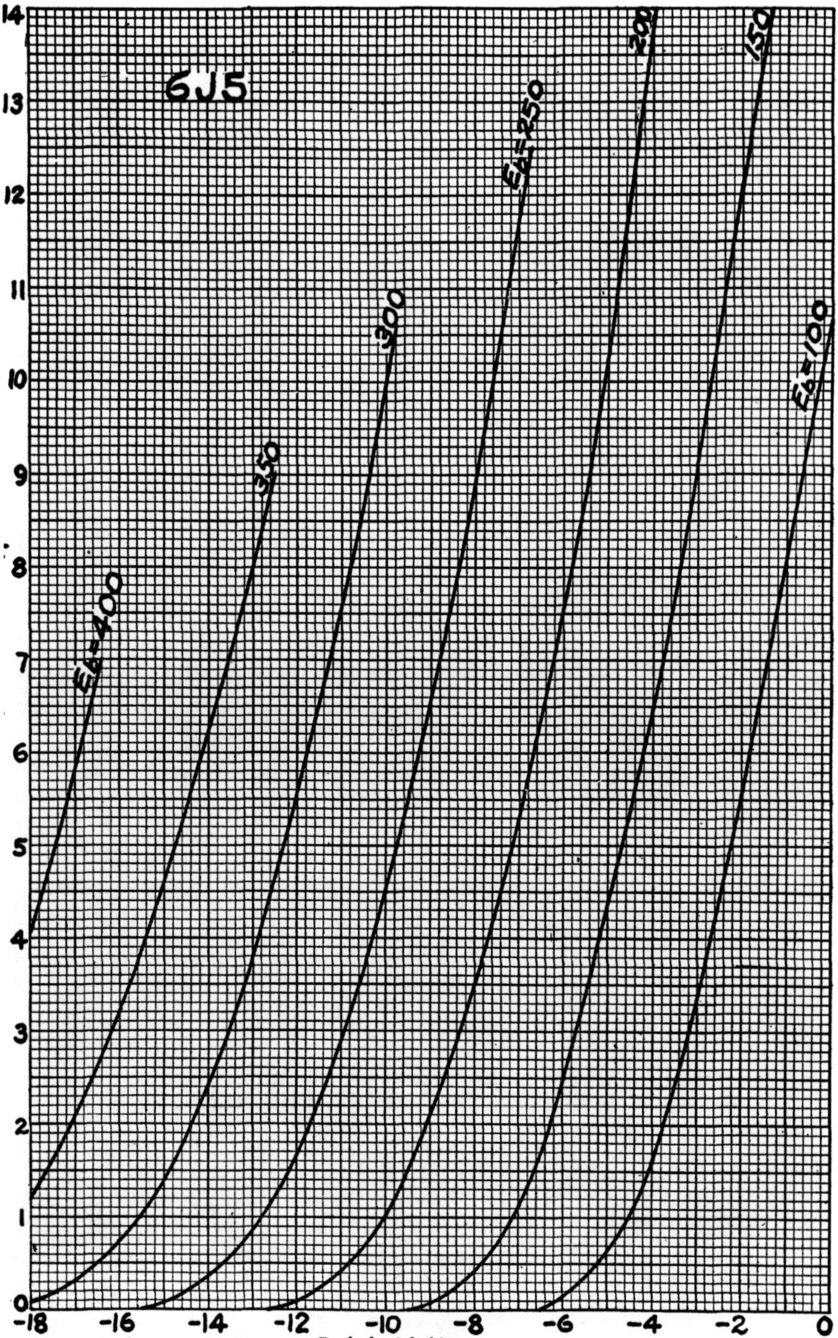

FIG. 8-5. The grid family of characteristic curves for the 6J5 triode.

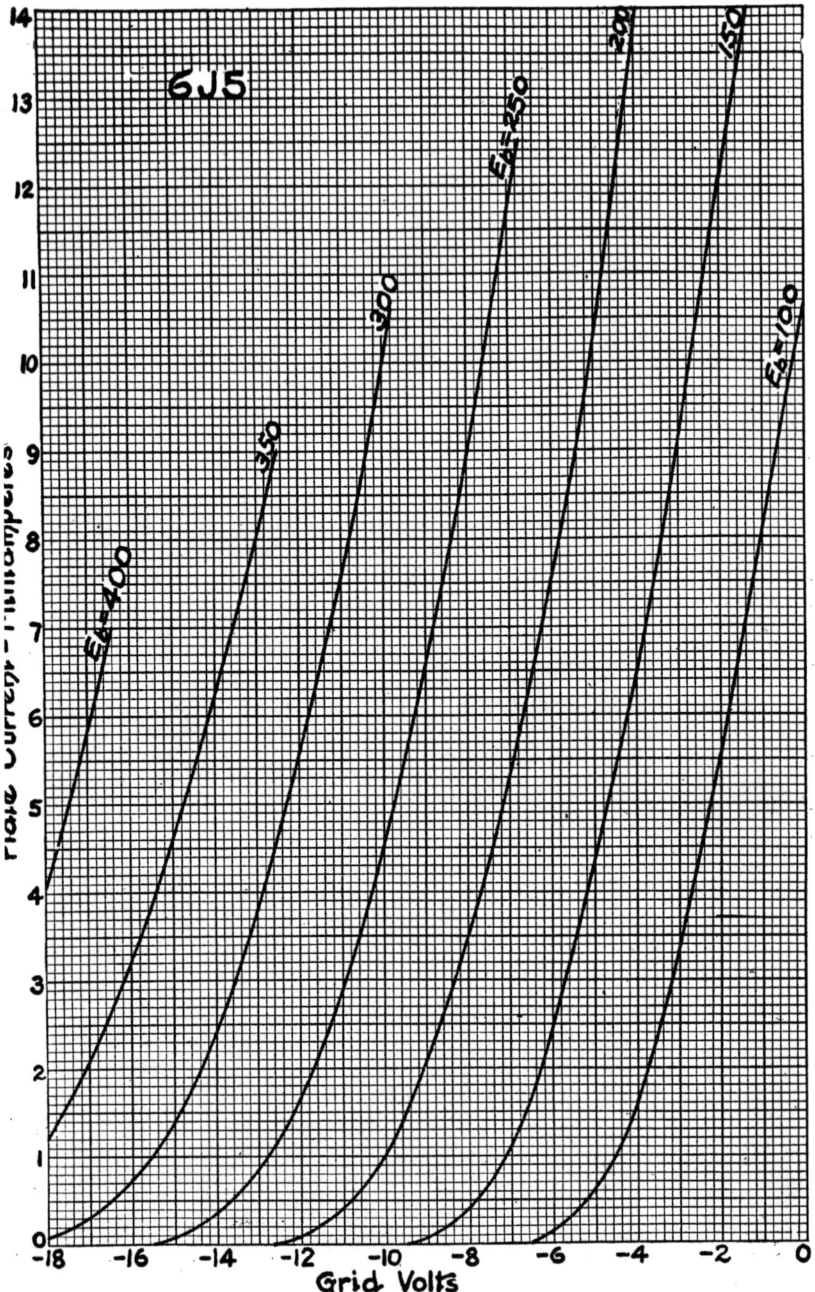

FIG. 8-5. The grid family of characteristic curves for the 6J5 triode.

Here then is the first numerical example of the interaction between the various electrostatic fields existing within the triode. That which was spoken of in the past as being effects, is now put into an actual relationship. Thus, if we desire to see the influence of grid voltage and plate voltage upon plate current, all we have to look for is the amount of plate-current change caused by a change in plate voltage and the corresponding change in grid voltage to cause the same change in plate current. For example, 200 volts on the plate and −8 volts grid bias is productive of 3.5 milliamperes of plate current. Suppose that the plate voltage alone is increased to 250 volts, while the grid voltage is held constant. What happens?

According to Fig. 8-5, the plate current would rise to 8.9 milliamperes. This is an increase of 5.4 milliamperes. How much increase in grid bias from the original −8 volts would be required to decrease the plate current again to 3.5 milliamperes with the same 250 volts on the plate? Looking down along the 250-volt plate-current curve, we find that 3.5 milliamperes corresponds to −10.6 volts. Having a family of grid voltage—plate current curves, we can establish details which were impossible with the single curve of Fig. 8-2.

What the Grid Family of Curves Shows

Such a grid family will indicate not only how much grid voltage in actual volts will offset a certain change in plate voltage, or vice-versa, but also the exact change which will take place in the plate current for any change in grid voltage. This same information was available in Fig. 8-2, but by having the family of curves we know the condition for *all* plate voltages within the operating range of the tube, and also the manner in which this condition varies at different points along the curve. For example, with 100 volts on the plate, it is evident that changing the grid voltage by 1.0 volt, namely from −2.0 volts on the grid to −3.0 volts, causes a reduction in plate current from 5.6 milliamperes to 3.4 milliamperes, that is, a change of 2.2 milliamperes. On the other hand, for the same plate-voltage curve, changing the grid voltage by the same amount, but this time from −4.0 volts to −5.0 volts on the grid, causes a change in plate current from 1.6 milliamperes to 0.65 milliampere that is, a change of 0.95 milliampere. It is therefore evident that the point of operation upon the plate-current curve has a profound effect upon the change in plate current for a certain change in grid or plate voltage.

The grid family also shows the great effect that plate voltage has upon the internal d-c resistance of the tube. The internal d-c resistance decreases rapidly as the plate voltage is increased. With −5.0 volts grid bias, the plate current is 0.65 milliampere with 100 volts on the plate and rises to 10.3 milliamperes with 200 volts on the plate. This is equivalent to a d-c resistance of 154,000 ohms for the lower plate voltage and about 19,400 ohms for the higher plate voltage.

This also shows, as was evident in Fig. 8-2, that increasing the negative bias voltage has the effect of increasing the internal resistance of the tube, for increasing the grid voltage in the negative direction with the plate voltage constant causes a reduction of the plate current, and hence must be increasing the internal d-c resistance of the tube. Similarly, making the grid voltage less negative has the opposite effect, namely, that of reducing the internal d-c plate resistance, for when the grid voltage is decreased the plate current increases.

The grid family of curves also illustrates the fact that a degree of non-linearity exists in all of the plate-current curves. The lower end of all of these curves regardless of the plate voltages, tends to bend. This indicates that under static conditions the change in plate current is not uniform throughout the range of grid voltage, for equal changes in grid voltage.

The information that can be gained from the grid family is associated with the grid voltage, and it therefore stands to reason that whatever final conclusions are derived from these facts must have some connection with the grid voltage. This is so, and although we have not yet discussed any special ratings which are associated with the triode or any special identifying constants, we can say that whatever they will be, they are under the influence of the control grid voltage.

You may wonder why we have not developed the various operating ratings for triodes during this discussion, for it is true that every item we mentioned in connection with Fig. 8-5 has some bearing upon these ratings. We purposely refrained from doing that because we desire first to acquaint you with the plate family of curves.

Plate Voltage—Plate Current Characteristics (Static Plate Family)

Earlier in this discussion of characteristic curves, we stated that both the grid voltage and the plate voltage were variables in the vacuum tube. Assuming that the emitter temperature was held constant, both the grid voltage and the plate voltage controlled the plate current. Having dealt with the grid voltage as the variable quantity, we shall now make the plate voltage the variable factor while holding the grid voltage fixed at different values. The circuit used to make

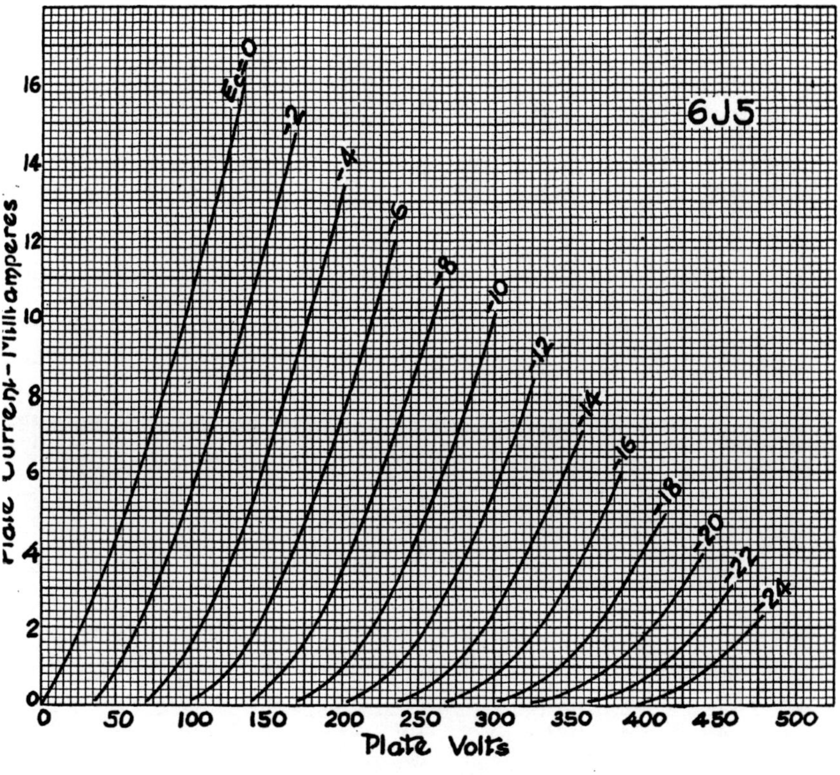

FIG. 8-6. The plate family of characteristic curves for the 6J5 triode.

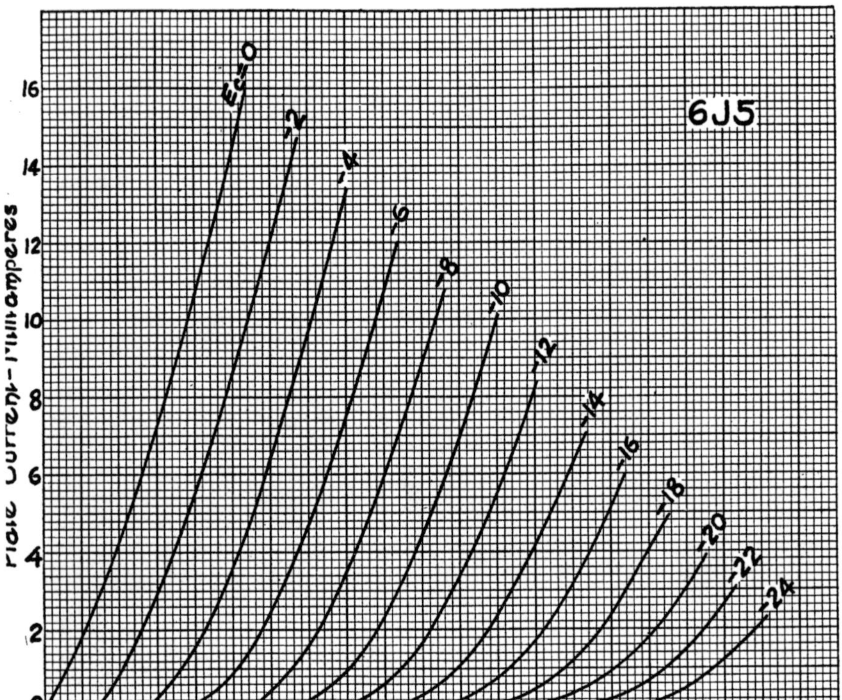

FIG. 8-6. The plate family of characteristic curves for the 6J5 triode.

such measurements is the same as that employed for the curves of Fig. 8-5 and is that shown in Fig. 8-2. The final result of the measurements appears as the plate family of curves shown in Fig. 8-6.

In this group of curves, the grid voltage is varied in steps of 2.0 volts, and plate-current measurements are made over a range of plate voltage, while the grid voltage is held constant at the different fixed values. For example, with the grid voltage at zero, the plate voltage is varied from zero to about 135 volts and the plate current noted. Then the grid voltage is increased to −2.0 volts and the plate current is noted for a range of plate voltage, starting with that value which is just offset by the grid voltage, in this case 35 volts, up to 170 volts. In every case the range of plate current shown for a certain fixed grid voltage starts at that point along the plate-voltage axis near a point for which the grid voltage would cause plate-current cutoff.

The upper limits of plate voltage for each value of grid voltage is determined by what the normal range of plate current is for that tube with the various values of grid voltage. You will see later, when this same family of curves is employed to show the behaviour of this tube under actual operating conditions in an amplifying system, that the value of plate-current flow is less than the maximum values shown in this family of curves. In other words, it is not necessary to establish the plate current at say—12 volts bias for plate voltages in excess of 330 volts, because the tube covered by this family of curves (6J5) is not operated at more than 330 volts upon the plate with—12 volts on the grid. If higher values of plate voltage are used, you can see that higher values of grid bias also are applied.

When speaking about grid and plate families of curves, it is customary to say that given one set of curves, it is a simple matter to plot the other family, for both furnish the same information except in a different form. This can be done, but should not be attempted in Figs. 8-5 and 8-6, because all of the information which would enable this operation is not given upon either of these graphs. Rather, the data shown upon the respective charts have been chosen for their value in an elementary discussion. Moreover, since the plate family is the more important of the two, and since that family is the one which is usually furnished in tube manuals, the need for replotting is obviated; given the plate family, the need for conversion into the grid family is infrequent.

Much more need not be said about the plate family, for it would only be a repetition of what was said about the grid family of Fig. 8-5. About the only item that justifies further comment is the fact that the grid-bias values which are associated with the various curves, indicate that the voltage applied to the grid is held constant at that value during the time that the plate voltage was varied. Values of grid voltage other than those indicated may be applied, of course, but were not applied during the time that the curves of Fig. 8-6 were developed. Had intermediate values of grid voltage been used, the family would have consisted of many more curves, but these are unnecessary for the proper application of the entire family. The number of grid-voltage values selected are sufficient. Let us now consider the first application of the grid family and plate family of curves.

Tube Constants

Having introduced you to the two basic families of curves, we are now ready to deal with the specific information which these curves can supply. Up to the present we have spoken only in generalities; now we shall convert these general remarks into more specific statements. The foundation for these specific facts has already been laid, although we did not so identify them when we brought them to light. We shall now do so.

One significant detail associated with all vacuum tubes which are capable of amplification, such as the triode and other multi-electrode tubes, is that they have three identifying ratings. Each rating bears a different name, but in general each is known as a *tube constant.* Now it is necessary to distinguish between these tube constants and tube characteristics. We defined tube characteristics as being graphical representations of tube behaviour, each characteristic being identified in accordance with the conditions it depicts. A well-known example of this is the grid voltage—plate current characteristic, which identifies the behaviour of the plate current under various conditions of grid voltage for a fixed plate voltage. As you have seen, there is nothing theoretical about such characteristics; instead they are definite and their exact form depends upon existing conditions.

Tube constants, on the other hand, are individual ratings, theoretical in nature and are figures that depend upon the geometric structure of the tube. We say that they are theoretical in nature because some of their values are determined before the tube is built and represent the goal to be achieved in the final tube. Of course, these ratings have a practical value in that they determine the performance of the completed tube and provide the means for selecting tubes which will fulfill specific requirements.

Each of these tube constants is associated with the characteristics of the tube, for tubes possessing certain constants will likewise possess certain characteristics.

Amplification Factor

The first of these constants, which is known as *amplification factor* or *amplification constant,* expresses numerically how much greater the effect of the electrostatic field set up by a voltage applied to the grid is upon the space charge, than the electrostatic field set up at the plate by the same voltage. In other words, this is a numerical expression which is based upon the action of the grid and plate fields upon the space charge with respect to the distance between the grid and plate. By definition, *amplification factor* μ *is the ratio between a change in plate voltage and a change in grid voltage for the same change in plate current.*

Otherwise expressed, $$\mu = -\frac{de_b}{de_c}.$$

It requires a much greater change in plate voltage than a change in grid voltage to cause the same change in plate current. Usually the amplification factor is symbolized by the Greek letter μ, which is pronounced "mu." It is a pure number without any reference to units. Thus a tube may have a "mu" of 10, which means that a change in grid voltage is 10 times as effective in controlling the plate current as the same change in plate voltage.

Judging by the fact that the word "amplification" appears in the name, you probably gather that this constant must have something to do with the property of amplification possessed by the tube. That assumption is correct, and not only does this apply to the triode, but also to many other multi-electrode tubes.

The reference to amplification occurs in the following manner: with a steady voltage applied to the control grid and a steady voltage applied to the plate, a steady value of current flows in the plate circuit. Now a variation in the voltage upon the grid, with the voltage applied to the plate of the tube held constant, causes a change in plate current. Since the current in the plate circuit has been varied, the effect is equivalent to a change in the plate voltage, namely, that change which would be productive of an equal variation in plate current if the plate voltage had been altered while the grid voltage was held fixed. Because of the greater effectiveness of a change in grid voltage than a corresponding change in plate voltage in influencing the plate current, we are justified in considering the equivalent change in plate voltage as being equal to the change in grid voltage multiplied by that number which identifies the number of times a change in grid voltage is more effective than a change in plate voltage in influencing the plate current. In effect, therefore, the change in grid voltage has been transferred to the plate circuit and *the change in plate voltage is equal to the grid voltage times the amplification constant of the tube.*

Determining the Amplification Factor

Let us consider a specific case of determining the amplification constant of a tube. We shall first use the grid family of curves and then the plate family. For this analysis we shall first use the grid family shown in Fig. 8-7. This is the same as Fig. 8-5 except that letters associated with the development of the amplification constant have been marked upon the curves.

Suppose that we desire to determine the amplification factor of the 6J5 when it is operated with 250 volts on the plate and −8 volts on the grid, corresponding to point *P*. We move down the $E_b = 250$-volt curve until we reach some convenient point *Q*. From point *Q* we project a line to the right until it intersects the next adjacent E_b curve, which in this case is point *S* on the $E_b = 200$-volt curve. From point *S* we then project a line vertically upward until it intersects the $E_b = 250$-volt line, thereby obtaining point *R*. (If it should happen that the upward projection of *S* did not result in an intersection with the $E_b = 250$-volt curve, as would have been the case if point *Q* had been chosen as 7.5 milliamperes, then another point *Q*, such as the one we have chosen, is selected.)

At point *R* the plate current is 11.75 milliamperes, while at point *Q* the plate current is 5.35 milliamperes. Also point *R* corresponds to a negative grid voltage of 7, while point *Q* corresponds to a negative grid voltage of 9.6 volts. Holding the plate voltage constant at 250 volts, we see that varying the grid voltage from *R* to *Q*, namely from 7 to 9 volts, results in a change of plate current from 11.75 to 5.35 milliamperes. On the other hand, holding the negative grid voltage constant at 7 volts, and changing the plate voltage from *R* to *S*, namely from 250 to 200 volts, a plate-current change from 11.75 to 5.35 milliamperes, is obtained. In each case there was a change in plate current of 6.4 milliamperes.

Referring to our former use of the letter *d* to mean "a change of," we observe that the change in plate voltage, $de_b = 50$ and the change in grid voltage $de_c = 2.6$ volts. Consequently,

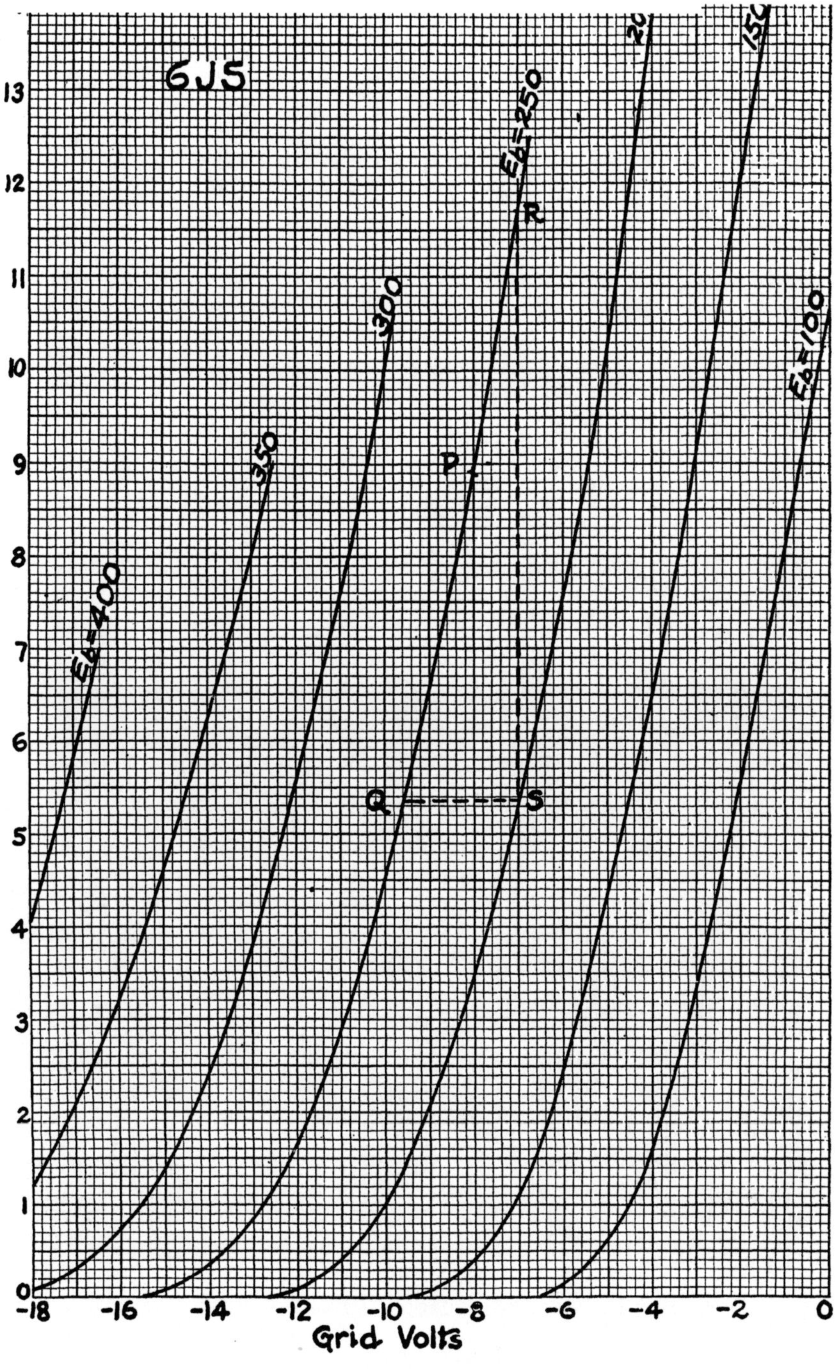

IG. 8-7. The amplification constant of the 6J5 triode at 250 volts on the plate and —8 volts id bias, point P, can be found by following the procedure in the accompanying text.

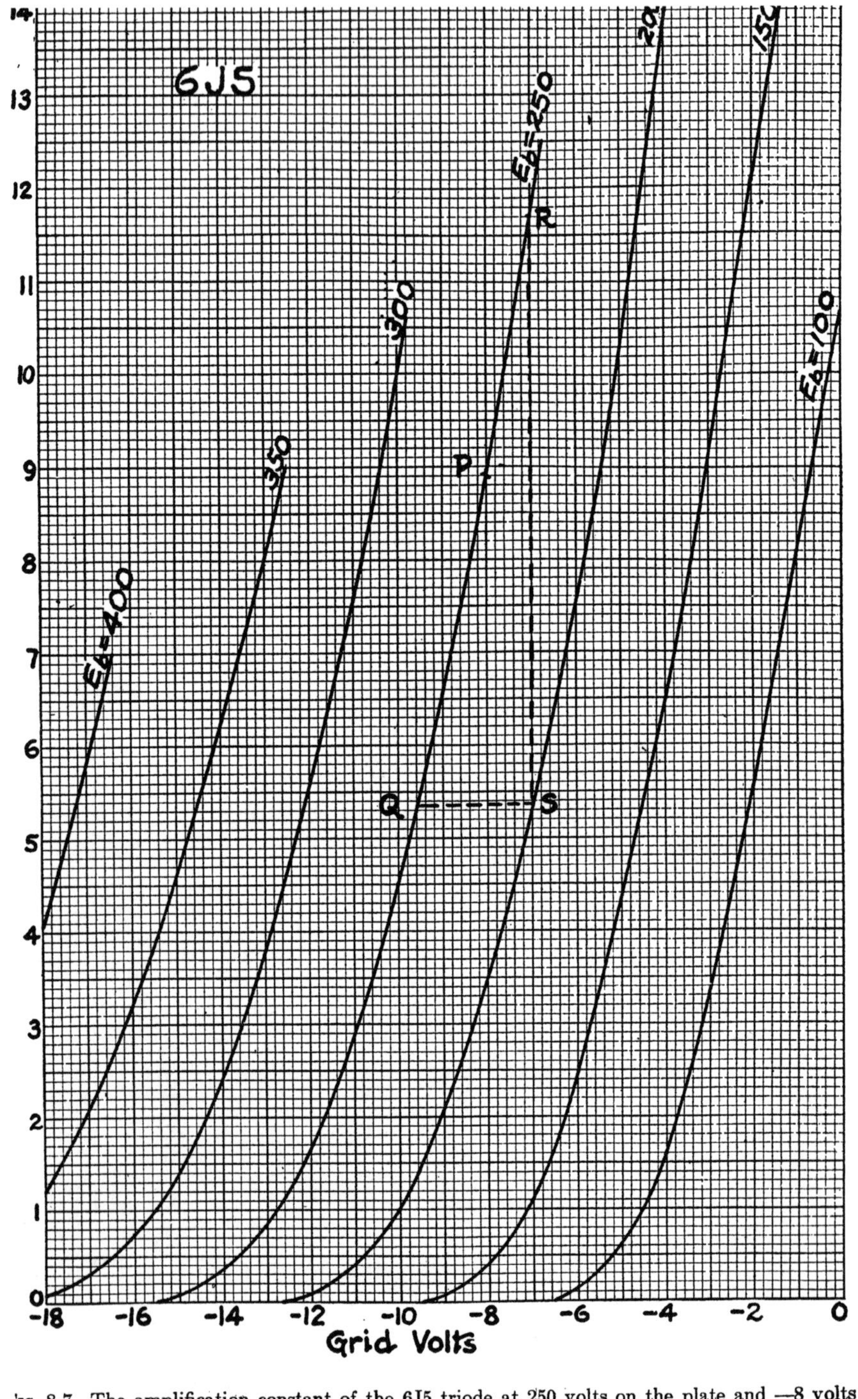

'IG. 8-7. The amplification constant of the 6J5 triode at 250 volts on the plate and —8 volts rid bias, point P, can be found by following the procedure in the accompanying text.

$$\mu = \frac{de_b}{de_c} = \frac{50}{2.6} = 19.2$$

This indicates that the grid voltage is 19.2 times as effective in controlling the plate current as the plate voltage is, and, theoretically speaking, the tube used at the plate voltage of 250 volts is capable of amplifying a signal fed into its grid by 19.2 times. The amplification factor thus obtained by graphical means is very close to the figure of 20 quoted by the manufacturer of the tube.

It is not necessary that such curves as are given in Fig. 8-7 be available to determine the amplification constant. All of the operations which were shown in the curves can be carried out by means of actual measurements. In other words, if tube curves such as those shown are not to be had, it is unnecessary that the grid family be developed before ascertaining the value of mu.

With regard to the accuracy of the "mu" figure established as being the amplification factor, the smaller the change of grid and plate voltages, the greater the accuracy, but inasmuch as a precision calculation is not needed, an error of 5% is perfectly permissible.

Determining μ from the Plate Family

It might be well to remark at this time that while the graphical method of determining the amplification factor as shown in Fig. 8-7, is the one usually described in textbooks, the general lack of availability of grid voltage—plate current characteristics makes the application of the aforementioned method somewhat difficult of execution in practice. This information can also be obtained from the plate family of curves, which we shall now employ. These curves contain the same information as the e_c-i_b (grid voltage—plate current) characteristics and we shall describe the method whereby the amplification factor may be graphically obtained from the plate family. The illustration to be used is shown in Fig. 8-8. As you can see this is the same as Fig. 8-6, except that the points upon the curves associated with the development of the mu of the tube, are indicated.

Let us assume that we desire to find the amplification factor approximately at point A on the $e_c = -8$-volt curve. Point A (which corresponds to a plate voltage $e_b = 216$ volts, grid voltage $e_c = -8$ volts, and plate current $i_b = 5.0$ milliamperes) is projected to the right until it strikes the next grid-voltage curve ($e_c = -10$ volts) at point B (which corresponds to a plate voltage $e_b = 257$ volts, grid voltage $e_c = -10$ volts, and plate current $i_b = 5.0$ milliamperes). We now project

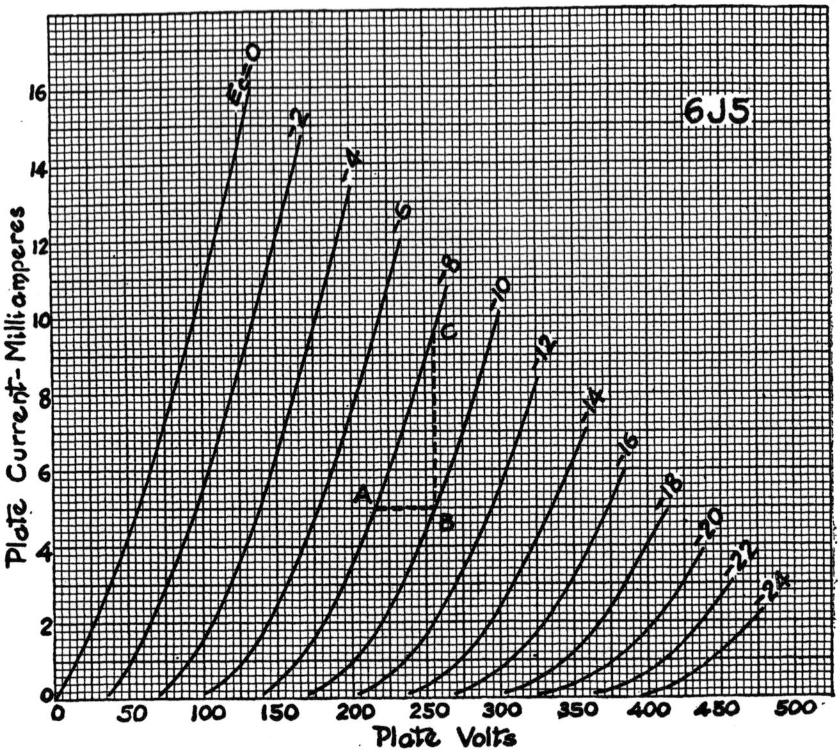

FIG. 8-8. The amplification constant can also be determined by using the plate family of characteristic curves as outlined in the accompanying text.

point *B* upward until it strikes the original grid-voltage curve ($e_c = -8$ volts) at point *C* (which corresponds to a plate voltage $e_b = 257$ volts, grid voltage $e_c = -8$ volts, and plate current $i_b = 9.6$ milliamperes).

It will now be seen that if we move from *A* to *C* along the $e_c = -8$-volt curve, we have maintained the grid voltage constant, but the plate voltage has changed from 216 volts at point *A* to 257 volts at point *C*. This corresponds to a plate-voltage change $de_b = 257 - 216 = 41$ volts. On the other hand, if we moved from point *B* to point *C*, the plate voltage would remain constant, but the grid voltage would change from −10 to −8 volts. In other words, the grid-voltage change is $de_c = 10 - 8 = 2$ volts.

According to the figures quoted and which are indicated on Fig. 8-8, a change of 41 volts on the plate is equivalent to a grid-voltage change of 2 volts. Therefore,

$$\mu = \frac{de_b}{de_c} = \frac{257 - 216}{10 - 8} = \frac{41}{2} = 20.5$$

which is a value substantially in agreement with the previously obtained value as well as the one given by the manufacturer. You may note that the value of mu as established from the plate family, is closer to the manufacturer's figure than that derived from the grid family. This is not due to the use of different families of curves, but rather to the use of a smaller change in grid and plate voltages.

Thus in the use of the plate family of characteristics for the determination of the amplification factor, the process is essentially the same as that which is employed for the grid family. As a matter of fact, the important thing to bear in mind is the equation which expresses the relationship determining the "mu" of the tube. The use of such plate families or grid families of curves is purely a matter of convenience, in that someone else has already completed the task of making the measurements.

Amplification Factor of Triodes

With regard to the amplification factor or constant of triodes, there are three general classifications: *low-mu triodes, medium-mu triodes,* and *high-mu triodes.* A low-mu triode is one which has an amplification factor which is less than 10; for example, the 2A3, which has a mu of 4.2. A medium-mu triode has an amplification factor which is more than 10 and less than 30. Such a tube is the 6J5 with a mu of 20. High-mu triodes have amplification factors of more than 30, examples of which are the 6K5G with a mu of 70, and the 6SF5 with an amplification factor of 100. Of course, these are not offered as rigidly fixed limits, but in general these are the normal limits of these various classifications of amplification factors.

In connection with this comment concerning amplification factor, it is well to remark that later in this text we shall discuss tubes which have amplification factors which vary with the applied grid bias. Since this is dealt with later, as well as the factors which determine how much of the amplification factor is usable, it is unnecessary to devote further time to the subject at this moment.

Plate Resistance

Another tube constant is the *plate resistance.* The internal plate resistance of the tube is an item of consequence, particularly in the selection of components to work with certain tubes. Like the diode, the triode bears two values of plate resistance: a-c and d-c. If you will remember, we have made several references to the d-c plate resistance in earlier portions of this chapter. As in the case of the diode, the a-c plate resistance of the triode, as well as other multi-electrode tubes, is of far greater interest than the d-c plate resistance. In fact, it is customary when speaking about plate resistance without any other qualification to have the a-c plate resistance in mind. The reason for this —and this may be slightly premature at the moment—is that the usual application of the triode involves the presence of a-c voltages and currents in the grid and plate circuits. Hence, the use of a-c plate resistance, often mentioned as *plate impedance,* for it is this which tends to limit the plate current flow.

Concerning the plate resistance, a-c as well as d-c, either the grid family or the plate family of curves can be used. Therefore, we shall show the application of first, the grid family and then the plate family for the determination of this tube constant. In Fig. 8-9 are shown the curves comprising the grid family, which we have already discussed in quite some detail.

Suppose that we wish to determine the d-c plate resistance at point P of Fig. 8-9 namely, with 250 volts on the plate and −8.0 volts on the grid, these two quantities corresponding to a flow of 8.9 milliamperes of plate current. Now since the d-c plate resistance of a tube is defined as being the d-c plate voltage divided by the d-c plate current, we obtain

$$R_p = \frac{E_b}{I_b} = \frac{250}{.0089} = 28{,}100 \text{ ohms}$$

For any other value of plate voltage, the d-c resistance would be determinable by the application of Ohm's law, wherein the numerator would be the plate voltage and the denominator would be the corresponding d-c plate current. The very simplicity of this calculation does not justify more than just a simple comment to the effect that it is determined in exactly the same manner, using the plate family of Fig. 8-10. The various triangles shown on this graph can be neglected in establishing the d-c plate resistance. Their presence is needed for the development of the a-c plate resistance.

Finding the A-C Plate Resistance

The determination of the a-c plate resistance is not quite so simple, although it is not difficult. You may remember the definition of the a-c plate resistance of the diode as being the ratio between a change

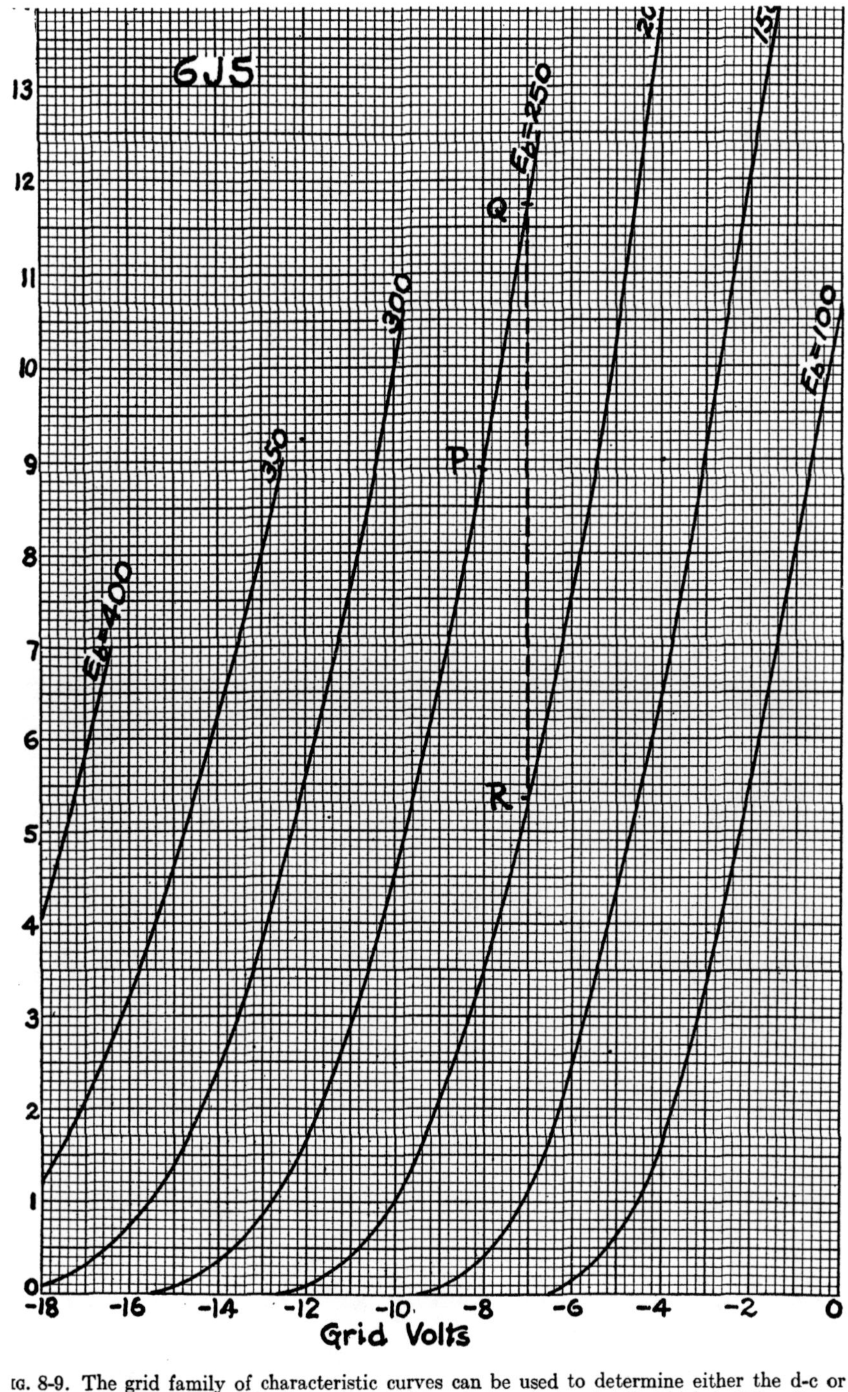

FIG. 8-9. The grid family of characteristic curves can be used to determine either the d-c or a-c plate resistance of a triode; here the resistance is found at point P, which is 250 volts on the plate and —8 volts on the grid.

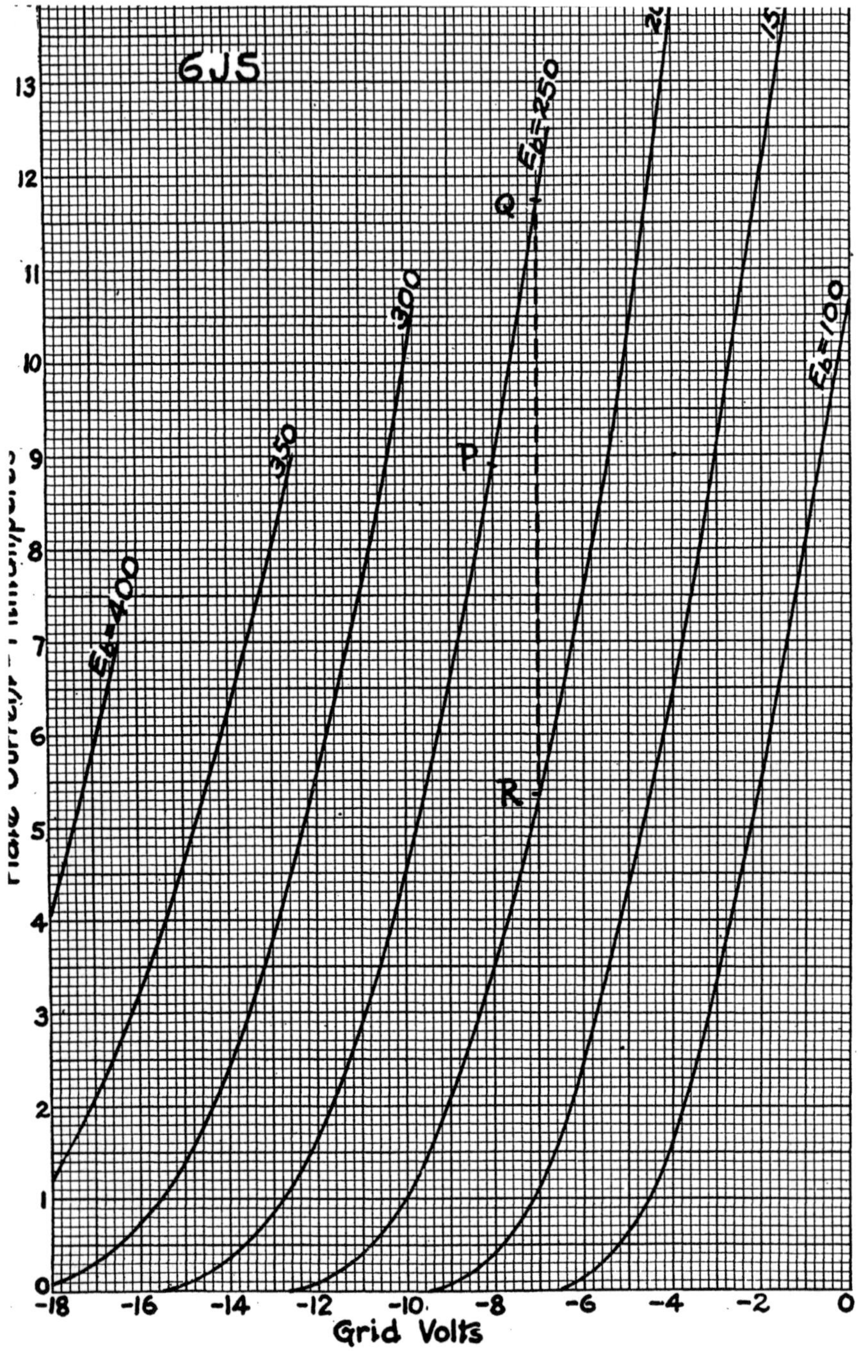

FIG. 8-9. The grid family of characteristic curves can be used to determine either the d-c or a-c plate resistance of a triode; here the resistance is found at point P, which is 250 volts on the plate and —8 volts on the grid.

in plate voltage and the resulting change in plate current. Expressed in the form of an equation, this is

$$r_p = \frac{de_b}{di_b} \text{ (with grid voltage constant)}$$

To establish the a-c plate resistance using the grid family of Fig. 8-9, we again refer to the previously used point *P*, namely 250 volts on the plate and −8.0 volts on the grid which results in the 8.9-milliampere flow of plate current.

As we wish to establish the plate resistance at this point *P*, we choose some point further up on the 250-volt curve where the curve is much straighter than it is at the lower end. The chosen point, *Q*, is on the −7-volt grid line and the intersection of this line and the 250-volt curve, shows a plate-current flow of 11.75 milliamperes. Since we desire a change in plate voltage with the grid voltage held constant, we follow down the −7-volt grid line until it intersects with the 200-volt curve at the point *R*. The plate-current flow for this new plate voltage is found to be 5.35 milliamperes.

We now have the two limits of plate voltage and two corresponding values of plate current with the same grid voltage. Applying these values to the equation for a-c plate resistance and converting milliamperes into amperes, we have

$$r_p = \frac{de_b}{di_b} = \frac{250 - 200}{.01175 - .00535} = \frac{50}{.0064} = 7810 \text{ ohms}$$

This value is in substantial accord with the manufacturer's figure of 7700 ohms.

Using the Plate Family in Finding A-C Plate Resistance

Determination of the a-c plate resistance from the plate family of curves is but slightly more detailed. Referring to Fig. 8-10, the initial point is *P* on the curve marked −8.0 volts. This corresponds to a plate voltage of 250 volts and a plate current of 8.9 milliamperes. The varying voltage representative of a "change of" plate voltage is accomplished by varying the plate voltage above and below the initial operating point *P*. This zone of operation included in the change in plate voltage is bounded by the small triangle drawn in dashed lines. In this drawing de_b is the total change in plate voltage equal to 20 volts, or a swing of 10 volts either side of the mean value of 250 volts, point *P*. The lower plate voltage limit is 240 volts, whereas the upper limit is 260 volts. The base of the triangle extends over the full voltage change of 20 volts.

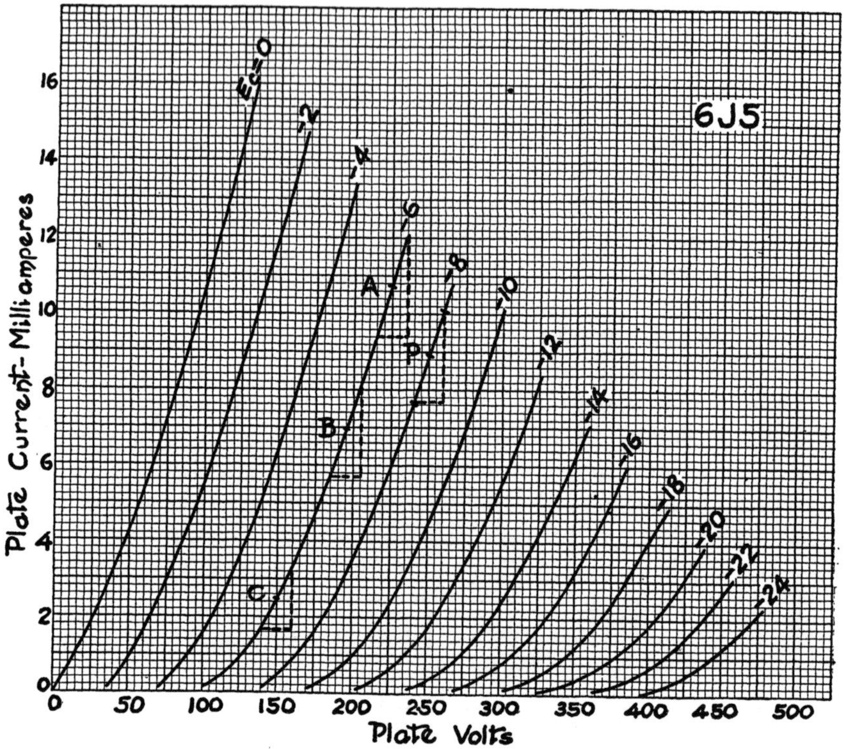

FIG. 8-10. The construction lines on this plate family of curves are used in finding the a-c plate resistance as described below.

The equation specifies that a change be made in plate voltage with the grid bias maintained constant. Accordingly, if the plate voltage is to be increased and decreased, say 10 volts from the average value, the upper point of the triangle lies where the —8.0-volt grid-voltage line intersects the 260-volt plate-voltage line. In turn, the lower limit is established where the 240-volt plate-voltage line intersects the —8.0-volt grid-bias curve.

The change in plate current, di_b, is established by noting where the —8.0-volt grid-voltage curve intersects the upper and lower plate-voltage lines and projecting these points towards the left to the plate-current axis. This establishes the upper plate-current value as 10.1 milliamperes and the lower value as 7.6 milliamperes. Converting these current values into amperes, the equation for the a-c plate resistance at the initial point is

$$r_p = \frac{de_b}{di_b} = \frac{260 - 240}{.0101 - .0076} = \frac{20}{.0025} = 8000 \text{ ohms}$$

This is not in exact agreement with the manufacturer's rating of 7700 ohms, but it is sufficiently close to meet our needs.

Discrepancies in a-c plate-resistance calculations of this kind are to be expected, unless all are made at the same point. For example, if we select −8.0 volts as the grid bias and determine the a-c plate resistance for three different values of plate voltage in accordance with the method just described, we find that at point A ($de_b = 20$ volts and $di_b = .0027$ ampere), the a-c plate resistance is 7410 ohms. At point B ($de_b = 20$ volts and $di_b = .0023$ ampere), the a-c plate resistance is 8700 ohms, and at point C ($de_b = 20$ volts and $di_b = .0016$ ampere), the a-c plate resistance is 12,500 ohms. The steeper the plate current curve, the greater the change in plate current for a given change in plate voltage, and consequently the lower the a-c plate resistance.

From the examples cited, it would appear as if several values of a-c plate resistance must be given in order to furnish the a-c plate-resistance data applicable to a tube. This is not the case, for it is general practice to give but one value, that which exists for the normal rated plate and grid voltages. Thus, in the case of our 6J5 tube with 250 volts on the plate and −8.0 volts on the grid, the a-c plate resistance, as determined from the plate family of Fig. 8-10, would be 8000 ohms.

Before concluding this subject of plate resistance, we should like to remark that very frequently the a-c plate resistance is referred to as *plate impedance*. Actually the word impedance signifies a term which includes both inductance and capacitance, neither of which appears in the determination of the a-c plate resistance, so that it would seem more proper to refer to the a-c plate resistance; however, both terms are used interchangeably.

Transconductance (Mutual Conductance)

We have shown the relationship existing between a change in grid voltage and a change in plate voltage to produce the same change in plate current. We have also discussed the subject of a-c plate resistance. Now we can suggest to you that the magnitude of plate-current change is definitely related to the grid- or plate-voltage change and to the a-c plate resistance. Forgetting for the moment that we are working with vacuum tubes and thinking solely in terms of simple electrical circuits, we know that any resistance in a circuit tends to limit the current. If we transfer this basic idea to the vacuum tube, it stands to reason that the value of a-c plate resistance, whatever it may be, must of necessity tend to limit the magnitude of the change in plate current which the change of grid voltage tends to produce.

It would, therefore, seem quite natural, considering the importance of the change in plate current caused by the change in grid voltage, if there were some expression which described this condition. In other words, it would appear from simple reasoning that the greater the change in plate current for a given change in grid voltage, the better the state of operation, for then the given input voltage to the vacuum tube would have the greatest effect in the output circuit. And if there is any such relationship, it should have some connection with the amplification factor as well as the a-c plate resistance, for both of these have a great bearing on the influence of the grid upon the plate current.

There is just such a relationship: one which expresses the amount of change in plate current for a unit change in grid voltage with the plate voltage held constant. This relationship is another of the "tube constants" and is known as *transconductance* or as *mutual conductance,* although the latter term is rapidly falling into disfavor. The definition of transconductance is the ratio of the change in plate current to the change in grid voltage which produced it, the plate voltage remaining constant. Symbolically, it is

$$g_m = \frac{di_b}{de_c} \quad \textit{(with } e_b \textit{ constant)}$$

Involving both the grid and plate circuits, the symbol for transconductance is g_{pg} and is expressed in micromhos. There are other types of transconductance between the *elements* of a tube, but with these we are not concerned. When the term transconductance is used without qualification, it is always the grid-to-plate transconductance that is meant, and this grid-plate transconductance beside bearing the symbol g_{pg} is usually symbolized as g_m. Another expression, although not so popular as the transconductance, is in direct conformity with the equation and is stated as ma/v or the number of milliamperes *change* per volt change on the grid.

Transconductance from Grid Family

Like the amplification constant and plate resistance, the transconductance of a triode can be determined from either the grid family or the plate family and as has been our custom, we will show both methods. To determine the transconductance of our sample 6J5 tube by using the grid family of curves, we shall refer to Fig. 8-11. Since the basic curves in this chart are the same as those for the several examples given before, we need not discuss them. The initial point is again chosen to be 250 volts on the plate and −8.0 volts on the grid. This is the point *P* on the 250-volt curve, where that curve intersects the −8.0-volt grid-bias line.

In accordance with the basic definition of transconductance, the grid potential is varied slightly above and below this operating point: down to −7.6 volts and up to −8.4 volts, or a swing of 0.4 volt in each direction. This is indicated by de_c, at the base of the triangle. When the grid voltage is −7.6 volts, the plate current is 10.0 milliamperes (plate voltage still 250 volts). When the grid voltage is changed to −8.4 volts with the plate voltage still at 250, the plate current falls to 7.9 milliamperes. Thus for a change of 0.8 volt, the plate current changes 2.1 milliamperes. Converting these current values into amperes and inserting the quantities in the basic equation, the transconductance then is

$$g_m = \frac{di_b}{de_c} = \frac{0.010 - 0.0079}{8.4 - 7.6} = \frac{0.0021}{0.8} = 0.00263 \text{ mho}$$

Since 1 mho = 1,000,000 micromhos, 0.00263 mho = 2630 micromhos. The value thus obtained is in close agreement with the figure given by the tube manufacturer, namely, 2600 micromhos. If this value of transconductance were to be expressed in ma/v, it would be 2.63 milliamperes per volt.

Importance of Transconductance

The importance of the transconductance now becomes obvious for it tells just how effective the grid is in effecting a change in the plate current. A large value of transconductance indicates that the grid is very effective, for it means that the change in plate current is high, whereas a small value of transconductance signifies a smaller degree of control. In fact, so important is the transconductance of a tube that it is frequently used as the basis of tube merit. When a tube is tested in any of the devices known as "tube testers" or "tube checkers," a low value of g_m as compared with a substantially higher normal value is accepted as conclusive evidence that the useful life of the tube is over. At the same time it is well to remember that the use of the transconductance as an indication of merit, is suitable only when working with tubes intended for similar service. It is preferable

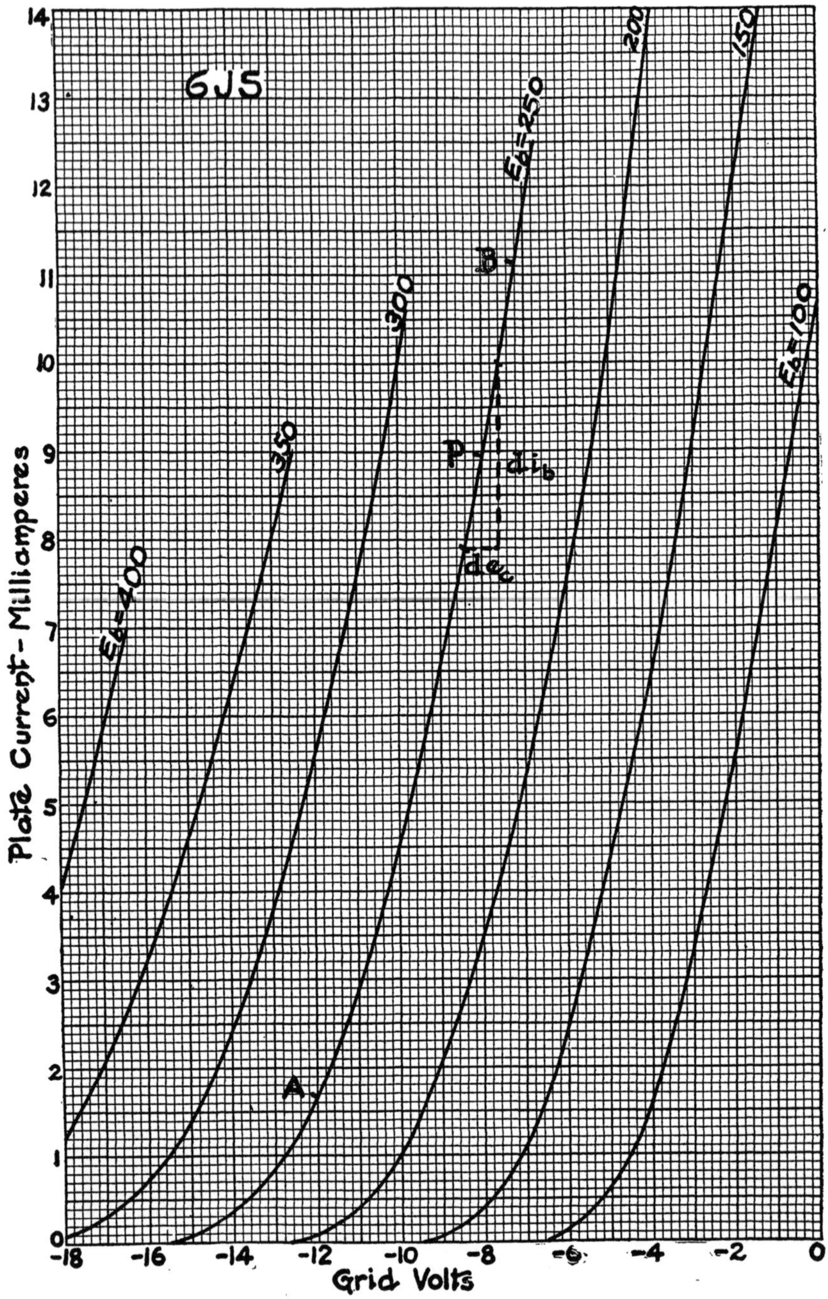

FIG. 8-11. How the grid family is used to find the transconductance of a 6J5 triode.

to consider transconductance in conjunction with the other constants of the tube.

Like the other factors already mentioned, the transconductance of a tube varies over wide limits depending upon the point of operation along the plate-current curve. For example, in Fig. 8-11, determination of the transconductance by changing the grid voltage around an operating point indicated by *A,* results in a value of only 1000 micromhos; whereas the same constant, by establishing the operating point at *B,* results in a value of 2830 micromhos. For ordinary use this variation can be neglected, for in practically all cases it is understood that the transconductance of a tube is given as that which would be established at the proper operating point by using the recommended grid and plate voltage. Thus operating point *A* is that with −12.0 volts on the grid instead of −8 volts and operating point *B* is for −7.2 volts instead of −8.0 volts.

These two examples are interesting, however, for they point to the manner in which the value of grid bias causes a variation in the transconductance of a tube. As you can see, increasing the grid bias in the negative direction tends to reduce the "g_m", while operating with lower-than-normal grid bias tends to increase the "g_m". This phenomenon is not entirely unexpected, for we have seen how an increase in the grid bias increases the a-c plate resistance, while a decrease in the grid bias decreases the plate resistance.

Transconductance from Plate Family

Since the grid family of curves is not always available, it is desirable to be able to determine the transconductance of a tube by using the plate family. This is done in Fig. 8-12. Remembering that the definition of the transconductance requires that the grid voltage be varied and the plate voltage remain constant, we draw a line upward along the 250-volt line from the point *A,* our initial point (250 volts on the plate and −8.0 volts on the grid). Ordinarily this line would intersect the adjacent curve, that for −6 volts, but in this case it does not. Therefore we choose a point on the −8-volt curve which does come under some point of the −6-volt curve. In this case we find that the 235-volt line intersects both curves; hence, we have the point *B* on the −8-volt curve giving a plate-current reading of 7.1 milliamperes. Following the 235-volt line upwards from *B,* we obtain a reading of 12 milliamperes where this line intersects the −6-volt curve at the point *C.*

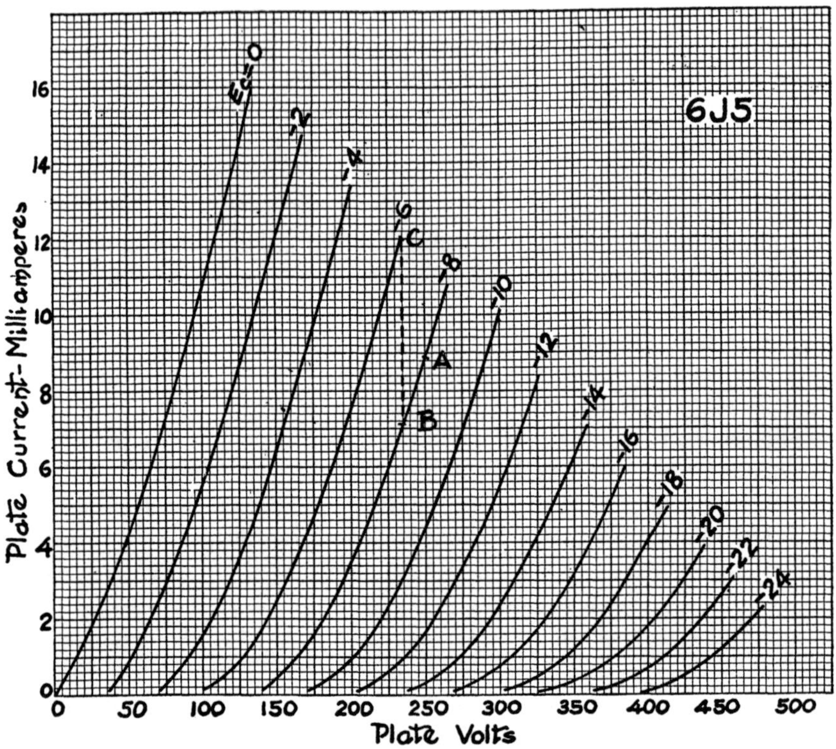

FIG. 8-12. The plate family of a 6J5 triode can also be used for finding the transconductance.

This gives us the two differences we require: that of the plate current and that of the grid voltage. By substituting these values in the above relationship, we have

$$g_m = \frac{di_b}{de_c} = \frac{0.012 - 0.0071}{8 - 6} = \frac{0.0049}{2} = 0.00245 \text{ mho} = 2450 \text{ micromhos}$$

You will note that this value is not the same as that given by the tube manufacturer, i.e. 2600 micromhos, but as it is within 10% of this value, it is sufficiently close.

This discrepancy in values brings out several important points. In the first place, we have again and again stressed the fact that in obtaining the various constants we were concerned with a *small* change in voltage or a *small* change of current. The smaller we make these changes, the more accurate will be our result. For example, in obtaining the transconductance from the grid family we varied the grid voltage by only 0.4 volt, and thereby obtained a transconductance of 2630 micromhos, which differed by only 1.15% from the manufacturer's rating. On the other hand, in making the determination of transconductance from the plate family, we varied the grid voltage by 2 volts. The transconductance of 2450 micromhos was lower than the manufacturer's rating by about 6%, largely due to this greater variation of grid voltage. If our variation had been still larger, the accuracy of our results would be even less.

Now, as you have probably observed, the family of curves does not permit us to make very small changes of voltage or current, and we are compelled to make a voltage or current change of a sufficiently high magnitude as to give us an answer which differs slightly from the manufacturer's rating. This circumstance is not of serious consequence, for in obtaining data from the charts we are concerned only with obtaining an approximate result, since this is all that is generally needed even in engineering computations. Moreover, variations in different tubes, although of the same type, will generally be of the order of 10 percent.

Relation Between g_m, r_p, and μ

We have completed our presentation of the development of the three major tube constants or ratings. We have seen that each varies for any particular tube, depending on the choice of the operating voltages. Yet for the same tube, all are interrelated and it is interesting to note the manner in which these vary with respect to each other. In order that a clear picture may be had of the relative variations of these three tube constants, they are shown upon a single chart and apply to the 6J5 tube. This chart is shown in Fig. 8-13 and the tube is that one which we have been using as our sample.

Before discussing this chart, we should like to say that while this chart applies to just one type of tube at certain operating potentials, the general behavior of these constants, that is, their mode of variation with respect to each other, will be found to hold true for all types of triodes. It seems almost impossible, due to the fact that these three factors are basically governed by the physical structure of the tube, to make tubes in which any one of the three constants will depart radically from the general pattern of Fig. 8-13.

Referring to the chart, the plate voltage was maintained at 250 volts, so that the horizontal axis could be drawn to represent either plate current or grid voltage, the former being chosen because it was

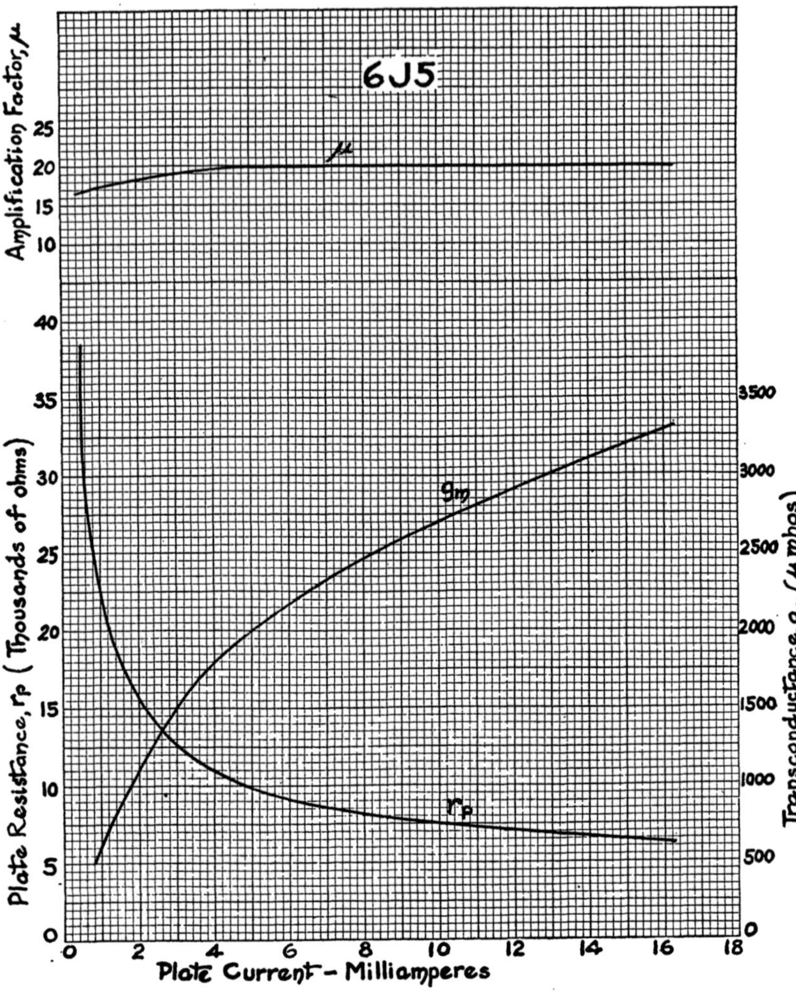

FIG. 8-13. These three curves show the relationship between the amplification constant μ, the transconductance g_m, and the plate resistance r_p for a 6J5 triode.

more convenient. You will understand that low values of plate current correspond to a more highly negative condition of the grid, while high values of plate current correspond to a less negative condition of the grid.

When reading the values indicated upon this chart, the values for the plate resistance are shown upon the lower left-hand vertical axis, the values of the transconductance are shown upon the lower right-hand vertical axis, and values of amplification factor are shown upon the upper left-hand axis.

Critical examination of this chart will give you valuable information relating to all triodes. For example, you will note that the amplification factor remains substantially constant over the full range of operating potentials applied to the tube. This is made evident by the fact that the "mu" does not change very much for plate current values from about 0.5 milliampere to about 17 milliamperes. Since the plate current is determined by both plate voltage and grid bias, it is evident that whatever conditions exist in the tube and result in plate current within the range stated, the "mu" is substantially the same. This is a general condition in all triodes, except those especially identified as being "variable mu" tubes, which will be considered later.

Concerning the "g_m" (transconductance or mutual conductance) of a tube, we can form some interesting conclusions. It is evident that the transconductance of any one tube does not vary with the amplification factor, for the latter is quite constant. The plate resistance does have a definite effect. Given any tube, the lower the plate resistance, the higher the transconductance. In fact, triodes with high transconductance ratings have a low plate resistance. The combination of high transconductance and high plate resistance is not available in triodes. If a high transconductance is desired, so that the grid-voltage change results in a large plate-current change, tubes with low plate resistance must be secured.

As to the plate resistance, there too, we find an interesting subject for discussion. While it is generally true in the design of a tube that the basic constants considered are the amplification factor and the transconductance, while the plate resistance is permitted to fall where it may, definite relationships do exist between the plate resistance and the other constants. For example, the separation between the grid and plate of a triode has a tremendous bearing upon the amplification factor, increasing as the separation between these two tube electrodes is increased. However, it also has a great bearing upon the plate resistance, for the longer the distance which must be traveled by the electrons between the cathode and the plate, the higher the plate resistance. Consequently, it is usually (although not invariably) the case that the higher the amplification constant of a tube, the higher is its plate resistance.

Although it is true that the amplification constant is controlled to a slight degree by operating potentials, it is essentially a property of the physical geometry of the tube rather than the operating potentials.

Plate resistance on the other hand, is a variable factor which is to a large extent controlled by the operating potentials, although for any given value of operating potential its basic value is determined by the structure of the tube, especially the spacing between the cathode and plate. Consequently, one tube designed to have a substantially higher value of amplification factor than another, will likewise have a higher value of plate resistance than the other, assuming that like values of operating potentials are applied to both. These are facts which you must bear in mind when comparing triodes.

Interdependence of g_m, r_p, and μ

Since we have said how the three major constants of the triode are related to one another, it should be possible to show the interdependence among the numerical quantities of these constants. Let us start with the two which are most closely related: the plate resistance and the transconductance.

According to the equations given

$$r_p = \frac{de_b}{di_b} \qquad g_m = \frac{di_b}{de_c} \qquad \text{and} \qquad \mu = \frac{de_b}{de_c}$$

Suppose that we multiply the plate resistance by the transconductance and see what happens. We then have

$$r_p \times g_m = \frac{de_b}{di_b} \times \frac{di_b}{de_c}$$

Since the term di_b appears in both numerator and denominator, we can cancel it from both, giving us the expression

$$r_p \times g_m = \frac{de_b}{de_c}$$

But we have seen from the equation above that this ratio equals the amplification constant, μ, and now we find that this same expression also applies for the product of the plate resistance and transconductance. Therefore, if the amplification factor and the product of the plate resistance and the transconductance are equal to the same thing, they must be equal to each other; so we can write

$$\mu = r_p \times g_m$$

Let us refer to Fig. 8-13 and see if this is so. Suppose that we select some point along the plate-current axis and call that our operating point. Imagine it to be the 10.0 milliampere point. Moving upwards we find that this line intersects the plate-resistance curve at 7500 ohms and the transconductance curve at 2700 micromhos. But since the transconductance is expressed in micromhos, we must convert the value of micromhos obtained from Fig. 8-13 into mhos (1 mho = 1,000,000 micromhos), so that 2700 micromhos equals 0.0027 mho.

Inserting these values of plate resistance and transconductance into our equation, we have

$$\mu = 7500 \times .0027 = 20.2$$

which is in very close agreement with the value of 20 shown in Fig. 8-13. Proceeding in the same manner you can select any point along the plate-current axis and establish the value of "mu."

Now since

$$\mu = r_p \times g_m$$

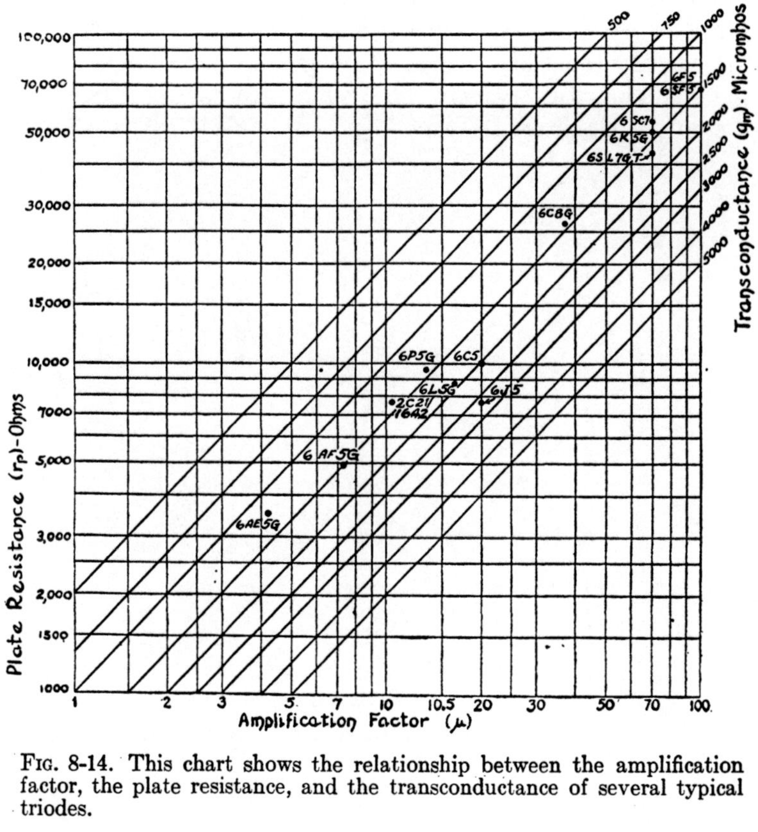

FIG. 8-14. This chart shows the relationship between the amplification factor, the plate resistance, and the transconductance of several typical triodes.

we can say that

$$r_p = \frac{\mu}{g_m}$$

and also that

$$g_m = \frac{\mu}{r_p}$$

thus showing all the relationships.

A chart showing the relationship of the amplification factor, the plate resistance, and the transconductance of a number of typical triodes is shown as Fig. 8-14. This chart enables the rapid selection of triodes having specific characteristics. It has already been pointed out that the higher the amplification factor of a triode, the larger the associated plate resistance, and the general tendency is graphically illustrated in Fig. 8-14.

Having established the three basic constants of the triode tube and the two basic families of characteristic curves, both of which are static in character, we can now advance to the development of the triode curves, which are typical of operating conditions, that is, *dynamic* characteristics, namely those which are developed with a *load* in the output circuit of the tube.

Chapter 9

TRIODE DYNAMIC CHARACTERISTICS AND LOAD LINES

IN THE FOREGOING CHAPTERS of this book, we have dealt with the various details associated with the fundamentals of diodes and triodes. The static characteristic curves and the three basic tube constants of these tubes were discussed. In this chapter we shall consider other characteristics of the triode, advancing from the static to the dynamic conditions.

Naturally if we use a piece of apparatus such as the vacuum tube, we rightfully expect it to be an efficient tool and to get a reasonable amount of work from it. As far as we have gone in our considerations of the triode, we have not seen it do any useful work; we have varied the voltages applied to the grid and plate and observed the amount of current flowing from the plate around to the cathode, but this current so far has done nothing productive, because there was nothing in the output circuit upon which it could work.

In order that a triode, or for that matter any vacuum tube, be of any practical use, a *load* must be inserted in its output circuit. The presence of a load in the output circuit of the tube makes possible the transfer of a signal applied to the input circuit from that circuit to an external circuit. In other words, the load is the means whereby the amplified voltage variations are made available for transfer on to the next tube in a series or to some other voltage-operated device. It has become common practice to say that a signal exists in the output circuit of a tube—that it exists across the load; and when you think it over you will agree that this is only right, as in a properly operating triode, the output should be an amplified replica of the input. Yet to make this statement true, as will be shown later, a load must be located in the output circuit.

From the foregoing you can see that the load and the output circuit of a triode are closely related—that if no load is in the output circuit, there will be no output signal voltage. Since all our considerations of the triode in the previous chapters were based upon a system which did not contain a load (that is, something in the output or plate circuit) you most likely wonder if the introduction of this load, whatever it may be, will not alter the facts already given about the basic triode.

The Effect of the Load

The load does indeed affect some of the things we learned about the sample triode. As a matter of fact, the load does not have very much effect on the basic constants, but it does materially affect the grid voltage—plate current relationship. It may seem strange to you that the grid family characteristics were mentioned and not the plate family; they have not been forgotten—in fact, we shall work with them, but we focus attention upon the grid family as that plays a very important part when we study the amplifying properties of a triode.

This study of the amplifying properties of the triode is the most important of all the capabilities of the tube, for it is an inherent property and as such, it has a great influence upon all the other tube functions. To appreciate the other actions of the tube, we must first of all understand its behavior as an amplifier and to achieve this end, we must use the grid family of characteristics. This is because this group of curves lends itself so well to illustrate the process of amplification, wherein the input signal voltage and the output plate-current variations are expressed with respect to time.

As you will see shortly, the presence of the load in the plate circuit of a triode has a tremendous effect upon the grid voltage—plate current relationship and the resultant curve or curves which display this relationship, are the *dynamic transfer characteristics.* These are in contrast to the static characteristics which are developed when the triode is studied without a load. Here then is the simplest answer to the question: what is the difference between dynamic and static characteristic curves? The *dynamic* characteristics express the relationship between the plate current and the controlling tube electrode potentials *with* a load, while the *static* characteristic shows the relationship between the plate current and the controlling tube electrode potentials *without* a load in the output circuit.

The Basic Triode Circuit

The circuit that we shall use for discussing the dynamic characteristic and the effect of the load is shown in Fig. 9-1. In order to have

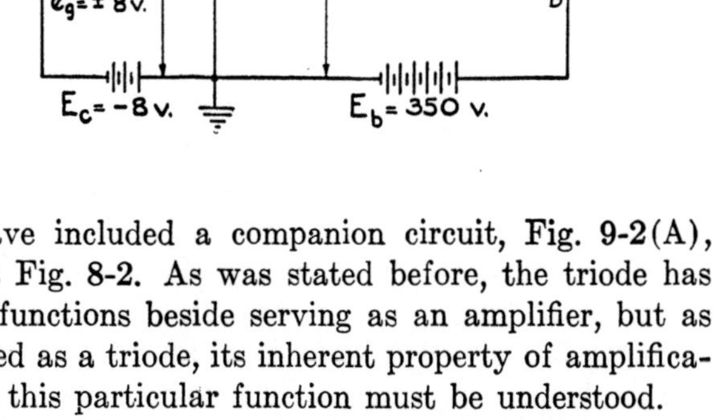

FIG. 9-1. The basic circuit for explaining the dynamic characteristics of a triode. The various symbols are explained in the accompanying text.

a comparison, we have included a companion circuit, Fig. 9-2(A), which is the same as Fig. 8-2. As was stated before, the triode has many other possible functions beside serving as an amplifier, but as long as the tube is used as a triode, its inherent property of amplification is employed and this particular function must be understood.

You will see that Fig. 9-1 is not very different from the basic circuit of Fig. 9-2(A); essentially two items have been added. The first is a 25,000-ohm load resistor R_b, in the plate circuit; the second is a source of input voltage, e_g, which is variable in magnitude as well as polarity. The latter is shown in Fig. 9-1 as being an a-c source, but for the present we can forget that and view it as being a d-c device which is capable of supplying the proper voltages. This is quite in order, because at any one instant a sine wave of a-c voltage can be replaced by an equivalent d-c voltage.

The value of 25,000 ohms for the load resistance was chosen arbi-

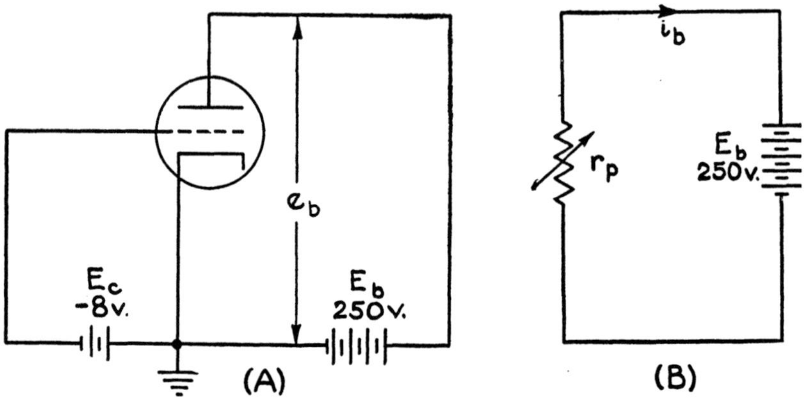

FIG. 9-2. The triode circuit in (A) has no load resistance in the plate circuit ($R_b = 0$) and no input signal; compare Fig. 9-1. The equivalent circuit of (A) is shown in (B) where the plate resistance r_p is shown as a variable resistor.

trarily and could just as readily be 50,000 ohms, 100,000 ohms, or any other value. In fact, later on we shall discuss what happens for higher and lower values of load resistance. The values of plus and minus 8 volts for the peak input signal voltage, e_g, were chosen so that the lowest voltage between the grid and cathode would be zero so as to keep the grid from ever becoming positive. This variable voltage, e_g, acts in conjunction with the grid bias, either to aid or to buck the bias, depending entirely upon the polarity of e_g. When e_g is positive and at its peak value, it offsets the grid bias E_c, and the resultant grid voltage e_c equals (+8 volts) + (−8 volts) = 0. When e_g is at its negative peak value, the resultant grid voltage, e_c, equals (−8 volts) + (−8 volts) = −16 volts.

Another difference between Figs. 9-1 and 9-2(A) is in the value of the plate supply voltage, E_b, which is now 350 volts instead of 250; also several different voltage symbols appear in Fig. 9-1, which do not show in Fig. 9-2(A).

Because of all the symbols which appear on Fig. 9-1 compared to the few in Fig. 9-2(A) the former looks quite a bit different from the latter, but, as we said before, only two new items have really been added. Furthermore, only Fig. 9-1 is of any practical value, because Fig. 9-2(A) is useful only insofar as it tells what is happening *inside* of the tube. First let us analyze the triode having a load as shown in Fig. 9-1.

Let us start with the grid circuit. Very little need be added to the statements made a few paragraphs back, for the references to E_c, e_g, and e_c are self-explanatory, particularly if you understand that at all times $e_c = E_c + e_g$.

Now examine the plate circuit. Note that the load resistor R_b is directly in the path of the plate current i_b. The movement of electrons representative of the plate current must be through this resistor, because the complete circuit of the plate current is from the cathode to the plate, through the resistor R_b, then to the positive terminal of the battery E_b, while electrons leave the negative terminal of battery E_b and return to the cathode.

Plate and Load Resistances in Series

Do you recall that during the discussion of the static characteristics of our sample triode, it was mentioned that a value of internal resistance, which we identified as plate resistance, r_p, existed within the tube? The internal plate resistance is rarely shown upon a diagram, for it is taken for granted. Realizing its existence, you should have no trouble in understanding the statement that two values of resistance exist in the output circuit which tend to control the flow of plate current. One of these is the plate resistance, r_p which is not shown but which is acknowledged to exist, and the other is the load resistance, which is shown as R_b.

Now we have to remind you of some simple d-c relationships. The plate circuit in Fig. 9-1 is a simple series circuit, for there is but one path for the plate current to follow. It is a rule in a d-c series circuit that the voltage divides in proportion to the resistances in the circuit. If only one resistance is in the circuit, then the voltage across this resistance is equal to the supply voltage. This is the case in the circuit of Fig. 9-2(A), as well as the circuit used to develop the static grid and plate families by measurement. In that circuit, there is only one resistance, that of the tube itself. (We are neglecting the internal resistance of the plate voltage supply E_b.) The voltage drop across the plate resistance of the tube is then equal to the supply voltage E_b no matter what the value of the plate current may be. That this is so can be seen by examining Fig. 9-2(B), which is the equivalent circuit of Fig. 9-2(A). The plate resistance r_p is shown as a variable resistance, since its value will be different for different values of grid voltage.

When two resistances are in the circuit, as in Fig. 9-1, and current flows in the system, the total supply voltage is then equal to the sum of the voltage drops which develop across the two resistances, i.e. that across the plate resistance r_p and that across the load resistance R_b. Assuming a steady value of plate current, i_b, the drop across the plate resistance r_p is $i_b r_p$ and the drop across the load resistance is $i_b R_b$.

The Plate Voltage, e_b

Suppose for the sake of illustration that the plate current is 5.25 milliamperes. Since $R_b = 25{,}000$ ohms, the voltage drop $i_b R_b = 0.00525 \times 25{,}000 = 131.25$ volts, which to three significant figures is 131 volts. Since the total supply voltage is 350 volts and the drop across the plate load is 131 volts, the difference between the supply voltage and the voltage drop across the load must equal the voltage drop across the plate resistance. In other words, the plate voltage $e_b = E_b - i_b R_b = 350 - 131 = 219$ volts. This relationship between the supply voltage, the voltage drop across the plate load resistor, and the voltage across the plate-to-cathode or the plate resistance of the tube, prevails no matter what the value of plate current or the specific individual values of voltage and resistance.

For example, let us assume that 10 milliamperes flows in the plate circuit. The increase in plate current does not change the value of the plate-supply voltage, E_b, for that remains constant at 350 volts. In this instance, the voltage drop across the load resistor R_b is $0.01 \times 25{,}000$ or 250 volts ($i_b R_b = 250$ volts) and the voltage e_b between the plate and the cathode is $350 - 250 = 100$ volts ($e_b = E_b - i_b R_b$). Next suppose that the plate current is reduced to 1 milliampere. Then the voltage drop across the load resistor R_b is $0.001 \times 25{,}000 = 25$ volts and the voltage across the plate resistance is $350 - 25 = 325$ volts.

Now visualize the flow of zero plate current. The load resistance will have no voltage drop across it and the full plate-supply voltage of 350 volts will appear across the internal plate resistance of the tube. At first thought this may be a difficult situation to appreciate, for you may think, "How can there be a voltage across the tube resistance, that is between plate and cathode, without the flow of current?" In a tube circuit, unlike a conventional continuous circuit, this is quite simple, for if the grid bias is such that it prevents the movement of electrons to the plate, the internal plate resistance will be so high that it is *effectively* infinite and there is no flow of current anywhere in the circuit.

The Effect of the Load

A study of these illustrative figures brings some very interesting and important conditions to light. First, we see that the presence of the load resistance has an enormous effect upon the conditions which exist in the plate circuit. In the simple circuit Fig. 8-2 that was used to obtain the static curves, the voltage at the plate of the tube was the same regardless of the value of the plate current; this is not the case when the load is present in the circuit. For now, the controlling influence is the amount of plate current: the *higher* the plate current, the greater the voltage drop across the load resistor and the *lower* the plate voltage across the internal resistance of the tube, i.e. between the plate and the cathode. On the other hand, the *lower* the plate current, the *less* the voltage drop across the load and the *higher* the voltage between the plate and the cathode.

From all this it is evident that a new kind of relationship exists between the plate current and the voltage between the plate and the cathode. Now, the plate voltage is not equal to the plate-voltage supply unless the plate current is zero.

When it is understood that the voltage between the plate and the cathode can vary as a result of the plate current flow, you can also see that the drop across the load resistance can fluctuate in accordance with the variation in plate current flow; in this case, however, the higher the plate current flow, the greater is the drop in voltage across R_b and vice versa. This leads to the statement that although the plate-voltage supply E_b is *constant,* we do have present in the tube plate circuit, when there is a load, two voltages which vary as the result of changes in plate current. One of these is the voltage e_b across the tube resistance r_p, and the other is the voltage i_bR_b across the load R_b.

Varying the Grid Voltage

It is now necessary to reintroduce the grid circuit. In all the discussion about variations in plate current, nothing was said about what can be considered as the reason for these variations. We merely stated that they took place. It is evident that the changes in plate current are not due to changes in the plate voltage supply E_b, for we have said that this voltage remains constant. The only other source of plate current changes is the grid voltage.

Let us then correlate changes in grid voltage with changes in plate current. For the moment we need not mention specific values of grid voltage, for it is sufficient to think in generalities. Since the minimum voltage e_c on the grid is zero while all the other voltages up to the maximum of -16 volts are negative, we must think of the grid as always being negatively charged to a greater or lesser degree. We know from past facts presented that the greater the negative charge on the grid, the less the plate current and vice versa. We also know that the grid bias voltage source E_c is fixed in value and that it is the variable e_g which is the determining influence in establishing the final voltage e_c, which is effective between the grid and the cathode. It is this voltage e_g which we consider as being the *signal voltage or the excitation voltage;* it is this that is responsible for the variations in plate current, because it is this voltage in the input circuit which is varying.

Try to visualize for a moment the presence of such a signal voltage in the circuit of Fig. 9-2(A). Imagine E_c to be variable in value rather than fixed as shown. Varying the input voltage to the grid will cause variations in the plate current, but it will not cause any variations in the voltage between the plate and cathode, because there is no load resistor in the plate circuit.

Now look again at the conditions in the circuit of Fig. 9-1. Varying the grid voltage now does cause a change in the voltage between the plate and cathode because of the voltage drop across the load resistor. For every momentary change in the total grid voltage e_c, a new value of plate voltage is established between the plate and the cathode, because every change in grid voltage causes a change in the plate current and hence a different value of voltage drop across the load. The only time that an unvarying voltage exists at the plate is when $e_g = 0$, which is the equivalent of no-signal voltage input; that is, when the grid voltage does not change. If we now plot the grid voltage—plate current characteristic we obtain a curve that is definitely different from the static curve. This new curve is the *dynamic transfer characteristic,* wherein the plate voltage e_b (not the plate supply voltage) is a variable depending on both the grid voltage and the plate current.

Before we put these data into graphical form and develop the dynamic characteristics let us mention one more point. Making the grid less negative than the grid bias E_c is the equivalent of applying a positive voltage from e_g, while making the grid more negative than the grid bias, is the equivalent of applying a negative voltage from e_g. If we forget about the bias and just consider the grid circuit from the viewpoint of changes in e_g, we note something very significant. When

e_g is *positive* (although the grid does not swing positive because of the grid bias E_c) the plate current increases and this means a *decrease* in the voltage at the plate of the tube *because the drop across the load is increased.*

On the other hand, when e_g goes negative, the plate current decreases, and since a decreased plate current means a lower drop across the load resistor, the plate voltage *increases.* Thus, the action of the grid voltage and the plate voltage is in the opposite direction. This condition is extremely important in its effects and will receive more attention later on. Having laid this groundwork for the dynamic characteristic, as well as the behavior of the load, let us now consider these in more detail.

Introducing the Load Line

To develop the grid voltage—plate current characteristic for our sample tube, we must perform certain operations. It would be possible to develop this curve from a series of measurements on the circuit of Fig. 9-1, but a much simpler method is available. This calls for the use of the static plate family to which is added a diagonal line known as the *load line.* The addition of this load line to the plate family furnishes us with the information which enables us to develop the grid voltage—plate current characteristic, otherwise known as the *dynamic transfer characteristic.*

Resistances Graphically Represented

Before going into this, it is necessary for us to digress for a while in order to explain how a resistance can be graphically represented. Let us consider Fig. 9-3, in which the horizontal axis represents volts and the vertical axis represents current in *milliamperes.* The relationship between the two can be shown by a straight line which represents a resistance having a fixed value. If we desire to show the relationship between a voltage (from zero to any finite value) and current, we can employ a straight line representative of resistance, and thereby indicate the value of current which will flow in a circuit containing that resistance for any value of voltage. Such a line representing resistance would start at the intersection of the voltage and current axes, since zero voltage applied to a circuit would result in zero current. The other limit of the line would appear at any point for which Ohm's law would be true.

For example, the line connecting the origin with the point corresponding to 150 volts and 30 milliamperes represents a resistance of 5000 ohms, since according to Ohm's law

$$R = \frac{E}{I} = \frac{150}{0.030} = 5000 \text{ ohms}$$

In a similar manner the line joining the origin with the point corresponding to 200 volts and 10 milliamperes represents a resistance having a value of 20,000 ohms since

$$R = \frac{E}{I} = \frac{200}{0.010} = 20{,}000 \text{ ohms}$$

You will observe in Fig. 9-3 that the more nearly horizontal the line becomes, the greater is the value of the resistance represented. A line which is horizontal, that is one which coincides with the horizontal axis, would indicate a resistance of infinite magnitude. This can be seen in Fig. 9-3, since the application of any voltage, however great, results in zero current. In the same way, the more vertical the line becomes, the lower is the value of resistance that it represents. A line coinciding with the vertical axis represents a resistance of zero ohms.

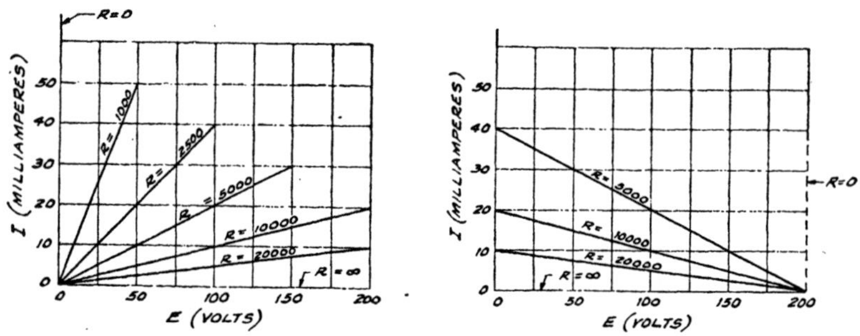

FIGS. 9-3 (left), 9-4 (right). A resistance having a fixed value can be represented by a straight line; those in Fig. 9-3 are for resistance when the voltage is unknown, and those in Fig. 9-4 are for resistance when the supply voltage is known.

The lines which start from the zero voltage-zero current point as origin, as shown in Fig. 9-3, may be used to indicate resistances when the voltage is not determined in advance. If the supply voltage is given, as in Fig. 9-4, then this point corresponding to the supply voltage may be used as the origin. In this case, the values of both the voltage and the resistance are given, the problem being to find the corresponding current.

Suppose, for instance, that the fixed voltage value is 200 volts and the given resistance is 10,000 ohms. Applying Ohm's law, we obtain

$$I = \frac{E}{R} = \frac{200}{10{,}000} = 0.020 \text{ ampere} = 20 \text{ milliamperes}$$

Accordingly, the line representing 10,000 ohms is drawn connecting the 200-volt (and zero current) point with the 20-milliampere (and zero voltage) point. The other lines in Fig. 9-4 are obtained in the same manner.

As before in the case of Fig. 9-3, you can also see in Fig. 9-4 that when the line representing resistance coincides with the horizontal axis, a value of infinite resistance is indicated. On the other hand, a vertical line starting from the 200-volt point on the horizontal axis, indicates zero resistance.

Now let us see how we can extend this information so that we can investigate the properties of a circuit which has two resistors in series connected to a voltage source. In Fig. 9-5(A) are shown two lines representing two resistors connected in series to a voltage source of 300

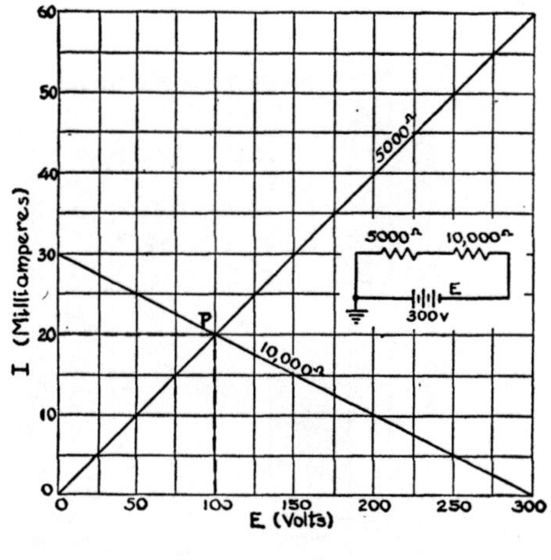

FIG. 9-5(A). Two resistors in series can be represented graphically. The intersection P indicates the current flowing in the circuit.

volts. We have assumed here that the resistors have values of 5000 and 10,000 ohms respectively, the lines in Fig. 9-5(A) being drawn by the methods outlined in connection with Figs. 9-3 and 9-4. For example, the 5000-ohm line is obtained by drawing a straight line between the zero voltage—zero current point and the 300 volt—60 milliampere point. The 10,000-ohm line is drawn with its origin at the 300 volt-zero current point. By Ohm's law, the resulting current is 30 milliamperes (0.030 ampere) and this determines the other end of the 10,000-ohm line, i.e. the 30-milliampere point on the vertical axis.

You will note that even though these two resistors are connected in series, their positions on the graph of Fig. 9-5(A) are determined as though they were alone in the circuit, i.e. when the 5000-ohm line was constructed, the other was ignored. Then in the determination of the 10,000-ohm line, the first was disregarded. This procedure can be followed inasmuch as the two straight lines intersect at the point P in Fig. 9-5(A) and this intersection gives the condition existing for the entire circuit. A dashed line from P to the vertical current axis shows that 20 milliamperes flow through the entire series circuit. A vertical dashed line dropped from P to the horizontal axis indicates 100 volts, this 100-volt point serving as a point of demarcation to indicate the voltage distribution across the two resistors. Here the voltage drop across the 5000-ohm resistor is 100 volts, and the voltage drop across the 10,000-ohm resistor is $300 - 100 = 200$ volts.

There was no necessity to have the origin of the 5000-ohm line start at zero volts and that for the 10,000-ohm line start at 300 volts. The reverse procedure could have been followed, as is illustrated in Fig. 9-5(B). At the intersection of the two lines point P, we again drop

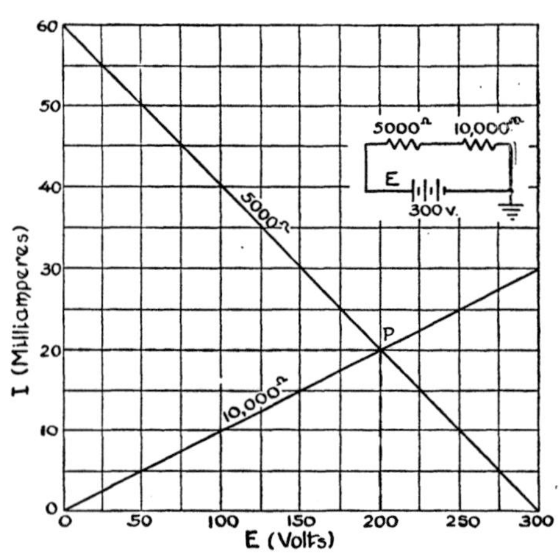

FIG. 9-5(B). The same two resistors as were represented in Fig. 9-5(A) are shown here, but in this case the line representing the 10,000-ohm resistor was drawn before the other, thus reversing the procedure followed in the preceding figure. The results are the same.

a vertical line to the volt-axis. It is again seen that there is a 200-volt drop across the 10,000-ohm resistor, and a voltage drop of $300 - 200 = 100$ volts across the 5000-ohm resistor, these results being consistent with our former results.

If both the 5000-ohm and 10,000-ohm lines had their points of origin at zero volts, there would have been obtained two non-intersecting lines, as shown in Fig. 9-3, each making different angles with the horizontal axis. The same is true if both lines had their point of origin at the 300-volt point, as shown in Fig. 9-4, a similar case but where 200 volts are used instead of 300 volts.

Another way of viewing the lines representing resistance is as follows: Two resistors are connected in series and the combination to a battery, one resistor having a resistance of 5000 ohms and the other 10,000 ohms and a 300-volt battery. According to Ohm's law the current I through the circuit will be

$$I = \frac{E}{R} = \frac{300}{15{,}000} = 0.020 \text{ ampere} = 20 \text{ milliamperes}$$

Since 20 milliamperes flow through each resistor, the point of intersection of the lines representing the resistor will intersect somewhere on the 20-milliampere line. That this is so can be seen by examining Figs. 9-5(A) and 9-5(B).

Linear and Non-linear Resistances

Thus far we have considered the two resistors to be linear circuit elements, i.e. the value of each resistor remained the same regardless of the amount of voltage applied or the amount of current that flowed through them. Now suppose that one (or both) of them was a non-linear element, which means that the resistance of one (or both) would vary as the impressed voltage or current flow was changed. In connection with vacuum tubes, we are concerned with the case where just one of the resistors is changed in value, i.e. it is non-linear, and the other remains constant. This is because the resistance of a tube varies with the applied voltage and so is a non-linear element, whereas the other element is linear, for instance the load resistor, the value of which remains unchanged no matter what voltage is applied to it. As far as the graphical method outlined above is concerned, that applies equally well to non-linear as well as linear elements and so can be used for our purposes.

A simple combination of linear and non-linear elements is shown in Fig. 9-6: a diode and a fixed resistor are connected in series with a battery. We desire to find the current flowing through the entire circuit, the voltage drop across the diode, and the voltage drop across the fixed resistor. Let us suppose that the voltage-current characteristic of the diode is given by the curved line, starting at the origin and going up to the right, this line being curved since the resistance of the diode R_d varies for different values of impressed voltage. We shall assume further that a 10,000-ohm fixed resistor R is in series with the diode, the line for this resistor being determined in the manner described previously. To make it more specific, we shall use a 300-volt supply, E. These three factors—the supply voltage, the size of the fixed resistor, and the diode characteristic—determine the point of operation P of the

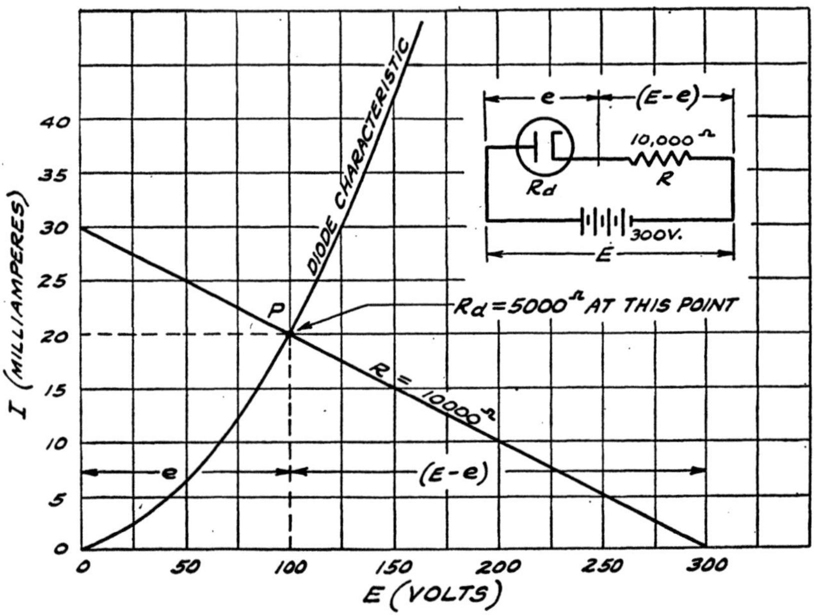

FIG. 9-6. When one or more of the resistances in a circuit are non-linear, the current in the circuit and the voltage drops across each resistance can be found graphically. In this graph these factors are determined for a diode (non-linear) in series with a fixed resistance.

diode. This point is found by the intersection of the resistance line with the diode characteristic, and at P the resistance of the diode is 5000 ohms, since

$$R = \frac{E}{I} = \frac{100}{0.020} = 5000$$

We now see that 20 milliamperes flows through the entire circuit, that the voltage drop e across the diode is 100 volts and the voltage drop across the fixed resistor is $E - e = 300 - 100 = 200$ volts.

Load Line Construction

Now we can investigate the load line as it is actually employed in practice in conjunction with the plate family of the triode.

The static plate family of our sample 6J5 tube is shown in Fig. 9-7. As you can see, the various plate-current curves are identical to those which have been shown a number of times before. The difference between this and the previous illustrations of the static plate family is the addition of the load line which graphically represents the 25,000-ohm load shown in the schematic of Fig. 9-1. This load line is surrounded by interesting details which must be brought to your attention.

Due to the apparently close association between this load line and the static plate family of curves, the first conclusion is that the load line is in some way related to the plate family of the vacuum tube for which it is drawn; that is, the plate voltage and the plate current relationships within the tube have a major bearing upon the construction of the load line. This, however, is *not* the case. The construction of the load line has nothing to do with the tube constants or the characteristics of the tube for which it is drawn. True, once the load line is drawn upon the characteristics, it furnishes us with valuable information about the tube, but its actual construction is independent of the tube characteristics.

The reason for this, as you will see, is that the load line is a graphic representation of the conditions set up by the circuital elements *outside* of the tube. It indicates in graphic form the fact that when a load is used in the plate circuit of a vacuum tube, the plate voltage e_b that is effective between the plate and cathode is less than the supply voltage E_b. The construction of the load line is determined only by the magnitude of the supply voltage and the magnitude of the load resistor, as we pointed out in our discussion of Fig. 9-5.

Accordingly, if we choose a load resistor of 25,000 ohms and a supply voltage of 350 volts, we may draw the load line so that one end-point is at the 350 volt-no current point, while the other end-point is at the 14.0 milliampere-no voltage point.

Meaning of the Load Line

Now that we have the load line, what does it indicate? A great deal, to put it mildly! First of all, the entire length of the load line represents a drop of 350 volts, namely, that of the entire voltage supply. A point of intersection of the load line with any plate-current curve represents a division of the plate-supply voltage between the load resistor and the tube plate resistance. Just what we mean by this is illustrated in the following.

The operating point that we have selected for our sample 6J5 tube has a grid bias of −8 volts. This value of grid-supply voltage E_c was chosen because with a signal-voltage input of 8 volts we can cover practically the whole working range of the tube, i.e. from 0 to −16 volts on the grid. Looking at the load line in Fig. 9-7 we find that it intersects the −8-volt grid line at 5.25 milliamperes, indicated by *P*. If we drop a line from this point of intersection *P* until it meets the plate-voltage axis, we find that it strikes the 219-volt point.

With 5.25 milliamperes of current flowing through the 25,000-ohm load, the drop across the load is $0.00525 \times 25{,}000 = 131.25$ volts, or $i_bR_b = 131$ volts, which is sufficiently close for all practical purposes. Now we see that the line (dashed) dropped from the recommended operating point *P*, divides the plate-supply voltage scale in exact accordance with the division that we determined mathematically. The distance along the plate-supply voltage axis to the left of this line represents the voltage across the plate and cathode and equals 219 volts. The distance along the plate-voltage axis to the right of this dividing line, is the voltage drop across the load resistance and is equal to $350 - 219 = 131$ volts.

This is not an accidental condition, for one purpose of this load line is to show the division of voltage and to show that the plate voltage decreases as the plate current increases. Thus for zero grid voltage (the $E_c = 0$ curve), the load line intersects at point *A*, which corresponds to a plate current of 10.1 milliamperes, a plate-cathode voltage e_b of 96 volts, and a drop of $350 - 96$ or 254 volts across the load resistor. With −4 volts on the grid, the plate current is decreased to 7.6 milliamperes (point *C*), the plate-cathode voltage is increased to 160 volts, and the drop across the load resistor is reduced to 190 volts.

With −6 volts on the grid, the intersection (point *D*) between the load line and the corresponding plate-current curve indicates that the voltage between the plate and the cathode is 190 volts. The difference between this figure and the total supply voltage E_b of 350 volts is the voltage drop across the load resistor or 160 volts. In the same way, the other points of intersection *B, E, F, G, H, I* and *J* indicate both the plate voltage e_b and the voltage drop e_{Rb} across the load resistor.

From the preceding paragraphs we can gather some more important information. First, the "swing in plate voltage." This is nothing more than *the full range of voltage effective at the plate.* We see from Fig. 9-7 (point *A* equal to zero grid voltage e_c) that the lowest value of the voltage e_b which is effective between the plate and cathode is 96 volts. In turn, the highest value of voltage which is effective between the plate and cathode, is 350 volts, namely that value existing when the grid voltage equals approximately −21 volts. The entire plate supply

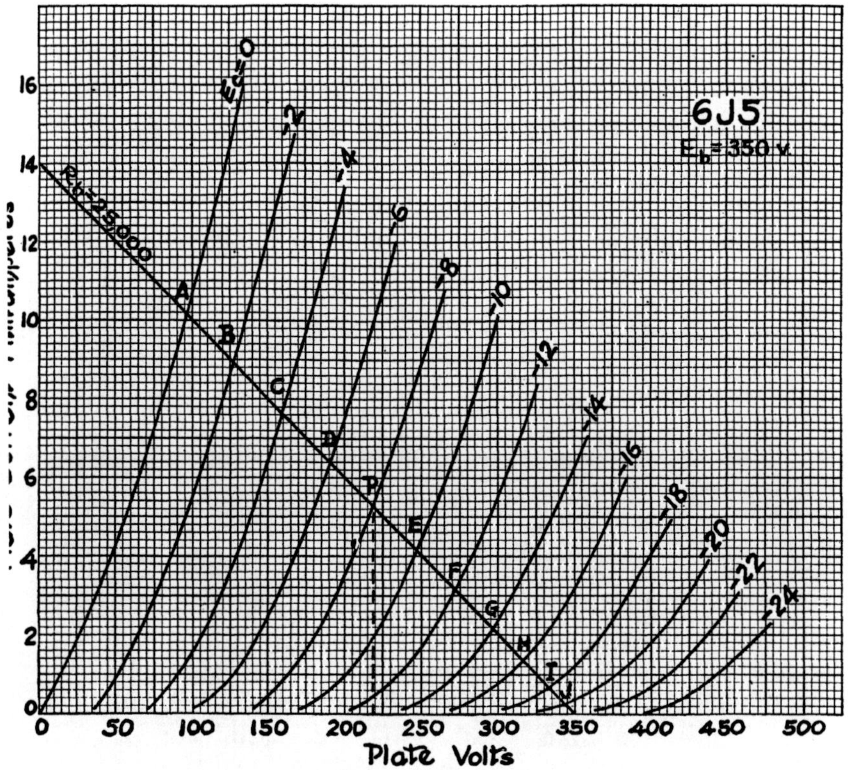

G. 9-7. The line superimposed on the plate family of characteristics of the 6J5 represents a ,000-ohm load in the plate circuit, as shown in Fig. 9-1.

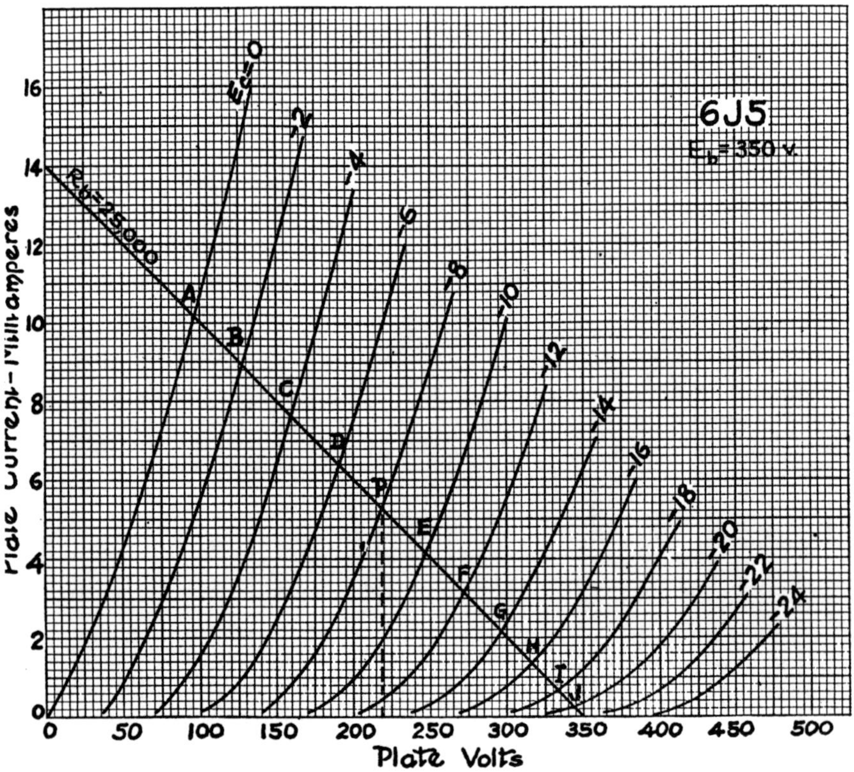

Fig. 9-7. The line superimposed on the plate family of characteristics of the 6J5 represents a 25,000-ohm load in the plate circuit, as shown in Fig. 9-1.

voltage $E_b = 350$ is now acting across the plate and cathode, and since no plate current is flowing, there will be zero voltage across the load resistor. This value of grid voltage is known as the plate-current *cutoff* value.

From this you see that the swing in voltage across the load resistor, i.e. the total change of voltage drop across the load resistor, is from 0 volts when the grid voltage equals -21 volts, to 254 volts, when the grid voltage is zero. Both of these facts will prove to be of importance later on.

Although it is true that the values given in Fig. 9-7 are representative of fixed values of grid voltage ($E_c = 0, -2, -4, -6$, etc.), these need not necessarily be fixed values secured from a battery, but may instead be considered *instantaneous values e_c of an a-c voltage between grid and cathode.* By taking the minimum (0) and maximum (-21) voltage limits at the grid of the tube, we have in effect the full range of -10.5 to $+10.5$ volts of signal-voltage swing. Therefore, the load line for 25,000 ohms drawn on the plate family, as shown in Fig. 9-7, gives *the dynamic behavior of the tube.* The load line indicates that conditions occur in the plate circuit for certain specific conditions of the grid circuit at every instant.

Dynamic Transfer Characteristic

Granting the importance of the load line when it is added to the static plate family, it does not tell the entire story as conveniently as it might. It has become a popular practice, when illustrating how the process of amplification takes place within a triode, to make use of a dynamic representation of the grid family, although this information may be indicated on the plate family. When the dynamic grid voltage —plate current characteristic is used, it lends itself most handily to the development of the graphic representations of the plate changes caused by the application of a varying grid voltage. Accordingly, it will prove highly advantageous if we add the dynamic characteristic resulting from the use of a 25,000-ohm load to the static grid family, such as that shown in Fig. 8-5. It is possible to show such dynamic characteristics for any different number of load resistors, but for the present we shall use the 25,000-ohm load as in Fig. 9-7.

No doubt you realize that such a dynamic grid voltage—plate current characteristic can be developed by measurements, but this seems unnecessary when we can develop the curve by transposing the information shown in the plate family of Fig. 9-7 onto the grid family of curves.

In Fig. 9-8 we have incorporated both the grid family and the plate family of curves on the same graph. The plate-current scale is the same for both families, but the voltage scales (horizontal) are different. That of the grid family (on the left-hand side of the combination graph) indicates the total negative voltage that is applied to the grid of the triode and that of the plate family indicates the plate voltage, as explained in the preceding section of this chapter. The 25,000-ohm load line has been drawn across this family of curves and the dynamic transfer characteristic curve has been drawn in the grid family in the left-hand side.

Construction of Dynamic Transfer Characteristic

Let us see why this transfer of data can be made. In order for this to be possible, the two families must have something in common, which is the total voltage, e_c, that is applied to the grid. In other words, the curves of the plate family, $E_c = 0, -2, -4, -6$, etc., correspond to certain divisions along the horizontal scale of the grid family. Hence, we can indicate equally well on either family the amount of plate current flowing when a certain voltage is impressed on the grid and if we choose certain points on the different curves in the plate family, we can show corresponding points on the grid family.

For example, take the point P in the plate family as a starting point, which represents a grid voltage of −8 volts and a plate voltage of 219. We can show this same condition in the grid family by first determining the plate current, which is 5.25 milliamperes as indicated on the plate-current scale. Then the −8-volt point is found along the horizontal scale of the grid family and the intersection of the vertical line from this point and the horizontal line from 5.25-milliampere point on the current scale, determines the location of P', which corresponds to P. In other words, first the plate current is found for the various points of intersection between the load line and the grid-voltage curves of the plate family; then these current values are measured up from the horizontal scale of the grid family on the e_c vertical lines corresponding to the same curves in the plate family.

Now let us refer to the load line of the plate family and note these points of intersection between the load line and the various grid-voltage curves, establishing the plate current for each point as outlined above. The list follows:

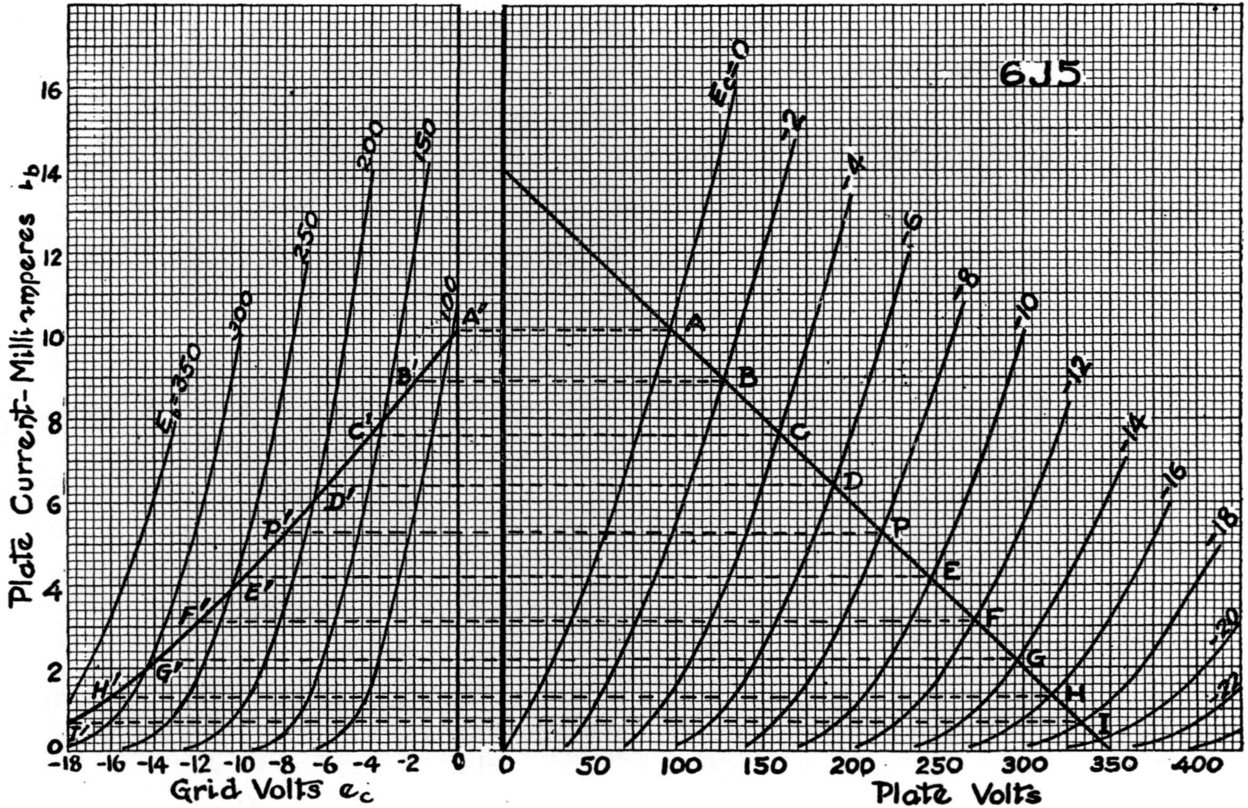

FIG. 9-8. The 25,000-ohm load line drawn on the plate family is projected onto the grid family of characteristics, as described in the accompanying text, in order to construct the dynamic transfer characteristic.

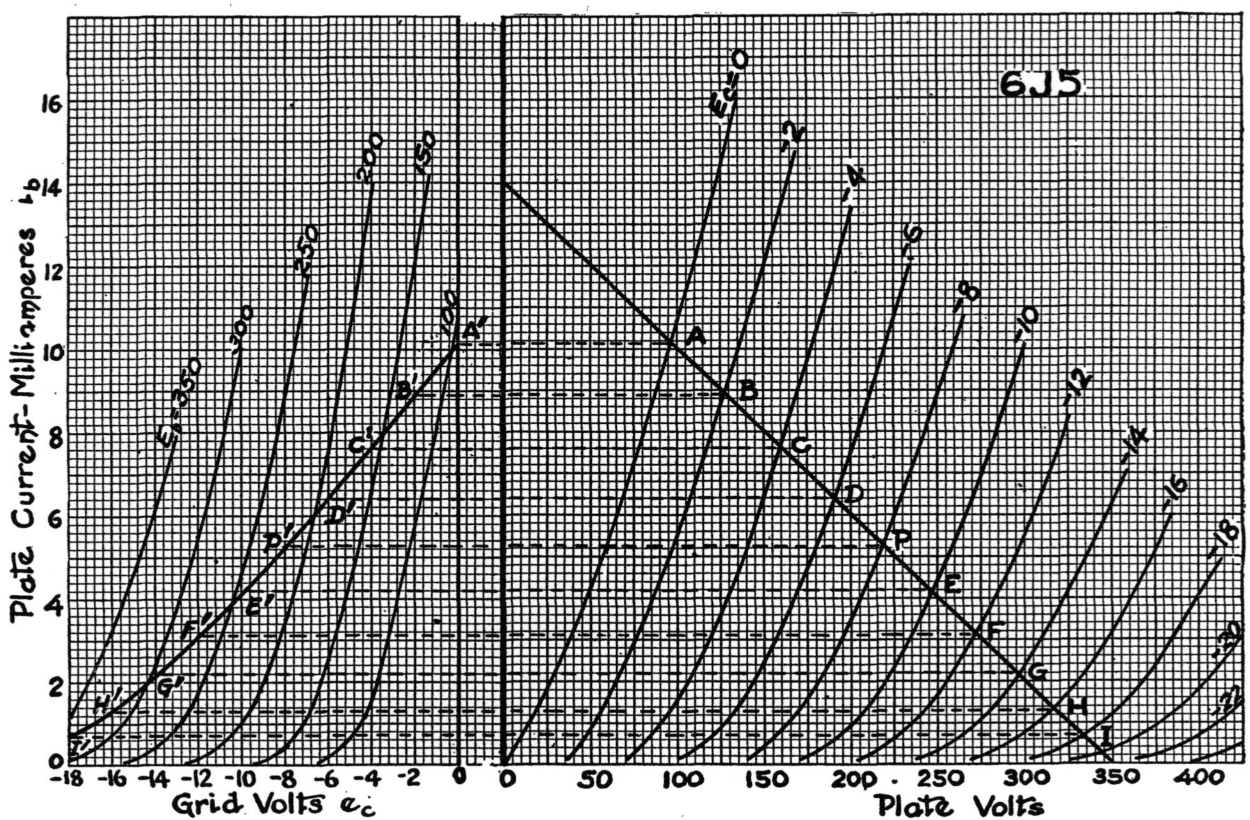

FIG. 9-8. The 25,000-ohm load line drawn on the plate family is projected onto the grid family of characteristics, as described in the accompanying text, in order to construct the dynamic transfer characteristic.

Point	e_c (volts)	i_b (milliamperes)
A	0	10.1
B	−2	8.9
C	−4	7.6
D	−6	6.4
P	−8	5.25
E	−10	4.1
F	−12	3.1
G	−14	2.2
H	−16	1.4
I	−18	0.7

We can now transfer these values of current on the proper grid-voltage lines of the grid family and so establish the dynamic transfer characteristic. Point A', which corresponds to point A on the plate family, is marked at the intersection of 0 grid voltage and the 10.1-milliampere line on the grid family. Point B', corresponding to point B, is established at the junction between −2 volts grid voltage and 8.9 milliamperes. Point C' is marked at the intersection of the −4 grid voltage line and 7.6 milliamperes. In the same way the locations of points D', E', F', G', H', and I' are found. When the points A' to I' inclusive are joined, the resulting curve represents the *dynamic transfer characteristic* or *dynamic grid voltage—plate current characteristic curve* for the 6J5 triode which is operated with a load resistor of 25,000 ohms in the external plate circuit.

The dynamic curve which has been drawn at the left of Fig. 9-8 enables you to see the transformation that is caused by the introduction of the 25,000-ohm load resistance in contrast to the static characteristic curves on the same grid family. Notice that the dynamic curve is much straighter throughout its entire length than the other static curves. It is true that it is not as straight as it might be or as straight as other curves which you will see later, but it is straighter than the static curves which correspond to a load resistor of zero ohms. In addition, this curve also shows *the exact change in plate current for all grid voltages between 0 and −16 volts,* which represents the range of e_c in Fig. 9-1. This information is lacking from the plate family, for that family of curves shows grid voltage variations only in steps of 2.0 volts.

The Value of the Load Resistor

Being in possession of these facts about the 25,000-ohm load line, some questions more than likely have arisen in your mind. Why was a 25,000-ohm load used instead of one 50,000 ohms? Why was not a higher or lower value employed? What effect would a higher or lower load have on the operation of the triode?

The answers to these questions are quite important. The selection of the value of the load resistance is related both to the application to which the tube is to be put as well as the amplification that is available from it. Although we are not ready to discuss these data in detail as yet, we can comment that in the process of amplification it is important that the variations in plate current conform, with respect to time, with the variations in grid voltage. If this is done, then the final amplified voltage output from the tube will be an enlarged version of the input voltage, and its operation will be free from distortion. Whether or not such a condition is obtained depends to a large extent upon the load applied to the tube.

The second consideration—that pertaining to the amplification of the triode—is that the value of the load resistance with respect to the tube plate resistance determines how much of the amplified signal available within the tube is taken out of the tube. This description is very meagre, but must suffice for the present.

Meanwhile, let us examine what happens when various values of load resistors are used with our sample tube. Suppose that instead of 25,000 ohms, we use 50,000 ohms. The construction of the new load line is carried out exactly as it was before, and this is shown as curve C, along with several others, in Fig. 9-9, which is, of course, the plate family.

Effects of Different Loads

What are some of the effects of increasing the value of the load resistor as compared with the results obtained with other loads? Table A shows these variations. First using the value of plate current which prevails at zero grid voltage as the basis of comparison, we find that the total swing in plate current is less with the 50,000-ohm load than with the 25,000-ohm load. With the latter it was from zero to 10.1 milliamperes, whereas with the 50,000-ohm load it is from zero to 5.7 milliamperes, as shown on the load lines *B* and *C* in Fig. 9-9. Thus we can say in general that the plate-current swing is reduced for an increase in load resistance.

In the second place, you will note a significant shift in the distribution of the voltage drop across the load resistance for like values of grid voltage. For example, at 0 volts on the grid, the voltage e_b effective at the plate is 96 volts for a 25,000-ohm load; whereas with the

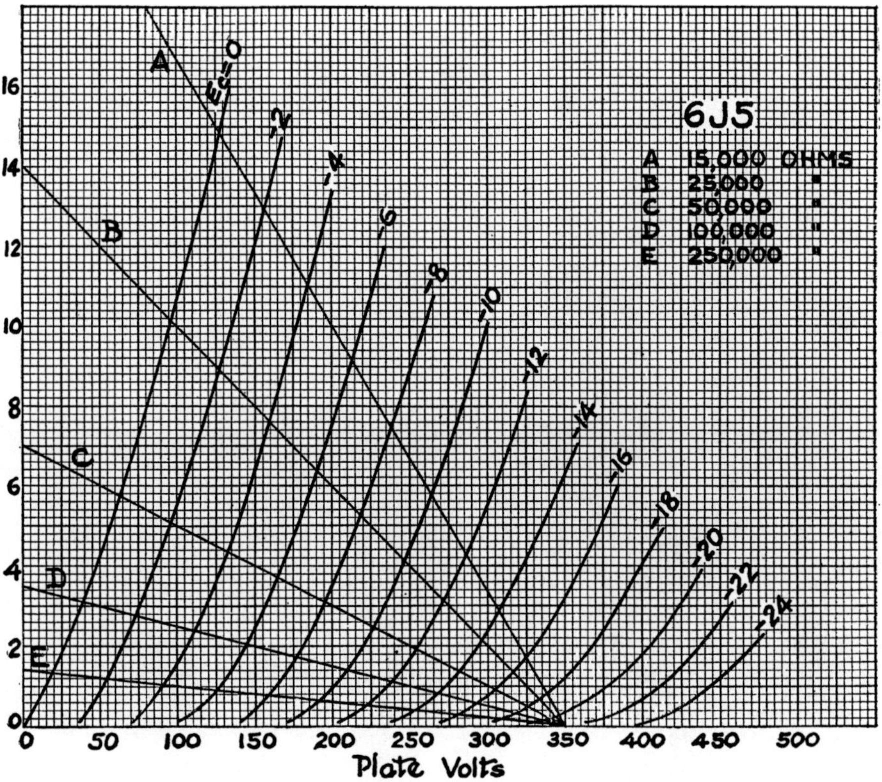

i. 9-9. The lines representing different values of load resistance are drawn on the plate ily of the 6J5 triode. Note that as the resistance of the load increases, the plate current ving in the circuit decreases.

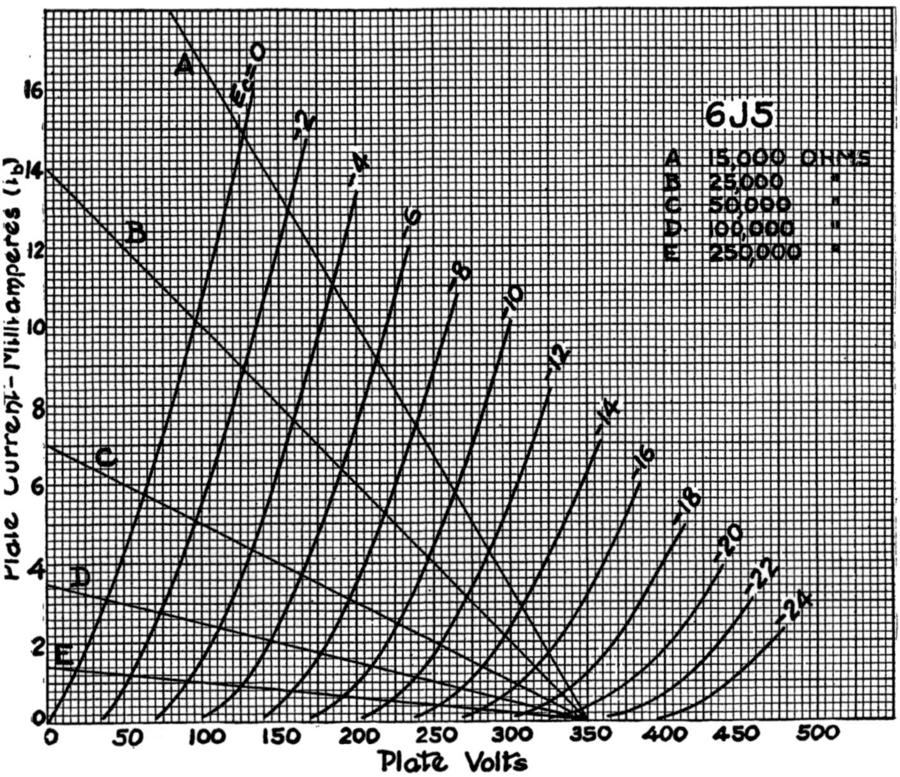

FIG. 9-9. The lines representing different values of load resistance are drawn on the plate family of the 6J5 triode. Note that as the resistance of the load increases, the plate current flowing in the circuit decreases.

50,000-ohm load the voltage e_b' effective at the plate is only 63 volts. These two voltage values e_b and e_b' are indicated on the general plate family of Fig. 9-10, wherein line *1* represents a load line of a resistor

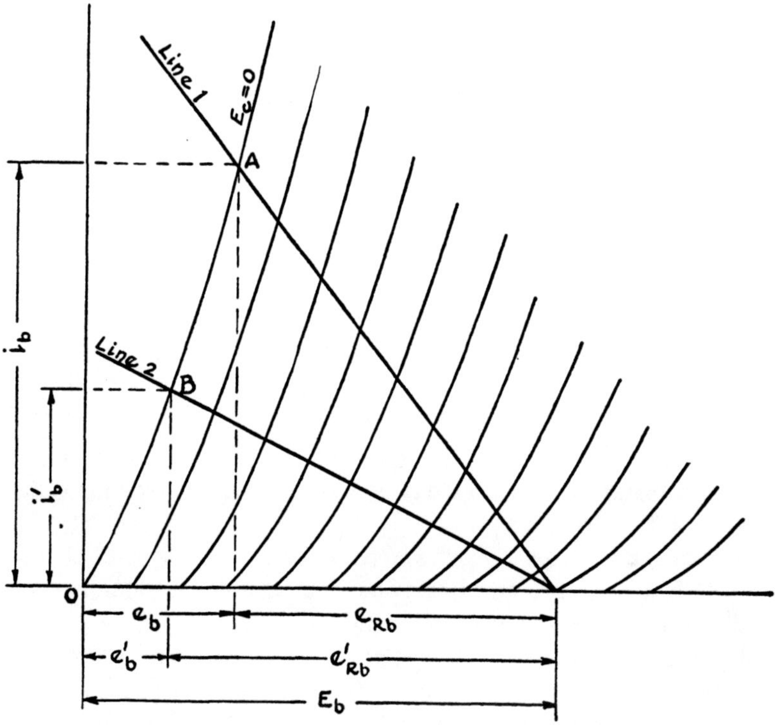

FIG. 9-10. A general plate family of a triode on which are drawn two load lines; line 1 represents a smaller value of load resistance than line 2. The various plate currents and voltage drops for each of these lines are shown in their relative positions.

which is *lower* in value than that used in line *2*. Lines *1* and *2* of Fig. 9-10 therefore correspond to lines *B* and *C* respectively of Fig. 9-9. As explained in connection with Fig. 9-7, the voltage e_b effective at the plate is found by dropping a line from the point of intersection (point *A* in Fig. 9-10) to the horizontal axis and reading the value of the plate voltage thus indicated. The voltage drop across the load resistance is then found by subtracting the value of e_b from the value of the plate-supply voltage E_b, giving e_{Rb} as shown in Fig. 9-10. In the case in point, we have a voltage drop e_{Rb}' across the 50,000-ohm resistor of 287 volts instead of the 254-volt drop e_{Rb} for the 25,000-ohm load.

TABLE A

E_c	Load Resistance in ohms		
	$R_b = 25{,}000$	$R_b = 50{,}000$	$R_b = 100{,}000$
$E_c = 0$	$i_b = 10.1$ $e_b = 96$ $e_{Rb} = 254$	$i_b = 5.7$ $e_b = 63$ $e_{Rb} = 287$	$i_b = 3.1$ $e_b = 38$ $e_{Rb} = 312$
$E_c = -8$ No-signal or Quiescent Value	$i_b = 5.3$ $e_b = 219$ $e_{Rb} = 131$	$i_b = 3.1$ $e_b = 196$ $e_{Rb} = 154$	$i_b = 1.7$ $e_b = 177$ $e_{Rb} = 173$
$E_c = -16$	$i_b = 1.4$ $e_b = 316$ $e_{Rb} = 34$	$i_b = 0.9$ $e_b = 304$ $e_{Rb} = 46$	$i_b = 0.6$ $e_b = 292$ $e_{Rb} = 58$

For your convenience in correlating these and other data which follow, we have arranged them in Table A, as well as the corresponding data for a load of 100,000 ohms.

When the grid voltage has been changed to −16 volts, the drop across the 25,000-ohm load is 34 volts and with the 50,000-ohm load it is 46 volts. (These values are found in the same way as outlined in the preceding paragraph except that lines are dropped from the intersection of the $E_c = -16$ curve of Fig. 9-9 and lines B and C to the horizontal axis from which the values are read.) The total swing across the 50,000-ohm load is from 287 to 46 volts, or 241 volts as against 254 − 34 or 220 volts for the 25,000-ohm load for the same 16-volt swing in grid voltage. Stated differently, for the same change in grid voltage, a greater change in voltage occurs across the load of 50,000 ohms than when the load is 25,000 ohms. This increased swing in voltage across the load and the decreased swing in voltage effective at the plate of the triode, will be found to be very important.

Making the value of the load resistor still greater, say 100,000 ohms, will result in a greater change in the distribution of voltages, namely, 312 − 58 = 254 volts, and a still further reduction in the swing of the plate current i_b, as shown by the load line $R_b = 100{,}000$ ohms drawn for that value in Fig. 9-9 and marked D. Following the same procedure as outlined above, at zero grid voltage, the voltage e_b effective at the plate is 38 volts and the voltage drop e_{Rb} across the load is 350 − 38 = 312 volts, which is also indicated in Table A. The fact that the major control of the plate current i_b is by the load resistance is also evident since the same range of 16 volts on the grid causes a change of only about $3.1 - 0.6 = 2.5$ milliamperes of plate current, whereas for a 25,000-ohm load, the swing of plate current was $10.1 - 1.4 = 8.7$ milliamperes. These values are included in Table A.

An extremely significant detail becomes evident when we compare the 25,000-ohm load with the 100,000-ohm load with respect to the change in plate current when the signal voltage is applied and the total grid voltage e_c is changed from the quiescent (no-signal) value of −8 volts first to 0 and then to −16 volts. With the 25,000-ohm load, changing the grid voltage e_c from −8 to −16 causes a change in the plate current i_b from 5.3 to 1.4 milliamperes or a change of 3.9 milliamperes. (These values are included in Table A.) On the other hand, when e_c is changed from −8 to 0 volts, the plate current increases from 5.3 to 10.1 milliamperes or a change of 4.8 milliamperes. Hence, it is plain that the increase in plate current is greater than the decrease notwithstanding the fact that the increase and decrease of grid voltage were uniform.

Let us see now what happens when the 100,000-ohm load is used. From Fig. 9-9 we find that the increase and decrease in plate current i_b around the no-signal input value of −8 volts on the grid is substantially the same when the grid voltage is decreased to 0 and increased to −16 volts, or a change of 8 volts in each direction. Such a condition of similar variation of the plate-current change for input signal-voltage changes is essential to amplification which is free from distortion. Mentioning this may be somewhat premature, considering the point reached in this text, but this relationship might as well be mentioned here even though it will be considered in more detail later on.

We have not referred yet to two of the load lines in Fig. 9-9. One of these is for 15,000 ohms and the other load line is for 250,000 ohms, these being marked *A* and *E* respectively. What we have discussed in connection with the other load lines is also evident in these, but rather than discuss the disadvantages of the use of either of these load lines with the tube in question, we prefer first to represent the information contained in Fig. 9-9 as dynamic transfer characteristics, and this is done in Fig. 9-11.

Effects of Resistance on Dynamic Transfer Characteristics

We have already described the procedure of transferring the load-line data to obtain the dynamic transfer characteristic in conjunction

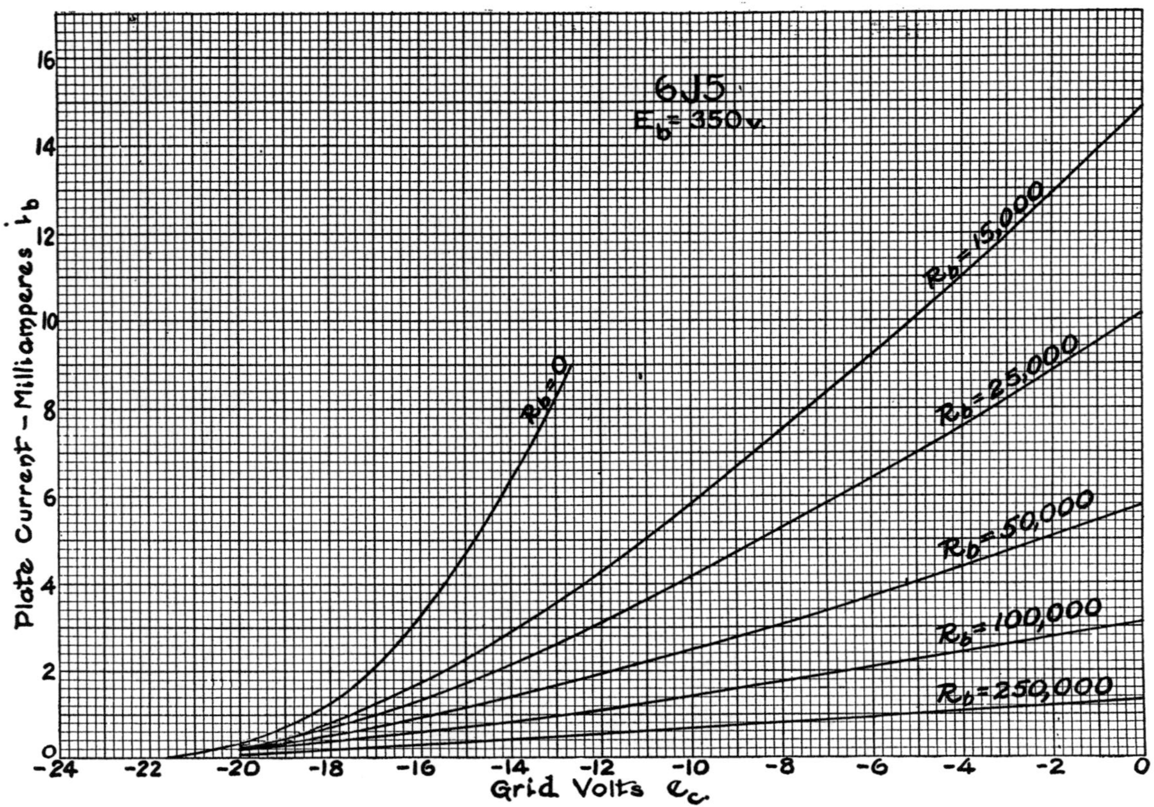

FIG. 9-11. These dynamic transfer characteristic curves correspond to the load lines of Fig. 9-9 for the same operating conditions.

with Fig. 9-8 and this same method is used to convert the load-line data of Fig. 9-9 to the dynamic characteristics of Fig. 9-11. That which will be said in connection with Fig. 9-11 applies equally well to Fig. 9-9, but it seems to be more easily understood when stated in connection with the dynamic transfer characteristic than when described with the load lines. You, of course, should understand that while dynamic transfer characteristics are developed from load lines, the load line is *not* the same as the dynamic transfer characteristic. The load line is always straight for an unvarying value of resistance, whereas the dynamic transfer characteristic becomes straight only as the resistance of the load is increased.

A casual examination of Fig. 9-11 reveals that the larger the magnitude of the plate load resistor, the straighter the dynamic transfer characteristic becomes. The resulting distortion which accompanies the process of amplification depends upon the amount of curvature in the dynamic transfer characteristic, for the greater the curvature, the greater the distortion. We appreciate that as yet you may not be familiar with the full significance of these references to distortion, but you will be very soon. Meanwhile, the details we are describing are given to you so as to form a slight foundation for the subject.

The ideal type of dynamic transfer characteristic would be, of course, an absolutely straight line, for under this condition no distortion would occur. Glancing at Fig. 9-11, that condition would seem to demand a load resistor of 250,000 ohms. For while the dynamic transfer characteristic with a 100,000-ohm load is almost straight, it is not as straight as the one resulting from the 250,000-ohm load. Unfortunately, such a high value of load is generally impractical for two reasons. The first is that with a fixed plate-supply voltage of say 350 volts, the voltage drop across the resistor would be excessive, and the resulting voltage effective at the plate would be insufficient to operate the tube properly. In addition, nothing worthwhile is gained in the way of amplification by using such high values. Furthermore, the amount of distortion introduced by the use of a lower value, such as 100,000 ohms with the tube in question, is not beyond the amount that can be tolerated.

A solution to this dilemma of high plate-load resistance suggests itself in the use of a higher plate-supply voltage. This is unjustified from either the viewpoint of economy or safety, as well as the fact that improvement in the performance of the tube as an amplifier is not materially improved.

Further examination of Fig. 9-11 and Table A shows the decrease in the amount of plate current as well as the range of this current covered by the dynamic transfer characteristic, as the resistance value of the load is increased. This is of no consequence, for this decrease merely indicates that the change in plate current over the full range of grid voltages is decreased, and also that the change in tube resistance with changes of plate voltage plays a lesser part in determining the overall change in plate current as the load resistance is increased. Since the swing in voltage across the plate load depends upon the plate current times the value of the resistance of the load, a high value of resistor and a low value of plate current may produce a greater voltage change across the load resistor than a higher value of plate current and a lower value of load resistance. Compare the plate current corresponding to various values of grid voltage for the 50,000 and 100,000-ohm loads; the voltage drop $e_{Rb} = i_b R_b$, as indicated in Table A, for the 100,000-ohm load is always greater than that for the 50,000-ohm load for like values of grid voltage.

In Chapter 10 we shall examine the process of amplification produced by the triode by utilizing graphic illustrations. During this discussion we shall have occasion to refer to distortion as well as to the variation in voltage across the load resistance for various values of grid voltage. Without doubt, this additional data will clarify whatever subjects presented in this ninth chapter that were not entirely clear.

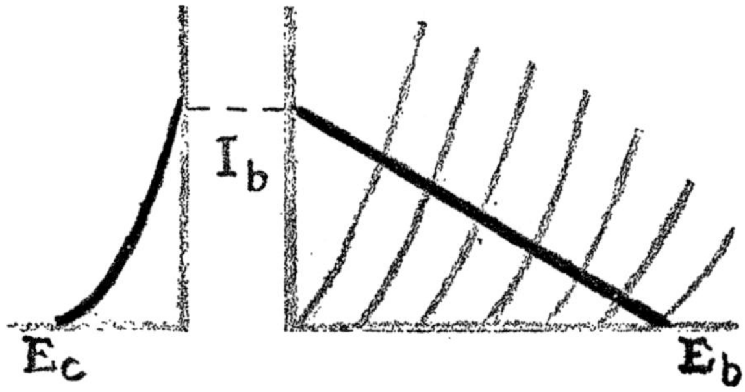

Chapter 10

DYNAMIC TRANSFER CHARACTERISTICS

WE NOW HAVE REACHED the most important portion in our discussion of the triode: the graphical illustration of the process of amplification and the various subjects which pertain to it. In previous chapters we have discussed several viewpoints of amplification, essentially those relating to the geometric structure of the tube. Now we shall consider the variations in voltage at the control grid of the triode and develop the signal which appears in the plate circuit; then the flow of plate current in all those portions of the vacuum tube system where its presence is significant.

Before starting with these considerations we would like to recall to you some of the facts presented in Chapter 3 concerning the fields existent between charged bodies and the behavior of test charges placed in these fields. It was established that a negative test charge (an electron) placed within the electrostatic field produced by the application of a voltage on two parallel plates, would be moved towards the positively charged plate by a force dependent upon the value of the voltage and the distance between the plates. If a third charged body, say a coarse wire mesh, be placed in the field, the progress of the electron towards the positively charged plate will be aided or hindered depending on whether the charge on the mesh is more or less negative than the charge on the original negatively charged plate. You can understand from what you have read in the chapters immediately preceding, that this is the action governing the electrons emitted from the cathode of a triode in their flight from the cathode to the plate. If the voltage on the grid is somewhat positive with respect to the cathode so that the field set up strengthens the field between the plate and cathode, an electron just being emitted from the cathode will be aided in going to the plate. On the other hand, if the voltage on the grid is negative with respect to the cathode so that the grid-cathode field bucks the plate-cathode field, then an electron between the grid and the cathode will be hindered in its passage to the plate.

Electrode Voltages Determine Fields

As we mentioned above, the force exerted on an electron by a field depends upon the proximity of the grid plate to the cathode and the value of the voltage establishing the fields. You are aware now that the grid of a triode is closer to the cathode than it is to the plate; hence if we desire to exert a controlling force on the electrons as they approach the grid, this force can be exerted by a field set up by a much smaller value of voltage applied to the grid than would be necessary if a similar field were set up by a voltage applied to the more distant plate. In other words, a smaller voltage on the grid is responsible for a field that exerts the same force on an electron between the grid and cathode as the field set up by a much larger voltage applied to the plate. It goes without saying then the voltages applied to the grid and plate of a triode are of the utmost importance as they determine the fields controlling the electron movements within the tube.

Let us see just how these voltages and the fields they set up bring about an amplification of voltage. You know that a fixed negative bias voltage is applied to the grid and a fixed positive supply voltage is applied to the plate in a triode. These set up their respective electrostatic fields within the tube and consequently a certain number of electrons will reach the plate per unit of time while the voltages remain unchanged. Now let us insert a source of varying voltage in series with the grid-bias voltage. When this varying voltage makes the total voltage on the grid less negative with respect to the cathode, the grid-cathode field will offer less hinderance to the movement of the electrons and more will arrive at the plate—in short, this is an increase in plate current. Because of the load in the plate circuit, the increased plate current will result in an increase in the voltage drop in the load.

Now assume that the varying voltage on the grid is changed so that the total grid voltage is more negative than it was originally. This results in a lessening of the grid-cathode field and a smaller number of electrons going to the plate. The decrease in plate current means a smaller voltage drop across the load.

You can understand from this discussion that the varying voltage impressed on the grid causes a varying grid-cathode field, which in turn influences the number of electrons (the plate current) that flow in the plate circuit. The amount of plate current determines the voltage drop in the load. You see that the voltage across the load resistor is really dependent on the grid voltage.

For a relatively small change in grid voltage, a correspondingly large change in voltage across the load resistor results. And this is the result we have been seeking: an enlarged replica of the voltage impressed on the grid in the plate circuit of the tube.

We have shown in numerous instances, examples of the variation in plate current caused by changes in the voltage applied to the grid, but in all of these we have considered this voltage to be essentially d-c in character. If you will recall, we did mention that this d-c voltage applied to the grid could be replaced with an a-c voltage. We shall do that in this chapter, although we must comment that the use of a grid voltage which varies with time as the condition set up for the illustration of amplification, does not mean that all amplifiers use such a-c grid or signal voltages. Just as you have seen the application of a d-c voltage to the grid of a triode and the consequent change in plate current for illustrative purposes, just so can the signal voltage, applied to the grid of an amplifier tube, be d-c in character. Such is the case for a direct-current amplifier, a system which is sometimes employed.

Influence of Grid Voltage on Plate Current

To illustrate the development of the signal in the plate circuit of a

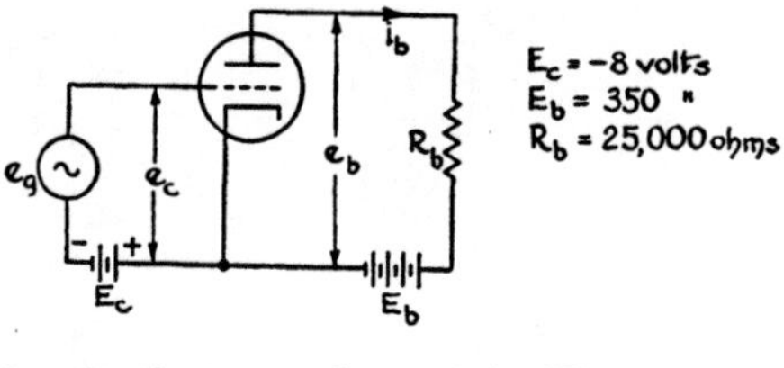

FIG. 10-1. This is the triode circuit employed to show how the signal impressed on the grid is developed in the plate circuit.

triode, we shall use the basic circuit that we showed in Fig. 9-1 and

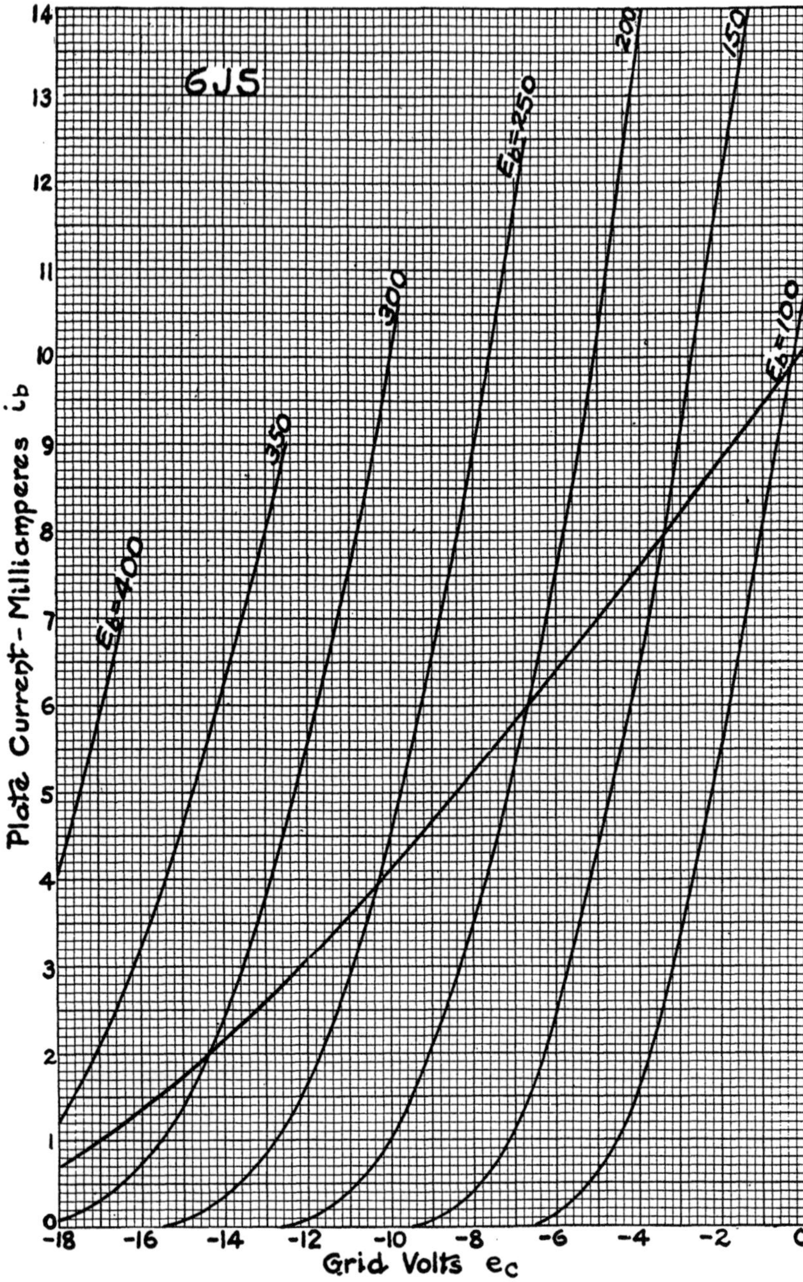

FIG. 10-2. The grid voltage—plate current characteristic curves (grid family) of the 6J5 triode on which is drawn the dynamic transfer characteristic for a load resistance of 25,000 ohms.

which is repeated here as Fig. 10-1 for your convenience. Since we desire to show how the plate current varies with changes in grid voltage, the logical characteristic with which to work is the dynamic transfer curve; therefore, we shall utilize the dynamic transfer characteristic we developed for our sample tube and showed in Fig. 9-8. This curve was for the 6J5 triode operated at a supply voltage of 350 volts and a load resistance of 25,000 ohms. That curve is reproduced as Fig 10-2.

Initial Operating Point and Quiescent Value of Plate Current

Before we can enter upon the discussion of the manner in which the plate current varies in conjunction with the changes in alternating voltage applied to the grid, a few significant details must be brought to your attention. One of these is most aptly described as the *quiescent value of grid voltage* or *operating point* which in turn gives rise to the *quiescent value of plate current.* Both of these terms should be thoroughly understood as they appear in connection with all vacuum tubes.

As you can readily understand a number of different conditions of voltage may exist in the grid circuit of a triode, particularly the bias

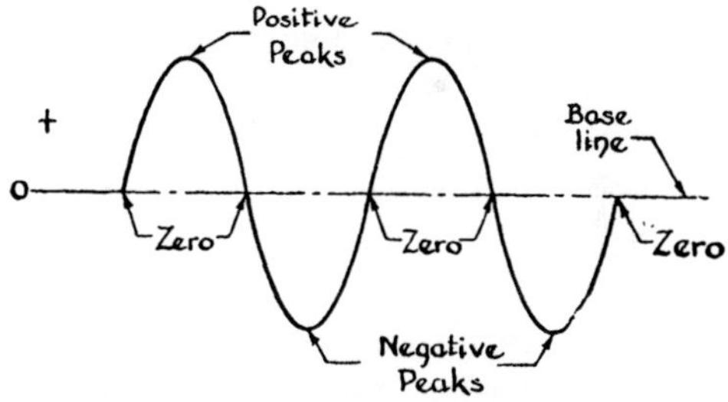

FIG. 10-3. In a wave of alternating voltage or current the instantaneous value of the voltage or current varies above and below zero, reaching positive and negative maxima or peaks each cycle.

voltage and the signal voltage. Assuming for the moment that the signal voltage e_g is of alternating character, you can see that in a wave of alternating voltage, as well as current, there are instants when the value of the alternating component of voltage or current is zero. These contrast with the other instantaneous values, which reach their maximum at the peaks of the positive and negative half cycles, as shown in Fig. 10-3.

As indicated, the zero values exist at the imaginary baseline. The representations of the voltage or current wave shown above the baseline are arbitrarily designated as being positive with respect to the zero value, and the representations of the wave shown below the baseline are also arbitrarily designated as being negative with respect to the imaginary baseline. With this in mind, you can see that with such a wave applied to the grid of a tube, at certain instants the voltage can be zero. At the same time, you can also understand the condition when a tube circuit is in an operative condition, yet no signal voltage is being applied to the input of the tube.

Thus we have a state which is identified as zero signal input, during which time plate current may or may not flow in the tube circuit, depending entirely on the relative conditions existing in the grid and plate circuits. The application of the signal voltage naturally alters the conditions in the grid circuit, but before the signal voltage is applied, a definite state of operation is created. In other words, if the operating capabilities of a triode are expressed by its dynamic transfer characteristic, a point on this characteristic is identified as the *operating point*. This point is created by the specific values of steady grid bias and plate voltage which exist upon the grid and plate elements respectively, when there is no signal input.

Since the tube is in an operative state, although no signal input is applied to the grid, a definite value of plate current will flow. This amount of plate current is designated as the *quiescent value* and a definite association exists between the initial operating point and the quiescent value of plate current.

Location of the Operating Point

This operating point must be understood if you are to comprehend what happens when the signal voltage is applied to the grid. After all, the signal voltage has a finite value and the voltage which will be effective between the grid and the cathode, (shown in Fig. 10-1 as e_c) depends upon what initial value of bias voltage E_c exists in the grid circuit. If we desire to illustrate the application of the signal voltage to the grid circuit, we must know where to locate the zero value of the signal-voltage waveform upon the dynamic transfer characteristic to show properly the change in conditions created in the grid circuit by the varying signal voltage. If we did not have an operating point with its corresponding value of plate current, it would be difficult to indicate graphically the manner in which the plate current is changed by the application of the signal voltage.

For example, in Fig. 10-4 a replica of the dynamic transfer characteristic of Fig. 10-2 is shown. We shall assume that the grid bias is —21 volts. This means that a steady flow of plate current flows in that tube system equal to about 0.03 milliampere, or 30 microamperes. As long as the grid voltage does not change, the no-signal value of plate

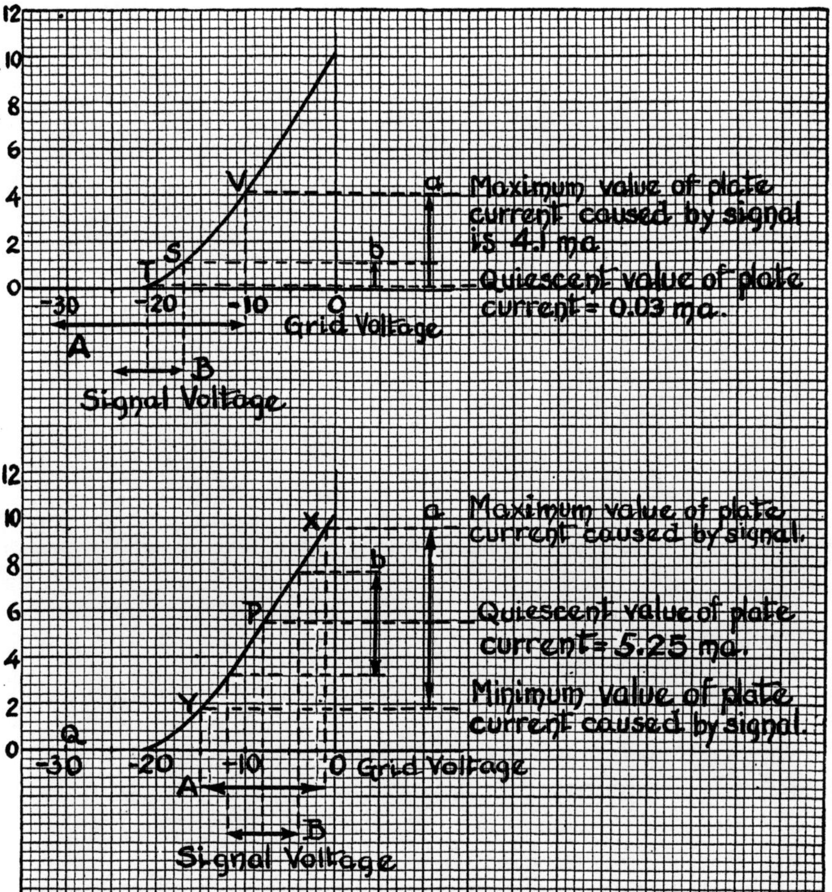

:G. 10-4 (above), FIG. 10-5 (below). The same dynamic transfer characteristic curve is used both cases, but different values of input signal voltage are used. In Fig. 10-4 the swing in gnal voltage is 21 volts (it varies 11 volts above and below its zero value) and the grid bias —21 volts. This results in plate current that is practically totally above the quiescent value. hen the grid bias is changed to —8 volts and the input signal reduced to a swing of 14 volts volts above and below zero) the plate current is practically the same above and below its iescent value.

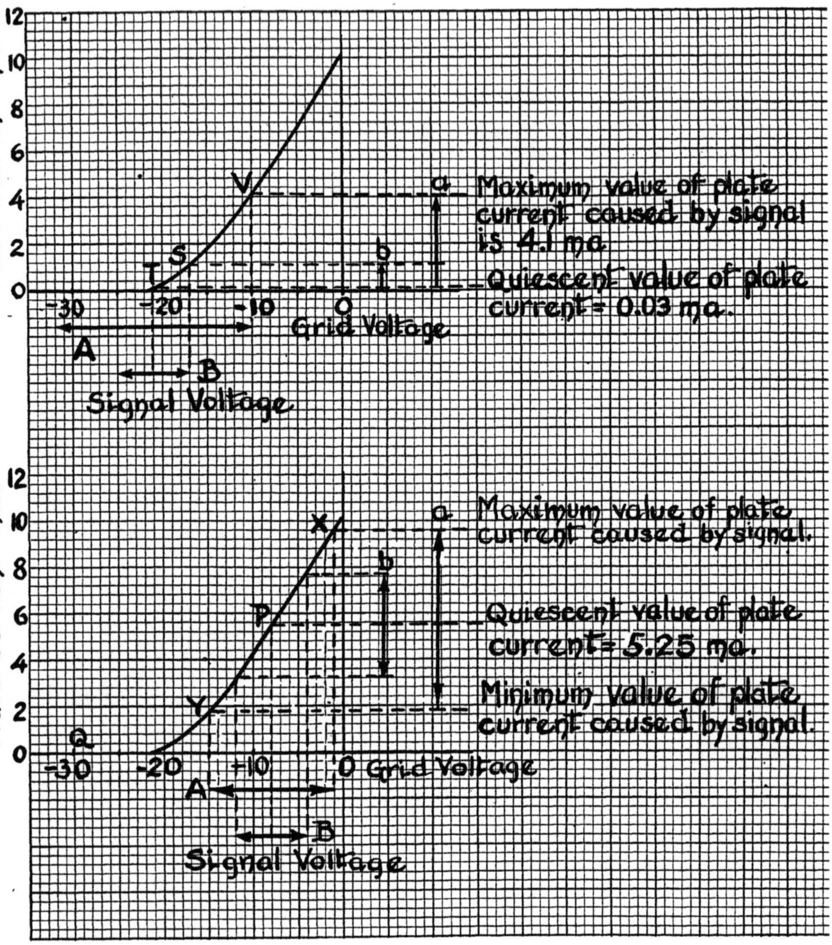

ɪɢ. 10-4 (above), Fɪɢ. 10-5 (below). The same dynamic transfer characteristic curve is used ı both cases, but different values of input signal voltage are used. In Fig. 10-4 the swing ir gnal voltage is 21 volts (it varies 11 volts above and below its zero value) and the grid bias —21 volts. This results in plate current that is practically totally above the quiescent value hen the grid bias is changed to —8 volts and the input signal reduced to a swing of 14 volts volts above and below zero) the plate current is practically the same above and below its ıiescent value.

current is 0.03 milliampere and the operating point on this dynamic transfer characteristic is point T.

Suppose we apply a signal voltage to the grid of this triode. How can we show this? For the moment, we shall not show this in the conventional way by means of the waveform of the signal voltage, but rather by a horizontal line with arrow heads, marked A, the length of which indicates the value of the voltage. Let us say that the positive and negative peaks of this voltage are 11 volts.

Where should we locate this signal voltage on the grid voltage axis of this graph? It must be placed so that zero signal voltage corresponds to the quiescent condition. To state this differently, the instant when the signal voltage is zero corresponds to the condition of no-signal input, and for that moment the tube conditions are such as if no signal at all were being supplied to the tube and the instantaneous value of plate current then is that which is determined by the operating point upon this dynamic transfer characteristic.

Therefore, we place the zero point of the representation of the signal to correspond with the quiescent operating point. This is indicated in Fig. 10-4 by the dotted line which joins the —21-volt point along the grid-voltage axis with the zero value of the input signal. Now when the signal voltage changes from zero and aids or bucks the fixed bias voltage, thus creating a series of instantaneous and different values of grid voltage e_c, we can follow the movement of the *instantaneous operating points* along the dynamic characteristic. For each of these instantaneous points upon this curve, there is a corresponding value of plate current.

For example, when the signal voltage e_g has reached its positive peak value of 11 volts, the voltage e_c effective between the grid and cathode is $(-21) + (+11) = -10$ volts, so that for that instant of peak signal voltage, the instantaneous point along the dynamic characteristic has shifted to point V, and here the corresponding plate current is 4.1 milliamperes. Another instant during the positive half cycle of the grid voltage e_g swing, the value is $+4$ volts. At that instant, the instantaneous point on the characteristic is at point S, which represents $e_c = (-21) + (+4) = -17$ volts. All during the positive half cycle of the signal voltage when the signal voltage is greater than zero, the instantaneous point is moving up and down along the dynamic characteristic, and at each instant the plate current corresponds to the amount which can flow for that instantaneous point on the curve.

The same condition prevails over the negative half cycle of the signal voltage, but due to the location of the *operating point,* as established by the fixed bias, there is but a small range of signal voltage over which any change can take place in the plate current. A very small change of negative signal voltage shifts the instantaneous operating point by adding sufficiently to the negative voltage being applied to the grid, to the extent that plate current cutoff takes place.

These changes in plate current are indicated in Fig. 10-4 by the vertical arrow *a* at the right-hand side of the illustration. Due to the location of the operating point, the full range of signal voltage during the positive alternation can cause a change in plate current, whereas only a small portion of the total signal voltage during the negative half cycle (which is too small to be shown), is effective in causing any change in the plate current.

A significant detail which you should bear in mind is this difference between the *operating point* and the *instantaneous point*. The former is set by the bias and the latter is a function of the signal voltage aiding or bucking the fixed bias.

Magnitude of the Signal Voltage

Changing the value or magnitude of the signal voltage is an important consideration in connection with the final results obtained from any vacuum tube. For example, assume that the signal voltage had a peak value of 4 volts instead of 11; this is indicated by the horizontal line *B* with arrow heads in Fig. 10-4. This would change the range along the dynamic characteristic over which the instantaneous points would shift for the full swing in signal voltage, but the zero point of the signal would still correspond to the quiescent operating point and the increase in plate current, shown by the vertical arrow *b* in Fig. 10-4, would be much greater than the decrease (which is too small to be shown), although the amplitude of the positive and negative halves of the input signal voltage are the same. This applies equally well to the 11-volt signal previously discussed.

Assume again that instead of applying a bias of —21 volts, we apply one of —8 volts and no signal. The new operating point or quiescent point is now at *P* along the dynamic transfer characteristic of Fig. 10-5 and this results in a plate current flow of 5.25 milliamperes. The application now of a signal voltage will swing the instantaneous points along the dynamic characteristic with the point *P* corresponding to the zero value of signal voltage. If we now show the signal voltage along the grid-voltage axis, the zero value of this signal must correspond with the quiescent operating point set by the —8-volt fixed grid bias.

Looking at Fig. 10-5, you can see that both the positive and negative alternations of the signal voltage, *A*, will create instantaneous points along the dynamic characteristic and cause changes in the plate current. Comparing this with the two signal voltages of Fig. 10-4, where only the positive half cycles affected the plate current to an appreciable extent, this change of the location of the quiescent operating point permits utilization of the entire signal voltage—both its alternations—to affect the plate current. This is illustrated by the instantaneous point *X*, Fig. 10-5, corresponding to the positive peak of the signal voltage and the point *Y*, corresponding to the peak value of the signal voltage during the negative half cycle.

The fact is significant that the changes in plate current, i.e. the rise from the quiescent value, is unequal to the decrease from the same value although the input is uniform; however, this will be left for discussion later on. The important thing to bear in mind now is that the location of the operating point does have a very great bearing upon how the input signal will cause the plate current to vary; also what portion of the total input signal is effective in causing a change in the plate current.

You will note that we are not recommending which of these two operating points should be used, the reason being that different requirements of tube operation call for the location of the operating point at different places along the dynamic transfer characteristic.

It is true, of course, that the magnitude of the signal voltage has a great bearing upon the location of the quiescent operating point, particularly when the function of the triode is that of a conventional amplifier, wherein uniform changes in plate current are desired for uniform changes in signal voltage. This is illustrated in Fig. 10-5, by comparing the two input signal voltages *A* and *B* and the corresponding changes in plate current. For instance, with a signal voltage input of 4 volts peak the changes in plate current are much more uniform than when the signal voltage has a peak value of 7 volts. You can see that all this depends upon the straightness of the characteristic curve as well as the location of the quiescent operating point upon the straight portion of the dynamic characteristic. Again we repeat, there is no rigid rule that all triodes are operated so that the operating point is located midway along the dynamic transfer characteristic. If an application so dictates, the operating point may be located at any point which will result in the desired type of operation.

For certain applications the operating point may be located at the zero grid-voltage point or, on the other hand, the operating point may be at some definite negative value and the magnitude of the signal voltage is so large as to drive the instantaneous total grid voltage into the positive region. A discussion of this condition will be reserved for a later chapter.

It is also possible to locate the operating point beyond the plate-current cutoff point, as shown by the point Q in Fig. 10-5, corresponding to a bias of -30 volts. This value has been arbitrarily selected and we say this because, if conditions demand, due to a certain required performance, the value of the negative bias can be several times the normal plate current cutoff value.

Therefore with a bias of -30 volts, the quiescent value of plate current is zero and the negative alternations of the signal voltage have no effect whatsoever on the plate current. In fact, during its positive half cycle the signal voltage must rise above 9 volts before it has any effect on the plate current. In other words, plate current flows in the form of pulses during a certain portion of the positive half of the signal voltage.

Summarizing these details shown in Figs. 10-4 and 10-5, you can see that the quiescent operating point, which is established by the bias applied, has a great influence upon how the signal voltage will affect the plate current. This is extremely important insofar as the instantaneous variations of signal voltage is concerned, because it determines whether or not any one type of amplifying system will provide distortionless amplification. Then again, it makes possible the attainment of a definite amount of distortion of the voltage wave, when such a form of operation is desired, which is the case in many instances.

Moreover, the operating point determines the length of life of the tube by establishing the quiescent value of plate current, since the quiescent operating point influences the operating capabilities of the tube inasfar as the dissipation of heat generated within the plate of the tube is concerned. You will recall that we mentioned the generation of heat at the plate of the tube due to the impact of the electrons. Each tube is designed to dissipate a certain amount of heat. If the quiescent value of plate current is too great, the heat generated within the tube will exceed the permissible rating and the tube may be damaged.

These details concerning the operating point are by no means all that may be given, but in an elementary text of this type, further elaboration does not seem necessary, for the above details embrace the most pertinent items. It might be well to mention, however, before entering another subject, that the general details about the operating point apply to *all* multi-electrode tubes as well. As you will see later, it is true that the value of the load upon the tube, the type of tube, and the operating voltages involved, have a decided bearing upon the magnitude of the change in plate current for any given change in grid voltage, but the fact that the grid bias establishes the quiescent operating point and that this operating point controls the magnitude of signal which may be applied to the grid of the tube, is not altered.

Output Plate Current and Input Signal Voltage

Now we are ready to deal with the development of the output wave of plate current corresponding in time to the input wave of signal voltage, which you may also call *excitation voltage* or even *grid voltage*. Now we are going to examine graphically the process of amplification: the manner in which the signal voltage applied to the grid appears in magnified form in the plate circuit.

We shall employ the circuit shown in Fig. 10-1, which has been reproduced again in Fig. 10-6, wherein the constants of the circuit are indicated. We shall assume that the fixed bias E_c equals -10 volts so that the operating point on the dynamic transfer characteristic of this typical tube is at point P in Fig. 10-6. This characteristic is the same as was used throughout this chapter and is the one developed in Fig. 9-4.

Instead of showing the input signal voltage as a horizontal line between certain limits of voltage as was done in Figs. 10-4 and 10-5, we shall use a sine wave, although this is not a requisite. What we intend showing can be illustrated by the use of any kind of an input wave of voltage, sinusoidal or otherwise; however, a sine wave lends itself more readily to a simple explanation. For the first example, we shall assume a peak value of 10 volts for e_g, the signal voltage.

Accordingly, a sine wave of voltage equal to 10 volts peak for the positive and negative half cycles is plotted on the chart with the zero value of this voltage corresponding to the quiescent operating point of -10 volts. Various points representing different instantaneous values of voltage of this wave are designated as *a,b,c,d,e,f,* etc.

Graphical Representation of Input Voltage and Output Current

Starting with zero signal voltage or *a* upon the input voltage wave, we note that the quiescent value of plate current is 4.1 milliamperes and by projecting point *a* to point *P* and then to point *A*, we indicate the start of the output plate-current wave at *A*. You will note in Fig.

10-6 that the input signal-voltage wave first increases in value in the positive direction, which means that the plate-current wave in the output also will increase in value from A in a positive direction. This is to be expected for if the input signal voltage increases in the positive direction, it means that it will oppose the grid bias, hence the plate current will increase.

The next point b on the input signal-voltage wave e_g represents a voltage of +3 volts and a lapse of time equal to 2 squares or divisions along the time baseline. As a consequence of the value of the fixed bias of −10 volts and the positive signal of +3 volts, the instantaneous point on the dynamic transfer characteristic corresponds to a resultant grid voltage, e_c, of −7 volts with a plate current of 5.9 milliamperes. Projecting along this current line of 5.9 milliamperes two divisions past the point A, we locate point B.

Point c on the input voltage wave e_g has a value of +5.9 volts and represents an advance of 2 more divisions along the time baseline. This value of signal voltage shifts the instantaneous point to $e_c = -4.1$ volts on the dynamic characteristic curve and this corresponds to a flow of plate current of 7.5 milliamperes. Projecting this value of plate current across to the fourth horizontal division to the right of point A we locate point C on the plate current wave.

Point d on the input signal-voltage wave represents a value of $e_g = +8.1$ volts and an instantaneous point upon the dynamic transfer characteristic of $(-10) + (+8.1) = -1.9$ volts. This means a plate current of 9 milliamperes. The projection of this value across to the second division beyond point C, establishes point D of the output current wave.

Point e on the input signal-voltage wave represents an input of 9.7 volts. This means that e_c, the instantaneous grid voltage, equals −0.3 volt, resulting in a plate current of 9.8 milliamperes. Projecting across this value of plate current from the characteristic curve to two divisions beyond point D, gives us point E of the plate-current wave.

Point f is the positive peak value of the input signal—voltage wave and equals +10 volts. Since the fixed grid bias equals −10 volts, the result of this bias and the input signal, is an instantaneous grid voltage equal to $(-10) + (+10) = 0$. For this instant, the operating point is along the zero grid-voltage axis. The plate current for the conditions existing here is 10.1 milliamperes. Projection of this value two divisions beyond the point E, gives the location of the peak of the positive half cycle of the output-current wave, point F.

Having reached the peak of the positive half cycle of the input voltage, the wave starts decreasing in magnitude although it still re-

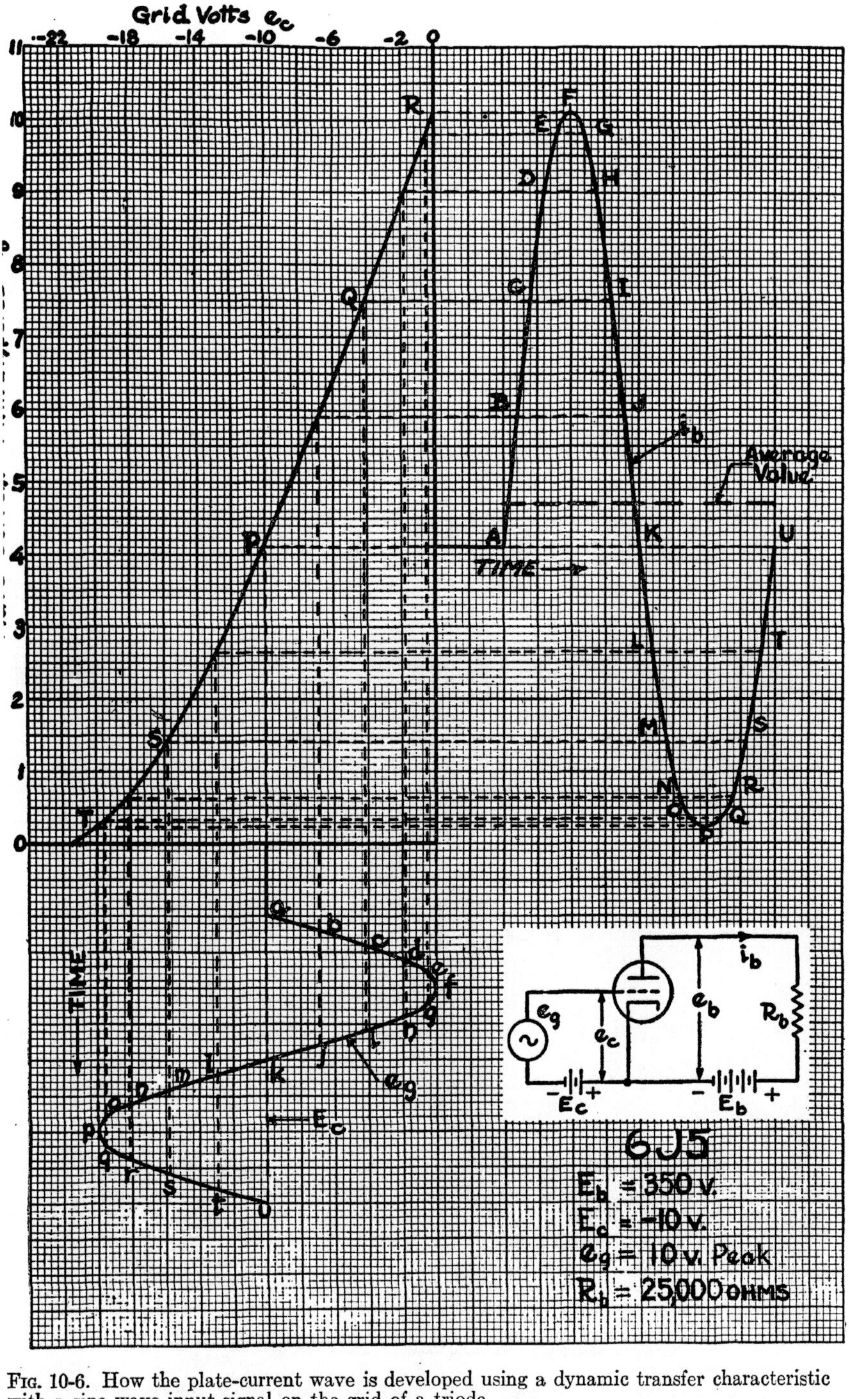

FIG. 10-6. How the plate-current wave is developed using a dynamic transfer characteristic with a sine wave input signal on the grid of a triode.

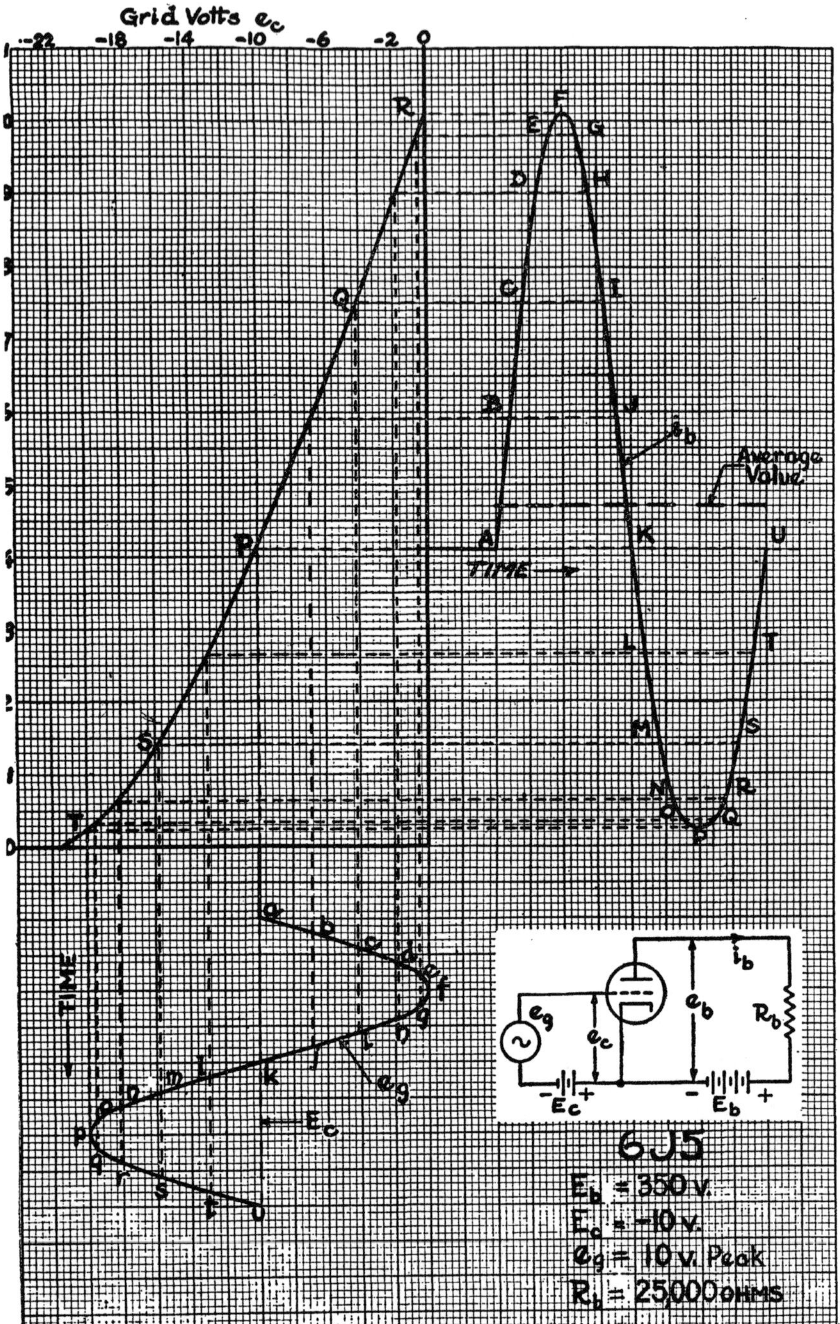

FIG. 10-6. How the plate-current wave is developed using a dynamic transfer characteristic with a sine wave input signal on the grid of a triode.

mains positive. As this part of the wave is symmetrical with respect to that which we have just considered, we can use the same values as we did in the first quarter. Thus points *g,h,i,* and *j* correspond with points *e,d,c,* and *b* and we can continue plotting the output current wave by simply extending the locations of *E,D,C,* and *B* until we locate the proper intersections with upward projections of the time base. In this way, points *G,H,I,* and *J* are located. During the time when the input signal voltage is decreasing in value, although it is still positive with respect to the baseline, the instantaneous point is retreating downward along the dynamic transfer characteristic, passing through the same points which were created during the time the input signal was increasing in value.

Finally the input signal again reaches its zero value, which is indicated by the point *k*. For this instant, the operating point coincides with the quiescent or no-signal point *P*, and the value of plate current is the same as at the start of the wave, which condition is indicated by the fact that points *K* and *A* on the plate-current wave are on the same line, both representing a flow of 4.1 milliamperes of plate current.

The signal voltage e_g now becomes of negative polarity so that it aids the fixed negative bias. The resultant instantaneous point must therefore move downward along the dynamic transfer characteristic so that it will correspond to values of grid voltage which are more negative than the fixed bias. For the sake of convenience, we select values of signal voltage which are the same as those used during the positive half cycle. Thus the first value of negative signal voltage e_g is that indicated by point *l*, representing a signal of −3 volts, making the resultant grid voltage e_c equal to $(-10) + (-3) = -13$ volts.

In exactly the same manner that was used to locate the previous points on the plate-current wave from points on the input signal-voltage wave, the remaining points *L,M,N,O,P,Q,R,S,T,* and *U* are located for points *l,m,n,o,p,q,r,s,t,* and *u* on the input signal-voltage wave. Points *u* and *U* on the input and output waves respectively correspond to zero signal input.

Comparison of Input and Output Waves

We have now completed the entire cycle of plate-current variations for the complete cycle of input signal-voltage variations. Let us now examine the output wave and compare it with the input wave. You will see several differences as well as several similarities and these are an interesting subject.

First considering the output plate-current wave i_b, note the absence of uniformity between the two halves of this wave. The curve above the time baseline, that is the positive half cycle, extends higher above this baseline than the other half extends below it. Yet the heights of the two halves of the input signal-voltage e_g wave above and below its baseline are the same. If we interpret the condition evident in the plate-current curve in values of plate current, we see that the positive half cycle of the plate current represents a greater change in plate current from the quiescent value than does the negative half cycle of this wave.

Starting with 4.1 milliamperes of current flow, the peak value of plate current in the positive half cycle corresponds to 10.1 milliamperes or an increase of 6 milliamperes. On the other hand, starting again at 4.1 milliamperes of plate current, the peak value of the plate current during the negative half cycle is 0.25 milliampere, a change of but 3.85 milliamperes. Considering only the shape of the input signal-voltage wave, you can see without even a critical examination that the plate-current wave is not an enlarged replica of the input signal-voltage wave. The lower half of the plate-current wave is much flatter than the upper half of this wave, which condition does not exist in the input wave. This relationship can be described by stating that having started with a sine wave of voltage, the plate-current output wave is distorted. Without concerning ourselves for the moment with the reason why the distortion exists, let us examine this relationship further.

Considered from the viewpoint of representations of alternating waves, although one is voltage and the other current, we note that both *vary in the same direction* with respect to time. The input signal-voltage wave, starting from zero, increases in value in the positive direction and so does the plate-current wave. Both start at the same point with respect to time and both reach their greatest values or peaks at the same moment. The same conditions prevail during the negative half-cycles. Granted that the height of the positive half-cycle above the plate-current baseline is greater than the height of the wave below the baseline during the negative half-cycle, the maximum or peak amplitude of the plate-current wave corresponds with respect to time to the peak of the positive half-cycle of the signal-voltage wave and the peak of the negative half-cycle of the plate-current wave is reached at the same time as the peak of the negative half-cycle of the input voltage wave. This can mean but one thing: *the output current wave and the input voltage wave are in phase.* The fact that the output current wave is distorted does not alter this relationship.

Similarity of Output Current to A-C Wave

But strangely enough, we now find cause for wonder. It is true that the illustration of the way in which the plate current varies with time and with the input signal-voltage variations, makes the plate current look like an alternating wave of current, but is it really that? Your idea of an alternating current is that the current reverses its direction of flow periodically. How can this plate-current wave be representative of an alternating current when it is a known fact, which we have stressed, that the direction of the electron flow comprising the plate current, always is in the same direction? How does the reversal of direction of this current come about?

The acceptance of the idea that an alternating current exists in the plate circuit of a vacuum tube comes about in a peculiar way. Of course, you are right when you think of the flow of plate current always in the same direction: the direction of the movement of the electrons. The introduction of a signal voltage to the grid does not change that condition. What it does, however, is to increase and decrease the value of the plate current *above and below the quiescent value.* During all this time the current, no matter how small its value, just as long as it is greater than zero, is moving from the plate and back to the cathode through the *external* circuit which joins the plate and the cathode. In our case, the external circuit is the load resistor R_b and the battery E_b. At all times the plate current i_b is moving through these two circuit elements in the same direction.

Because the plate current is varying in value with respect to time, a voltage drop occurs across the load R_b; this drop also varies with respect to time. For a definite period of time, controlled by the character of the a-c voltage applied to the control grid (the input signal voltage), the periodically varying voltage across the load resistor is *greater* than the voltage drop during the quiescent period. For another period of time, again controlled by the character of the input signal voltage, the periodically varying drop across the load resistor is *less* than that during the quiescent moment.

Such a periodically varying voltage, which rises and falls above some steady value, creates an effect which is similar to that of a flow of alternating current, so that it is possible to view the periodically varying voltage drop across the load resistor as being the equivalent of an alternating voltage. If the output plate-current wave varies in amplitude with respect to time exactly as the variation in amplitude of the input signal, then the *variations* in voltage drop across the load resistor are considered as being an enlarged version of the input signal voltage and it is said that amplification of the input signal has taken place; also that the input signal has been transferred in enlarged form to the load upon the tube.

It is customary to consider that the plate-current flow, shown in Fig. 10-6, as a combination of direct current and alternating current. This is so because if an alternating current of proper magnitude which varied exactly as the wave shown, were introduced into the tube circuit and superimposed upon the flow of a direct current, a resultant wave like that shown in Fig. 10-6 would be obtained. It then is assumed that during the positive half-cycle of the alternating current, the direct current and the alternating current would be additive and the result would be a value of current greater than the steady direct current. During the negative half-cycle, the *assumed* alternating current is flowing in the opposite direction and thus tends to buck the direct current and reduce the effective value of plate current, so that the resultant plate current would be less than the steady value of the direct current.

What we have said is by no means intended as a complete discussion of the process of amplification; rather it is quite a meagre explanation, but it will serve as a foundation upon which we can build. We have prepared you for three subjects which will receive further elaboration: the factors which determine the production of an output plate-current wave which varies in amplitude with respect to time in exact accordance with the input signal voltage; the phase relationships which exist in the triode and multi-element tubes; and the factors which determine the magnitude of the signal voltage developed across the load resistance. Incidentally, the last named subject should be understood as also being of value when the load is not a resistor, for there are numerous instances when this is true. You will remember the statement that unless a load is applied to a tube, it cannot be gainfully employed.

Why is the variation in amplitude with respect to time of the output plate-current wave not identical to that of the input voltage wave? Perhaps before we discuss this subject, we should include a few words about what is meant by the variation in amplitude with respect to time. Those of you who are familiar with a-c phenomenon understand the term fully, but those who are not so familiar can benefit from the following short discussion.

Variation of Amplitude of A-C Wave with Time

Time as applied to alternating currents can be expressed in fractions of a second, depending upon the frequency of the wave of current or voltage; or else in degrees, which is independent of frequency. Since the latter is employed in almost all texts, we shall follow this procedure. If the time base of a single cycle of voltage or current is divided into 360 equal divisions, each represents the lapse of 1 degree of time. One quarter-cycle is equivalent to 90 degrees of time; a half-cycle to 180 degrees of time; three quarters of a cycle to 270 degrees of time, and the complete cycle to 360 degrees of time, as shown in Fig. 10-7.

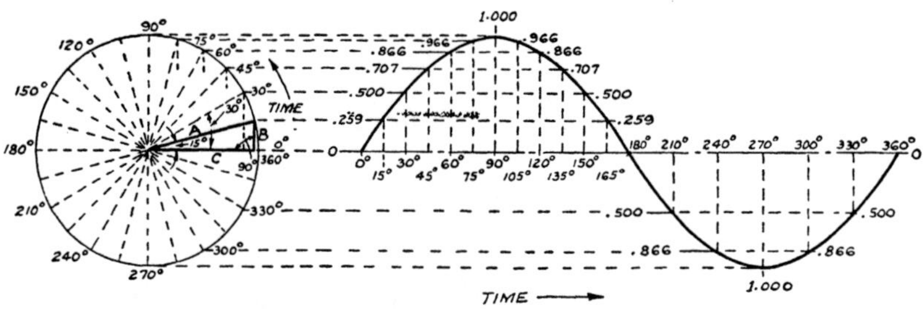

FIG. 10-7. The instantaneous values of a sine wave for every 15° are shown at each division of the positive half-cycle; although not shown in entirety, these values are the same for the negative half-cycle.

If this wave of current or voltage occurs 60 times per second so that 360 degrees of time represents 1/60 of a second, each division equals 1/21,600 of a second.

The instantaneous amplitude of a sine wave for various numbers of degrees, bears a definite relationship to the total peak value of the wave. As illustrated in Fig. 10-7, the sine wave shown is developed from a circle divided, let us say, into 24 equal angles of 15° each. If the peak value of the sine wave is equal to a certain number of units (which may be representative of volts, amperes, or fractions of volts or amperes) and the wave starts at zero, a lapse of 15 degrees means an instantaneous amplitude that is equal to this total peak value times 0.259; a lapse of 30 degrees equals the peak value times 0.500; 45 degrees equals the peak times 0.707; 60 degrees equals the peak times .866; 75 degrees equals the peak times .966; 90 degrees equals the peak value times 1.000, etc., for the remainder of the sine-wave cycle.

The decimal values 0.259, 0.500, 0.707, etc., corresponding to 15°, 30°, 45°, etc., respectively, are obtained from trigonometric tables and represent the multiplier used in the solution of a right-angle triangle where one side and one angle is given, that is, solving for the side B of the triangle ABC in the circle of Fig. 10-7. When the sine of the angle of 15° and the side A (the radius of circle and represented as 1.000) are given, we have the formula

$$B = A \times \textit{sine of angle of } 15^\circ$$
$$\text{then} \quad B = A \times 0.259$$

and the instantaneous amplitude B for 15° is therefore equal to the total peak value of 1.000×0.259.

Solving for a 30° angle, $B = A \times \textit{sine of angle } 30^\circ$

$$\text{then} \quad B = A \times 0.500$$

and the instantaneous amplitude for the 30° position is equal to the total peak amplitude times 0.500.

The remaining multipliers and instantaneous peak values can be determined in this way as the sine wave advances through its complete cycle of 360°. The instantaneous amplitudes will have plus values in the first half of the sine-wave cycle (the positive swing), and minus values in the second half of the cycle (the negative swing).

By variation in amplitude with respect to time is meant the instantaneous values of voltage or current at the various time intervals. When two voltage or current waves are said to be in time phase and one is an enlarged replica or version of the other, a lapse of say 45 degrees of time means that in both cases the instantaneous value of voltage or current is 0.707 of the peak value of the respective waves. This is substantially true for the positive half of the plate-current wave in Fig. 10-6, wherein for a total of about 60 squares for the peak of the half cycle, the height of the wave after a lapse of 45 degrees (⅛ cycle) is about 42 squares, namely 0.707 or about 71%. This compares favorably with the structure of the positive half-cycle of the input signal-voltage wave.

You will understand that these references to squares in connection with the height of the wave or its amplitude in Fig. 10-6 is entirely due to the type of graph used. Referring to the negative half of the plate-current wave, we find that the amplitude (31 squares) at the ⅝-cycle point (between M and N) is equal to approximately 74% or 0.74 of the peak value, whereas in the positive half-cycle of the input voltage wave, the proper ratio of 0.707 or about 71% (42 squares) exists at the corresponding point. The existence of such a difference in the variation of the amplitude with respect to time is why we say that the output current wave is *not* an enlarged replica of the input voltage wave.

As to the phase relationship between the input and output waves, you can see that what was stated about these two waves is still true even though distortion exists in the plate-current wave. Both waves pass through their zero and maximum points at the same instant and move in the same direction. It is this that establishes the phase relationship.

Non-linearity of Characteristic Causes Distortion of Output

Why is distortion present in the output plate-current wave? Although we said that the positive half-cycle of the plate-current wave is representative of a sine-wave condition, it really is not so because the variation of the amplitude with respect to time of that half-cycle departs slightly from the proper mode of variation. The basic reason for this condition is the non-linearity of the dynamic transfer characteristic.

The axis on which the signal-voltage wave is plotted is a straight line and the characteristic is slightly curved. Any attempt to extend uniform divisions from a straight line onto a curved line, will result in non-uniform distances along the curved line between the points of intersection. The dynamic characteristic in Fig. 10-6 is a curved line with minimum curvature between the points *P* and *R* and an increasing curvature downward from *P* towards *T*.

Because the dynamic transfer characteristic of Fig. 10-6 is almost straight between *P* and *R*, projection of the uniformly divided points along the input signal-voltage wave produces almost uniform distances along the dynamic transfer characteristic curve between the points of intersection. The wave, therefore, that is developed in accordance with these points of intersection nearly forms an amplified replica of the original. When the same procedure is followed for the negative half-cycle of the input signal voltage, the distances along the characteristic between points of intersection are not the same, so the wave developed in accordance with these points of intersection is unlike the original.

The distortion which appears in the plate-current wave of Fig. 10-6 is due entirely to the curvature of the dynamic transfer characteristic, *TSPQR*. The distortion in the negative half-cycle of the output current wave is more prominent than in the positive half-cycle because the curvature of the dynamic transfer characteristic between *P* and *T* is greater than between *P* and *R*.

The reasons why the dynamic transfer characteristic is curved and what can be done to straighten it, have been given before, but we shall repeat them. The characteristic is curved because the variation of the internal plate resistance of the tube, due to changes in operating potentials, plays an important role in controlling the plate current. In other words, the 25,000-ohm load resistance which we have considered in connection with this tube, is not high enough to nullify the action of the internal plate-resistance variation in determining the plate current. From the above you can see that by increasing the load resistance, the dynamic transfer characteristic will be straightened, and so the distortion which appears in Fig. 10-6 will be reduced.

Operation on Linear Portion of Characteristic

We mentioned previously that this dynamic transfer characteristic approached a straight line between the points P and R in Fig. 10-6. If we can arrange to operate along this portion of the curve rather than along the entire curve, it should be possible to obtain an output plate-current wave which would be reasonably free from distortion. It would not be wholly free from distortion because the characteristic is not an absolutely straight line, but the result would be much better than that obtained under the conditions existing in Fig. 10-6.

If we limit the zone of operations to that portion of the curve between P and R, we must take into account the range of input signal voltage which is covered by those limits of the dynamic transfer characteristic. We can see from the diagram that the permissible swing in voltage which will operate between P and R on the characteristic, is the range of voltage indicated by the projections from P and R down upon the grid-voltage axis. With point R corresponding to zero grid voltage and point P to -10 volts grid voltage, we must be limited to an overall swing of 10 volts on the grid, that is, of course, if we are to operate between P and R on the dynamic transfer characteristic.

It is quite worthwhile trying a reduced signal input e_g of 5 volts. This has been done in Fig. 10-8. You can see that this is an improvement over what was obtained in Fig. 10-6, but it is not all that we wanted. It does, however, point to a conclusion, which we do not yet want to make definite: the signal voltage fed into a tube is a factor in determining the ability to secure amplification that is free from distortion. Also that the signal voltage is associated with the load impedance, as far as distortionless amplification is concerned.

The first of these thoughts originates in the fact that reducing the magnitude of the input signal voltage creates an improvement, even though it is not as much as we desire. The negative half-cycle of the plate-current wave is still flat compared with the positive half-cycle,

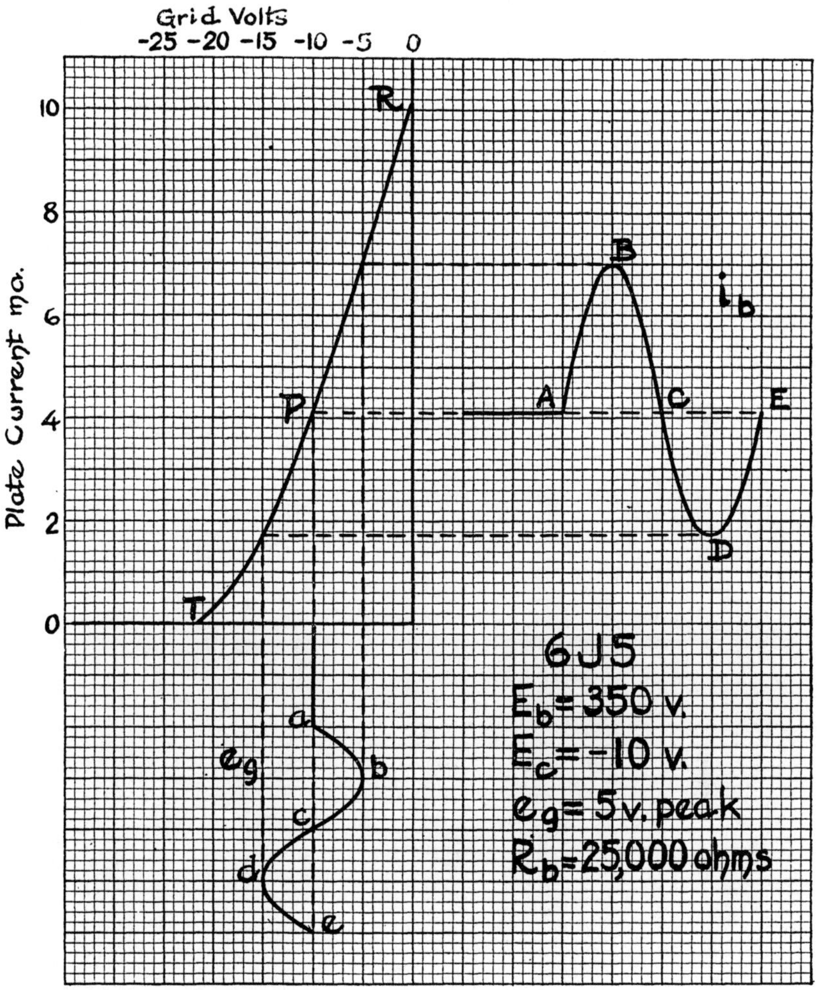

FIG. 10-8. In this case the input signal was reduced, thereby decreasing the distortion in the output current wave.

thus indicating that the characteristic below the point P has too great a curvature to be of much use.

Now let us again employ a signal of 5 volts peak, but this time so that it operates in the region between P and R of Fig. 10-6.

This has been done in Fig. 10-9, but another change has been made and this must be discussed. We cannot use the P position of Fig. 10-6 as the quiescent operating point, for if we do, then the negative half

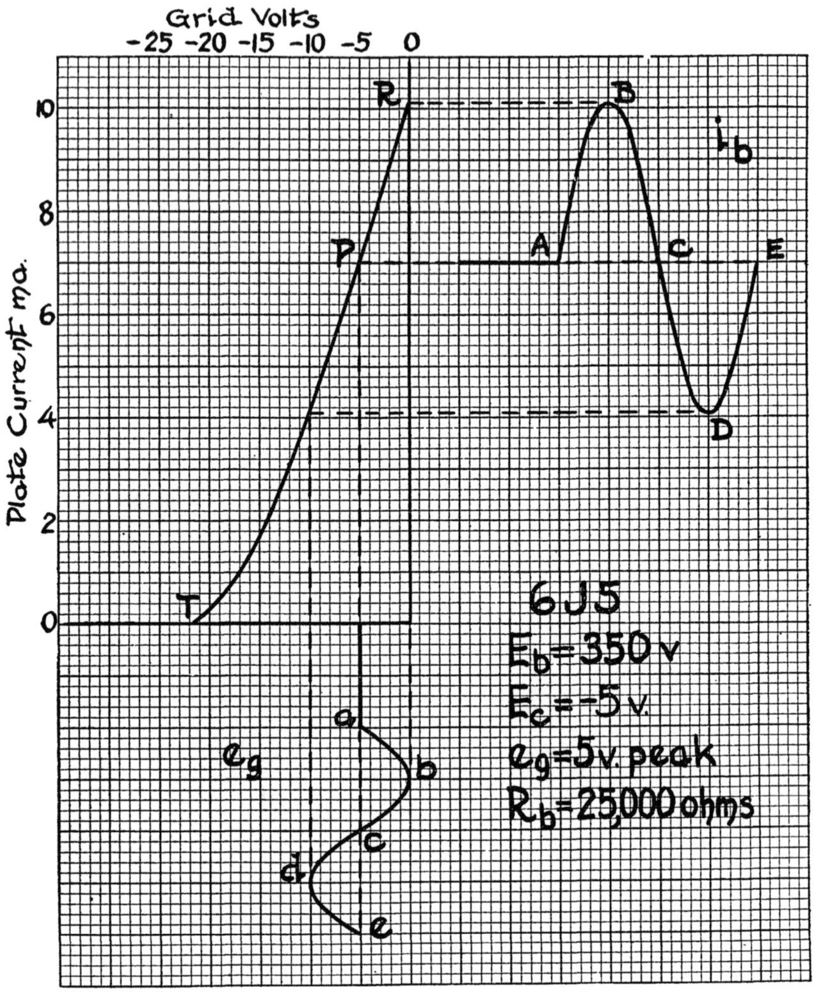

FIG. 10-9. Here the grid bias was so selected that the input signal voltage operates on the straight portion of the dynamic transfer characteristic, resulting in an output wave that is practically free from distortion.

of the input signal voltage will make the grid more negative than −10 volts. We must relocate the quiescent operating point *P* so that the range over which the instantaneous signal voltage will travel will be on the straight portion of the dynamic transfer characteristic. To relocate this operating point and still utilize this same tube, the same plate-supply voltage, and the same load resistance, we must shift the bias voltage.

In Fig. 10-9 the new operating point P corresponds to a grid bias voltage E_c of -5 volts and a plate current of 7 milliamperes. On the positive peak of the half-cycle of input signal voltage e_g, the resultant voltage e_c is reduced to zero and on the peak value of the negative half-cycle of the input signal voltage, the effective grid voltage e_c is increased to -10 volts. The swing in plate current i_b from the quiescent value of 7 milliamperes is up to 10.1 milliamperes and down to 4.1 milliamperes, an almost equal increase and decrease from the quiescent value. With a sine wave of input signal voltage, we now secure a sine wave of output plate current, and the shape of the output current wave is almost an exact amplified replica of the input signal-voltage wave.

Returning to the conclusions we mentioned several paragraphs back, we must supplement the comment concerning the reduction of the value of the input signal voltage in order to secure distortionless amplification, by the condition that it also depends upon the portion of the dynamic characteristic where operation is being carried out. This is borne out by the fact that the same value of input signal voltage, i.e. a peak of 5 volts applied to two different portions of the dynamic transfer characteristic, results in different operation; the output of one is distorted and the other is almost free of distortion.

Thus we find that we are confronted with three different conditions associated with distortion during amplification: the value of the load resistance, the magnitude of the input signal, and the location of the quiescent operating point.

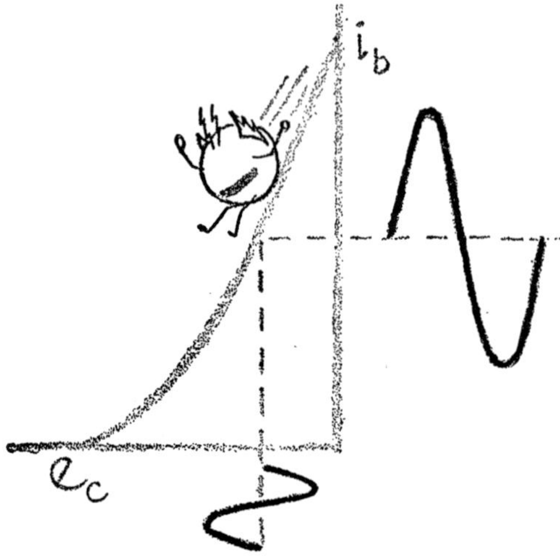

Chapter 11

VOLTAGE AMPLIFICATION

IT HAS BEEN MENTIONED heretofore in this text that in order for a triode to be of some use as an amplifier a load must be inserted in the plate circuit and *the changes in the plate current* flowing through this load from instant to instant, are what enable us to obtain an enlarged version of the signal voltage that is impressed on the grid. Up to this point, we have dealt almost exclusively with the flow of plate current and its effect upon the functioning of the tube; now we must consider the voltage developed in the plate circuit, for after all we are mainly interested in the triode as a voltage amplifier. Later on, of course, you will see that the current is the more important aspect when it comes to power amplifiers, but now let us think of the triode as essentially a means for amplifying voltages.

Let us examine the Fig. 11-1 schematic, which is essentially the same circuit as Fig. 10-1, but with some additions. The signal voltage e_g, here has a peak value of 8 volts; $E_c = -8$ volts, $E_b = 350$ volts, and

$R_b = 25{,}000$ ohms, as indicated on the diagram. The input signal e_g has a peak-to-peak value of 16 volts and, as has been mentioned before, this is the value of the total grid swing. When the plate current flows through the load resistor R_b, the resulting voltage drop, according to Ohm's law, is equal to the product of the plate current and the value of the resistance, which can be here written as

$$e_{Rb} = i_b \times R_b$$

wherein e_{Rb} = the instantaneous value of the voltage across the load in volts; i_b = the instantaneous total plate current in amperes; and R_b = the resistance of the load in ohms.

Consider for a moment the plate current. As indicated in Fig. 11-1 by the arrow, the plate current i_b is flowing away from the plate and down through the load R_b to the battery. This means that one end of the load resistor will be less positive than the other end, or, as has been shown in the schematic, end A of the resistor is more negative (or less positive) than the end B. The plus and minus signs at the resistor have been enclosed within parentheses so that the polarity relationship shown is with respect to the resistor itself. It should be remembered that both ends of the resistor are positive with respect to the cathode of the tube, point B being more positive than point A. The voltage drop developed across the load has such a polarity as to buck the plate supply; therefore, we can indicate the *voltage effective at the plate* by the expression

$$e_b = E_b - e_{Rb} = E_b - i_b R_b.$$

Let us consider some examples now, first by computation, and then by obtaining all data from the graph of Fig. 11-1.

Computing the Instantaneous Plate Voltage

In computation ($e_b = E_b - i_b R_b$), so that when the instantaneous signal voltage e_g applied to the control grid is zero, the instantaneous total grid voltage is thus $e_c = -8$ volts, the instantaneous plate current is 5.25 milliamperes, the voltage drop across the load resistor R_b is $e_{Rb} = i_b R_b = .00525$ ampere $\times$ 25,000 ohms or approximately 131 volts, and the instantaneous total plate voltage is

$$e_b = E_b - i_b R_b = 350 - (.00525 \times 25000) = 350 - 131 \text{ or approximately } 219 \text{ volts.}$$

When the instantaneous signal voltage $e_g = +8$ volts, the instantaneous total grid voltage is thus $e_c = 0$, the plate current $i_b = 10.1$ milliamperes (.0101 ampere), the voltage drop $e_{Rb} = i_b R_b = .0101 \times 25000$, or approximately 252 volts, and the instantaneous total plate voltage

$$e_b = E_b - i_b R_b = 350 - (.0101 \times 25000) = 350 - 252 \text{ or approximately 98 volts.}$$

Hence we can state that when the signal voltage e_g is at its maximum *positive* peak of +8 volts, making the instantaneous total grid voltage e_c equal to 0, the voltage effective at the plate is a minimum, or in this case, 98 volts. Similarly, when the signal voltage e_g reaches its negative peak and the instantaneous voltage e_c on the grid is −16 volts, $(-8) + (-8) = -16$ volts, the voltage drop e_{Rb} across the load becomes $.00135 \times 25000$ or approximately 34 volts, and we have $e_b = 350 - 34 = 316$ volts, approximately.

In other words, when the instantaneous total grid voltage e_c is at its maximum *negative* peak, the voltage effective at the plate is at a maximum, or in this case, 316 volts.

These results are tabulated in the following table:

e_g	e_c	i_b	e_{Rb}	e_b
−8	−16	1.35	34	316
0	−8	5.25	131	219
+8	0	10.1	252	98

Graphical Method of Finding Instantaneous Plate Voltage

In the graphical method ($e_b = E_b - e_{Rb}$) we refer to graph of Fig. 11-1 (which was discussed in Chapter 9 as Figs. 9-7 and 9-8 in conjunction with Fig. 9-10) where we have a load resistor R_b of 25,000 ohms represented as a load line drawn across a plate family of characteristic curves, and a plate supply voltage E_b of 350 volts. When the instantaneous signal voltage e_g applied to the control grid is zero, the instantaneous grid voltage e_c is thus the quiescent operating point or −8 volts, and the instantaneous plate current i_b is seen graphically to be 5.25 milliamperes by reading the vertical plate current scale at the intersection of the load line with the grid curve $e_c = -8$ volts.

By dropping a perpendicular line from this intersection of the load

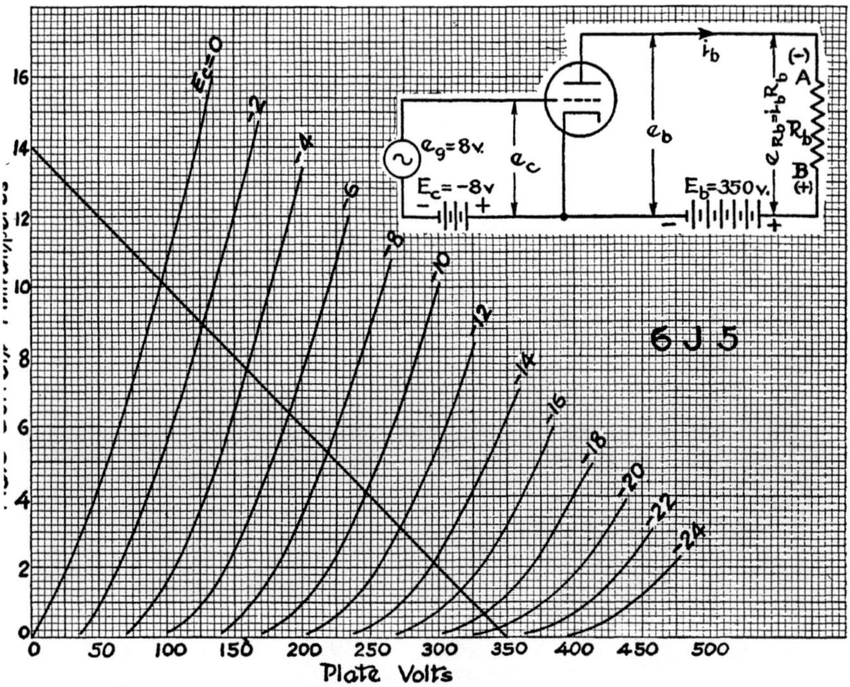

'IG. 11-1. The instantaneous plate voltage and other factors can be found by using this plate amily upon which has been drawn the 25,000-ohm load line. The various voltages and curents are indicated symbolically on the schematic diagram in the insert.

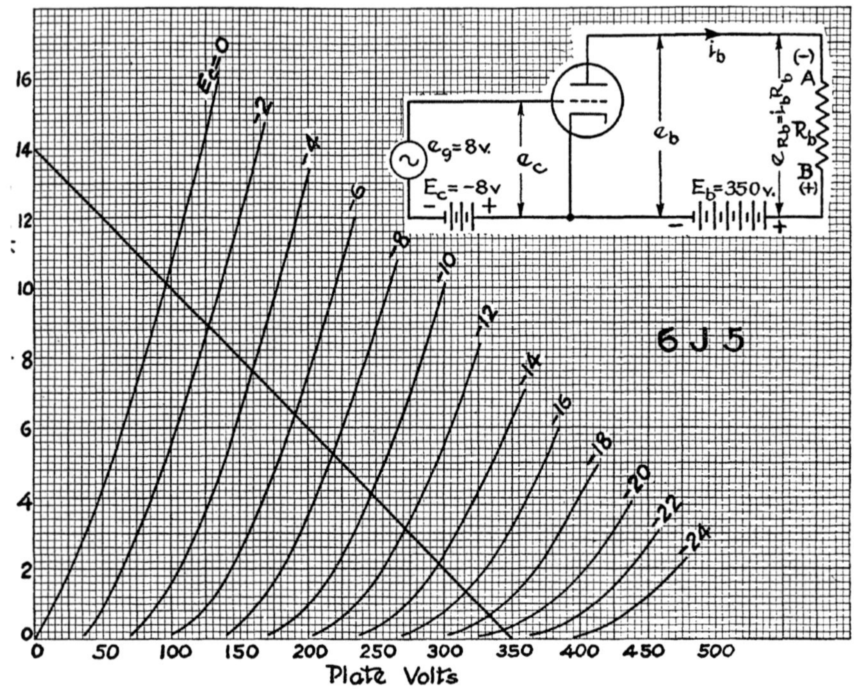

IG. 11-1. The instantaneous plate voltage and other factors can be found by using this plate .mily upon which has been drawn the 25,000-ohm load line. The various voltages and cur- nts are indicated symbolically on the schematic diagram in the insert.

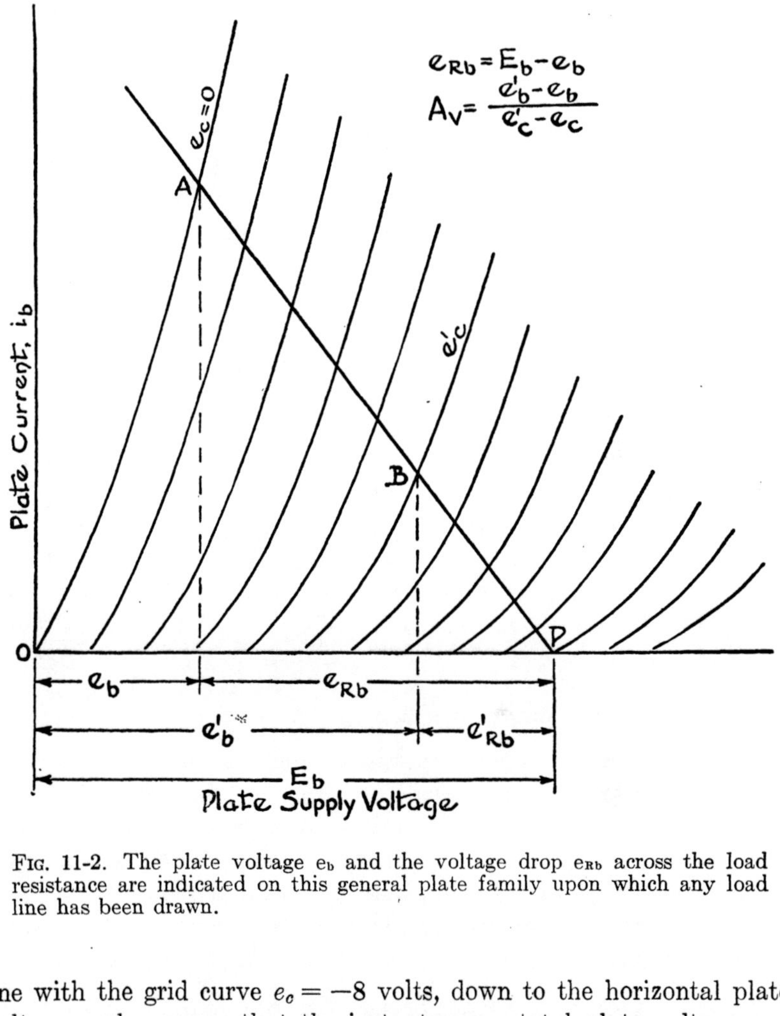

FIG. 11-2. The plate voltage e_b and the voltage drop e_{Rb} across the load resistance are indicated on this general plate family upon which any load line has been drawn.

line with the grid curve $e_c = -8$ volts, down to the horizontal plate-voltage scale, we see that the instantaneous total plate voltage e_b is 219 volts and the voltage drop e_{Rb}, across the load resistor R_b, is determined by subtracting 219 volts from 350 volts ($e_{Rb} = E_b - e_b$).

In the same way we find that when the instantaneous total grid voltage $e_c = 0$, that the instantaneous plate current $i_b = 10.1$ milliamperes, the instantaneous total plate voltage e_b is 97 volts (which is about the same as the computed value of 98 volts), and the voltage drop across the load resistor is $e_{Rb} = E_b - e_b = 350 - 97 = 253$ volts.

Any other values of voltage drop across the load for corresponding values of signal voltage can be determined in the same manner from the load line. This is shown in a general form in Fig. **11-2** wherein any load line is drawn on any set of plate characteristic curves. The intersection of the load line with the zero grid voltage curve at the point A determines the voltage effective on the plate e_b, which is the distance from the zero plate voltage axis to a perpendicular dropped from A. The distance from the vertical line dropped from A, to the vertical line at the value of the plate-supply voltage at point P, indicates the voltage drop e_{Rb}. Any other point along the load line, as B, corresponding to some other value of grid voltage e'_c, will provide similar information, the effective plate voltage being e'_b and the voltage drop across the load being e'_{Rb}.

Now you will recall that for a sinusoidal input grid voltage, we have in the plate circuit of our triode a voltage that is constantly changing sinusoidally, although its value is always positive. Moreover, this sine-wave voltage is an enlarged version, neglecting any distortion for the time being, of the signal voltage that is impressed on the grid of the tube. Hence in our particular example, as a result of the 16-volt total grid-voltage swing, the voltage swing across the load resistor is from 252 volts to 34 volts, or 218 volts. Correspondingly, the plate voltage varies from 98 to 316 volts, giving us a plate voltage swing also of 218 volts.

The Voltage Amplification, A_V

In Chapter 8 we considered the various factors that determined the amplification constant of a triode and, as you will recollect, it was stated that in effect the change in grid voltage de_c has been transferred to the plate circuit and that the change in plate voltage de_b is equal to the change in grid voltage de_c times the amplification constant μ of the tube. In the preceding paragraphs we have been considering changes of signal voltage or, as we have called them there, total grid-voltage swing, so in reality we have been referring to the same thing. Of course, the conditions which we were depicting in Chapter 8 were ideal and now we are considering practical uses of the triode; nevertheless, the same thoughts hold good as you shall see presently.

We have seen that a total grid swing of 16 volts in our sample triode, results in a plate-voltage swing from 98 to 316 volts or a total swing of 218 volts. What does this mean in terms of amplification? How can it be expressed? If we take the ratio of the amount of work obtained from a device to the amount put in, we obtain an idea of the efficiency of that device under certain working conditions. We can do the same thing in the case of a triode: the amplification may be considered to be *the ratio of the plate-voltage swing to the grid-voltage swing,* there being a resistive load in the plate circuit. In our particular case, we have therefore 218.75 ÷ 16 or a voltage amplification of 13.7. In other words, the output voltage is 13.7 times greater than the input signal voltage.

It was stated in Chapter 8 that the amplification constant μ for our sample tube, the 6J5, was 20 and you may well wonder why the discrepancy exists between this figure and the amplification of 13.7 derived in the preceding paragraph. We shall point out the reasons later on why this discrepancy does exist, but we can say here that *the voltage amplification A_v obtainable is always less than the theoretically maximum value given by the amplification factor μ of the tube.*

Determining Voltage Amplification Graphically

It goes without saying that the amount of voltage amplification for a given set of conditions can be obtained from the load line and its accompanying grid curves. Let us consider the general load line in Fig. 11-2. Assume that the grid swing in this instance is from $e_c = 0$ to some value e'_c, the latter being a more negative voltage. From each of the two points of intersection *A* and *B,* which the load line makes with these two grid-voltage curves, vertical lines are dropped to the horizontal axis. The distances e_b and e'_b indicate the amount of plate voltage obtained when the respective values of signal voltage are impressed on the grid. The difference between e'_b and e_b divided by the difference between e'_c and e_c, will give the voltage amplification for the conditions under consideration. This can be stated mathematically as

$$A_v = \frac{e'_b - e_b}{e'_c - e_c} = \text{Amount of voltage amplification.}$$

You must bear in mind that e_b, e_c, etc. are *any values at all* and that you can apply this system of determining the amount of amplification with any load line on any set of curves of any plate family.

Taking a specific case on Fig. 11-1, we have the limits of the grid swing from $e_c = 0$ to $e_c - 16$ volts with a load of 25,000 ohms. You see that the 25,000-ohm load line intersects the zero grid voltage curve at 97 volts and the other end of this load line intersects the −16-volt grid curve at 316 volts. Substituting these values in the above equation, we have

$$A_v = \frac{316 - 97}{-16 - 0} = \frac{219}{16} = 13.7$$

We have dropped the minus sign in the denominator in the result, as this is a ratio and the minus sign does not mean anything here. Of course, if the lower end of the grid swing (the positive peak) was some negative number, as $e_c = -4$ instead of $e_c = 0$, then the value of the denominator would be found by subtracting the two figures algebraically, thus: $(-16) - (-4) = -12$. (The sign within a parenthesis is changed in an algebraic operation, when the parenthesis is preceded by a minus sign).

The A-C Components

Thus far in our considerations of the voltages and currents in the triode we have dealt with *total* values: the instantaneous total plate voltage e_b; the instantaneous total plate current i_b, etc. These total values include not only the alternating-current component, but the direct-current component as well. Of course, the d-c components are of importance in establishing the operating point of the tube, but once this has been found, it is unnecessary to consider the d-c components and we need only concern ourselves with the a-c components.

It was mentioned previously that the quiescent value of the plate current was that at which the only voltage on the grid, was that afforded by the grid bias. If this condition is maintained, then the plate current and plate voltage remain unchanged, in other words, this is the equivalent of a direct-current condition. Now when a signal voltage is impressed on the grid in addition to the bias, and if the input signal is an a-c voltage, then the plate current and the plate voltage will be changing in accordance with the variations of the signal on the grid. Thus, you can see that we are justified in considering the total plate current and total plate voltage as consisting of two components: the a-c component and the d-c component.

If you will refer to Fig. 11-3, you will see that we have indicated the breakdown of the total voltages and current in which we are interested. In the input circuit we show that

$$e_c = E_c + e_g$$

where e_c = the instantaneous total grid voltage; E_c = the grid-bias supply voltage; and e_g = the instantaneous value of the varying component of the grid voltage. Here are two voltages coming from finite sources and so are easily understood; the bias voltage source and the source of signal voltage, whatever it may be. The plate voltage and plate current may perhaps be a trifle more obscure in the nature of their breakdown, but we feel sure you can understand them if you will

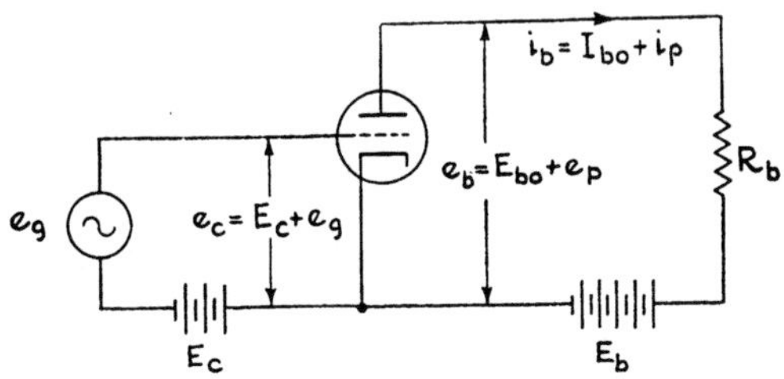

FIG. 11-3. The instantaneous values of the alternating components of the various voltages and current are indicated on the schematic of the triode, these are e_g, e_p, and i_p.

recall some of the data concerning the quiescent values which we have already covered.

As has been mentioned before, when e_g is 0, the plate current and the plate voltage are at their quiescent values, and we will call the values of the plate current and the plate voltage under this condition, I_{bo} and E_{bo} respectively. If you will refer to Fig. 11-4, you will find $e_g = 0$ and I_{bo} indicated graphically. When e_g is some other value than 0, that is, when the signal voltage is functioning, the plate current and plate voltage will be constantly undergoing a change, as has been explained before. We will call the instantaneous values of the varying component of this plate current and voltage i_p and e_p respectively, which are indicated in Figs. 11-3 and 11-4. Therefore, we can say that the instantaneous total plate current is composed of two parts: the quiescent value I_{bo}, and the varying component i_p, as shown in Fig. 11-4. As these two currents are both flowing in the same circuit, we can write their total value mathematically as

$$i_b = I_{bo} + i_p$$

and similarly we can write the expression for the plate voltage

$$e_b = E_{bo} + e_p$$

These two equations are indicated in their proper places in Fig. 11-3 and from the stand-point of the signal, it is this group of components in which we are chiefly interested: e_g, i_p, and e_p.

Incidentally, if you refer again to Fig. 10-6, you will observe that the plate current wave is shown both with a time axis and with an axis which is labeled *average value.* Although we shall not further concern ourselves with the distinction between these two axes in this book, and shall instead consider that they coincide, we shall just comment briefly on the existence of the average-value axis. Whenever a waveform occurs in which the top-half and bottom-half are not com-

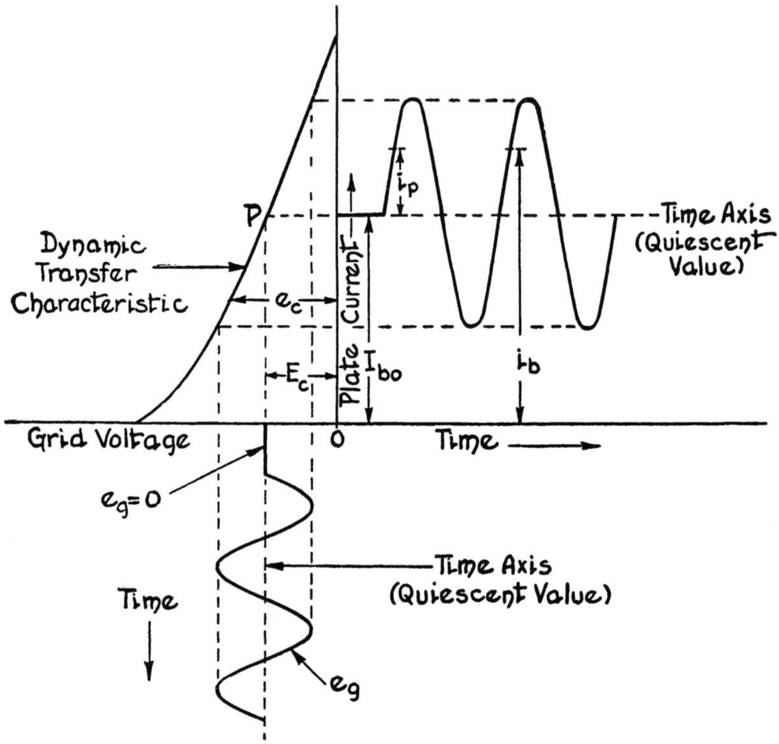

FIG. 11-4. The a-c and d-c components of the various voltages and current are identified in conjunction with a general dynamic transfer characteristic. Compare with Fig. 11-3.

pletely symmetrical, a difference will result between the time axis and the average-value axis. The average-value axis is located in such a manner that the area included in the top-half of the wave is equal to the area included in the bottom-half of the wave. In Fig. 10-6, when no signal is impressed upon the grid, the d-c plate current will have a value of 4.1 milliamperes, but when a signal is applied to the grid, the d-c plate current will jump to 4.7 milliamperes.

Plate Circuit Theorem

Once again we refer to the fact that the grid was μ (mu) times as effective as the plate in causing a variation in the plate current. Moreover, we know that the triode possesses the property of acting as a resistance. Putting these two facts together we can draw the circuit of a triode as an a-c source of voltage in the output of which are two resistances, as shown in Fig. 11-5. In other words, we can replace the

FIG. 11-5. In the equivalent circuit of a triode the tube is indicated as a source of a-c voltage being equal to the input voltage times the μ of the tube. The plate and load resistances are shown in series with the a-c voltage source.

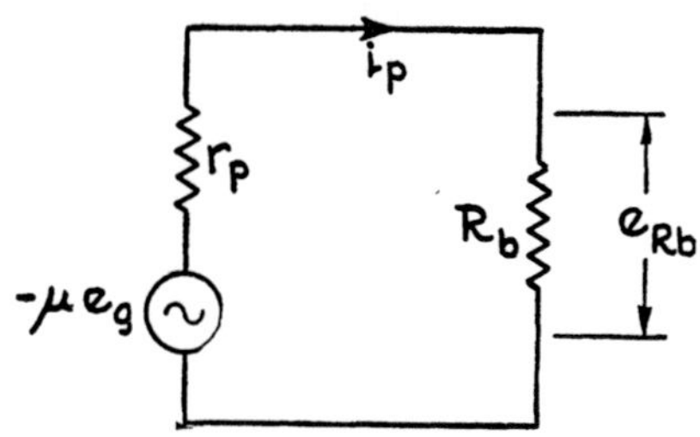

tube with an a-c generator having a voltage equal to the value of the signal voltage on the grid times the amplification constant of the tube μ. In series with this are two resistors, r_p, representing the a-c plate resistance at its point of operation, and R_b, which is the resistance of the plate load resistor. You will have noticed the minus sign preceding the expression for the voltage, $-\mu\ e_g$. The reason for this sign is that we are considering both the input signal and the equivalent generator to be acting from the cathode and also to the fact that phase reversal occurs within the tube, which will be explained later on.

Now according to Ohm's law, we know that the current in a series circuit is equal to the voltage divided by the total resistance of the circuit, namely,

$$\text{Current} = \frac{\text{Voltage}}{\text{Resistance}} \quad \text{(Ohm's law)}$$

In the equivalent circuit shown in Fig. 11-5, the voltage is $-\mu\ e_g$ and the total resistance is $r_p + R_b$, so that the current i_p may be expressed as

$$i_p = \frac{-\mu\ e_g}{r_p + R_b}$$

The negative sign that appears in the above equation does not mean that a negative current is flowing, but that it is used to show a difference in phase relationship between the a-c and d-c plate current components. If a minus sign is used before i_p, it indicates that the a-c plate current is flowing in the same direction as the quiescent direct current in the plate circuit, and conversely, if a plus sign or no sign at all is used in front of i_p, it means that the a-c plate current is flowing in the opposite direction to the no-signal quiescent d-c plate current. This system of designations is unfortunate, but it is necessary if we wish to indicate that the input and output signal voltages are out of phase, as will soon be explained.

Solving Ohm's Law for voltage, we obtain

$$\text{Voltage} = \text{Current} \times \text{Resistance}.$$

Consequently, the alternating voltage e_{Rb} appearing across the load resistor R_b due to the current i_p flowing through it, is

$$e_{Rb} = i_p R_b = \left(\frac{-\mu\ e_g}{r_p + R_b}\right) \times R_b = \frac{-\mu\ e_g\ R_b}{r_p + R_b}.$$

Determining Voltage Amplification

Now we are in a position to obtain the expression for the actual voltage amplification A_v available in a triode. Keeping in mind that the *voltage amplification* A_v is the ratio of the output voltage e_{Rb} to the input voltage e_g—the a-c components in each case—we have

$$A_v = \frac{e_{Rb}}{e_g} = \frac{\dfrac{\mu\ e_g\ R_b}{r_p + R_b}}{e_g} = \frac{\mu\ R_b}{r_p + R_b} = \left(\frac{R_b}{r_p + R_b}\right)\mu$$

In the above, we substituted the expression for e_{Rb} in the numerator of the fraction and then clearing the fraction, e_g was cancelled out. You will see that the plus and minus signs in front of the ratio have been omitted, since in this expression for voltage amplification we are not concerned with phase relations.

In this equation, let us substitute the values of the a-c plate resistance $r_p = 7700$ and that of the load $R_b = 25000$ for our sample triode, which you will recall has an amplification constant $\mu = 20$. We have the following:

$$A_v = \left(\frac{R_b}{r_p + R_b}\right)\mu = \frac{25000}{7700 + 25000} \times 20 = \frac{500000}{32700} = 15.3$$

If you will compare this value of the voltage amplification with that value which we derived earlier by means of the load line, you will see that these values of 13.7 and 15.3 agree quite closely, in fact, closely enough for all practical purposes. You must realize by now that none of these formulae are exact by any means, for several assumptions have been made in their derivation. In the first place, it was assumed that the amplification constant μ and the plate resistance r_p were constant, which is true only for very small input-signal voltages. Then it was further assumed that a sine-wave input signal resulted in a sine-wave output signal, which, while almost true, nevertheless is not absolutely exact. Be that as it may, these equations and the methods outlined in this chapter furnish us with data that are sufficiently accurate for many different purposes.

In the expression that we found for the voltage amplification of a triode, namely

$$A_v = \left(\frac{R_b}{r_p + R_b}\right)\mu$$

you can see that the actual voltage amplification depends upon three factors: the amplification constant μ, the load resistance R_b, and the a-c plate resistance r_p. Of these three, two are fixed in value for any given tube and for any given set of operating conditions, namely, the amplification constant μ and the plate resistance r_p. This leaves only the load resistance as a factor over which we have any control.

Let us consider the above equation for a moment. You can see that if we make the value of the load resistance R_b large in comparison with the plate resistance r_p, the effect of the plate resistance will become more and more negligible the larger the load is made. If R_b is made very large compared to r_p, we can neglect r_p and then the expression $\frac{R_b \ \mu}{r_p + R_b}$ can be thought of as being $R_b/R_b = 1$, eliminating r_p entirely. This makes the voltage amplification A_v equal to the amplification constant μ of the tube, as all the symbols for the resistance are eliminated in the above equation.

From this it would seem as if we should employ a load resistance that was very large in comparison to the plate resistance in order that we get a maximum amount of voltage amplification from the triode. Several other facts enter which make such a choice undesirable, as will be discussed a little later.

Considering the various factors thus far discussed, it may be stated that the best value of the load resistance lies somewhere between 3 and 7 times the value of the a-c plate resistance. This can be shown graphically in Fig. 11-6, in the case of our sample triode, the 6J5, which has an amplification factor of $\mu = 20$ and plate resistance of 7700 ohms. According to the reasoning given above, the maximum possible gain or voltage amplification from this tube is 20 and we have indicated this limit by a dashed horizontal line drawn from the 20 point on the

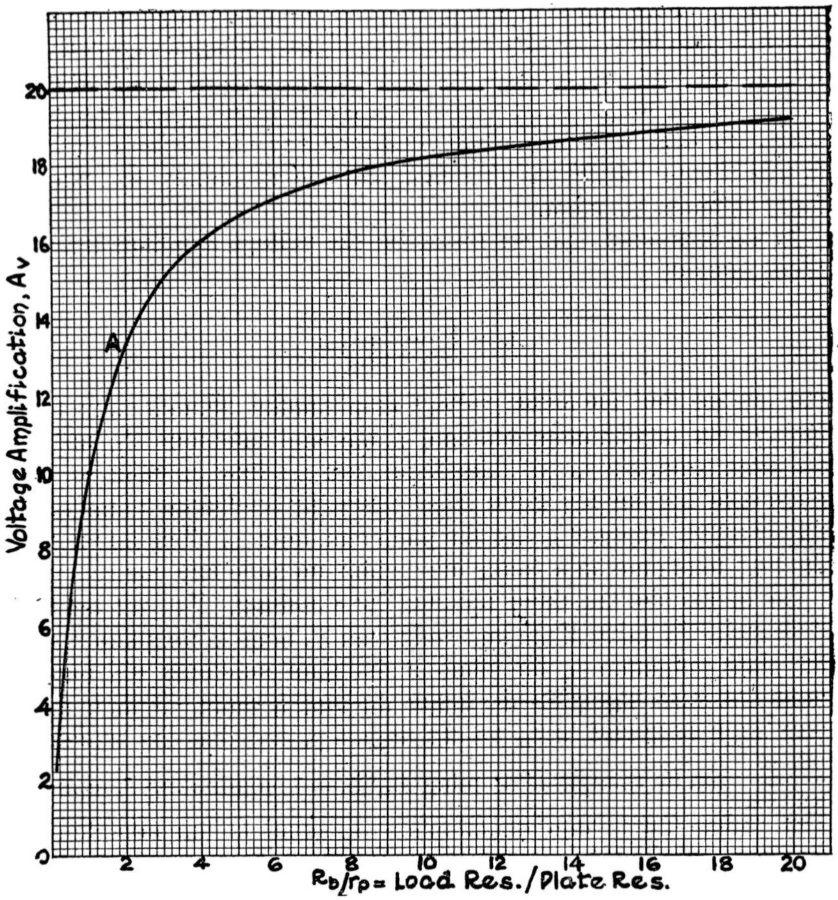

FIG. 11-6. The ratio of the load resistance to the plate resistance is plotted against the voltage amplification for the 6J5 triode. Note that when the ratio is above 8, the amplification does not appreciably increase.

vertical scale. The horizontal scale is the value of the ratio of the load resistance to the plate resistance of the tube, the latter being a fixed value. The vertical scale indicates the voltage amplification A_v.

Now consider two or three points on the curve. When the load is twice the plate resistance (that is, when R_b is 15,400 ohms, making $R_b/r_p = 2$), then the amplification is found by substituting in the equation for A_v, as follows

$$A_v = \frac{\mu\ R_b}{r_p + R_b} = \frac{20 \times 15{,}400}{7700 + 15{,}400} = \frac{308{,}000}{23{,}100} = 13.3$$

This is indicated as the point *A* on the curve and successive points are found in the same way by substituting in the equation for A_v. Now if you will follow up the vertical line from the point on the horizontal scale indicating a ratio of 4, you will see that the amplification is 16 and for a ratio of 8 the value of A_v is 17.8. Out at the end of the curve for a ratio of 20, that is, when the load is 20 times the value of the plate resistance, the amplification is 19.1.

Now this curve gives us a graphical confirmation of what we said previously when we were referring to the size of the load resistance. You can see from Fig. 11-6 that the curve is very steep for low values of the ratio R_b/r_p and then as this ratio increases the curve becomes less and less steep, flattening out and approaching more closely to the dashed line representing the μ of the triode. This informs us that beyond a certain ratio of the load resistance to the plate resistance, say around 6 to 8, the increase in amplification is so small for an increase in the value of the plate load resistance, that it is not worthwhile to increase the load resistance further. In fact, the upper limit of the load being 7 times the value of the plate resistance, which we gave previously, you can see is a very reasonable figure. This value of the ratio gives an amplification of 17.5, and above this very little if anything

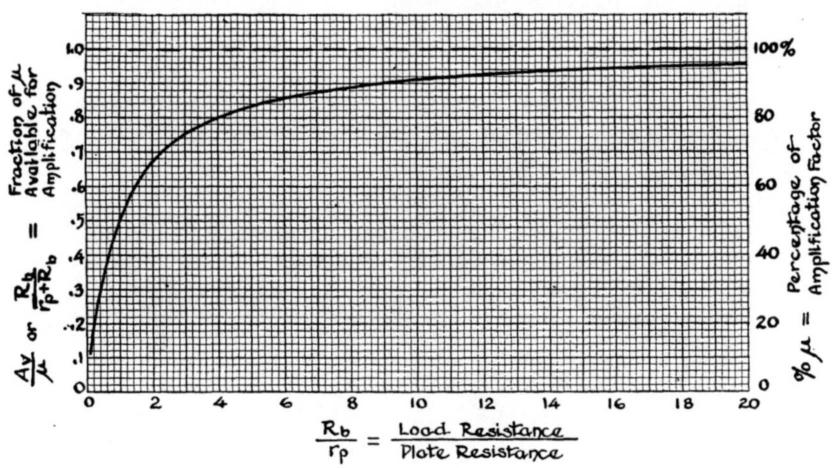

FIG. 11-7. This curve shows the relationship between the ratio of the load resistance to the plate resistance and the percentage of the amplification factor of any triode.

is gained when the load resistance is increased giving a higher value to the ratio of the load to the plate resistance.

We have dealt in the preceding paragraphs with the 6J5 triode, but *the same reasoning applies to any triode.* We can show by use of the equation for the voltage amplification stated above that it is unfeasible to employ a load resistance that is greater than 5 to 7 times the a-c plate resistance of the triode. Starting with the equation

$$A_v = \frac{\mu\ R_b}{r_p + R_b}$$

we can divide both sides by μ, thus

$$\frac{A_v}{\mu} = \frac{R_b}{r_p + R_b}$$

Furthermore, we can divide both the numerator and the denominator of the right-hand side of the equation by r_p without changing its value, thus

$$\frac{A_v}{\mu} = \frac{\dfrac{R_b}{r_p}}{1 + \dfrac{R_b}{r_p}}$$

Now the left-hand side of the above equation is a fraction indicating how much of the amplification constant is available for amplification, which can also be considered in terms of percentage of the amplification. The right-hand term of the equation is entirely in terms of the values of the a-c plate resistance and the load resistance. Hence, we can plot a curve showing percentage of amplification available against the ratio of the load resistance to the plate resistance, as has been done for the general case in Fig. 11-7.

Here the maximum amplification that is available from a triode is the value of the amplification constant, μ, which is taken as 100%; a horizontal dotted line is drawn as a reference for the curve at this value. Points on the curve are found by substituting different values of the ratio R_b/r_p in the right-hand member of the last equation as follows:

$$\frac{A_v}{\mu} = \frac{2}{1+2} = \frac{2}{3} = .667 = 66.7\%$$

where the value of the ratio is 2 or the load is twice the value of the plate resistance. Similarly a voltage gain of 80% is secured when the load is 4 times the plate resistance, and when the load is 20 times as large as the plate resistance, the amplification is 95%.

Here is further proof that in the case of any triode it is entirely unnecessary to have the load resistance any greater than 7 times the value of the a-c plate resistance. It is evident that as the value of the load is increased, the amount of amplification is small in proportion; hence, we can state that, in general, a load resistance that is 3 to 7 times the value of the plate resistance of the triode is quite satisfactory, since a load within this range effects an excellent compromise with respect to the possible available gain without introducing excessive distortion or excessive high-frequency attenuation.

Input and Output Phase Relationships

In the preceding pages you have seen how the signal applied to the grid of a triode causes the development of an amplified signal in the plate circuit of the tube. Also, references have been made to the phase relationships that exist between the input signal voltage, the plate current, and the output voltage in the plate circuit of the tube. Let us now examine this matter of phase more closely.

In Fig. 11-8 five sine waves are shown which illustrate the phase relationships that occur within the triode. The first wave A, represents the sine-wave signal-voltage input e_g to the grid. Once more we are considering the operation of our sample 6J5 triode and here we will not concern ourselves with distortion that might exist because of the grid bias used, but shall assume that the bias is $E_c = -8$ volts. Furthermore, we shall assume that the input signal voltage e_g has a peak value of 8 volts, thus varying the total voltage e_c applied to the grid between 0 and -16 volts, and never driving the grid positive, as shown in wave B, wherein the sine wave varies 8 volts above and below the grid bias E_c of -8 volts.

The sine wave at C in Fig. 11-8 represents the a-c component of the plate current. This wave has been plotted from the dynamic transfer characteristic of Fig. 10-6 and varies above and below the quiescent value I_{bo} of 5.25 milliamperes. The fourth sine wave D represents the total voltage drop e_{Rb} varying above and below 131 volts, which is the voltage drop when the signal voltage $e_g = 0$, the total grid voltage $e_c = -8$ volts, and the grid bias $E_c = -8$ volts. The fifth sine wave E represents the voltage e_b, effective at the plate of the triode, varying above and below the quiescent value of the voltage on the plate, 219 volts, E_{bo}. How this wave was developed will be explained in a moment; let us first consider the sine waves A,B,C, and D so that you will understand their phase relationships and see why they are drawn as they are in Fig. 11-8.

Assume that the input wave A is just about to begin the positive half of its cycle, that is, the value of the voltage e_g is 0, indicated at point a. As you have read before, this means that the total voltage e_c impressed on the grid is equal to the amount of the grid bias $E_c = -8$ volts, and so the point a_1 in wave B is located on the $e_c = -8$-volt line. At this same instant, the plate current i_b is at its quiescent point I_{bo} and so the corresponding point a_2 on the plate-current wave, is located on the 5.25-milliampere line. At this same instant the voltage drop e_{Rb} is at the corresponding point a_3, the 131-volt line of curve D. This simultaneous condition is indicated on the graph by the vertical dotted line connecting a, a_1, a_2, and a_3.

When the input signal e_g has reached its most positive peak, a value of +8 volts, (point b), the total voltage e_c applied to the grid is 0 ($e_c = e_g + E_c$) indicated at point b_1 in wave B. This results in an increase of plate current to its maximum value of 10.1 milliamperes, indicated at point b_2 and an increase in voltage drop to its maximum value of 252 volts indicated at point b_3. From b to c in wave A, the input signal is becoming less positive and the resulting voltage e_c on the grid is becoming more negative (from b_1 to c_1, wave B), until it finally reaches its quiescent value at c_1 of −8 volts. In accordance with this diminishing of the grid voltage, the plate current decreases in value to 5.25 milliamperes, indicated from b_2 to c_2 in wave C and the voltage drop e_{Rb} decreases in value to 131 volts, as indicated from b_3 to c_3 in wave D.

After point c in wave A, the input signal voltage e_g starts on its negative half-cycle, reaching its maximum negative value of −8 volts at the point d. Resulting from this increase in negative potential on the grid ($e_c = -16$ volts) as indicated at d_1, the plate current decreases from c_2 to its lowest value of 1.35 milliamperes at the point d_2 on wave C, and the voltage drop decreases from c_3 to its lowest value of 34 volts at the point d_3 on wave D. Between the points d and f on wave A, the input signal voltage e_g becomes less negative; consequently, the total voltage on the grid e_c increases from its minimum of −16 volts, point d_1, to the quiescent value of −8 volts at point f_1 on wave B. Correspondingly, the plate current increases from 1.35 milliamperes at point d_2 on wave C to its quiescent value of 5.25 milliamperes, point f_2, and the total voltage drop e_{Rb} increases from its minimum of 34 volts, point d_3, to the 131-volt line at point f_3 on wave D.

From each of the points a,b,c,d, and f on wave A, vertical lines have been drawn to the corresponding points on waves B, C, and D, and as each of these four sine waves pass respectively through their positive

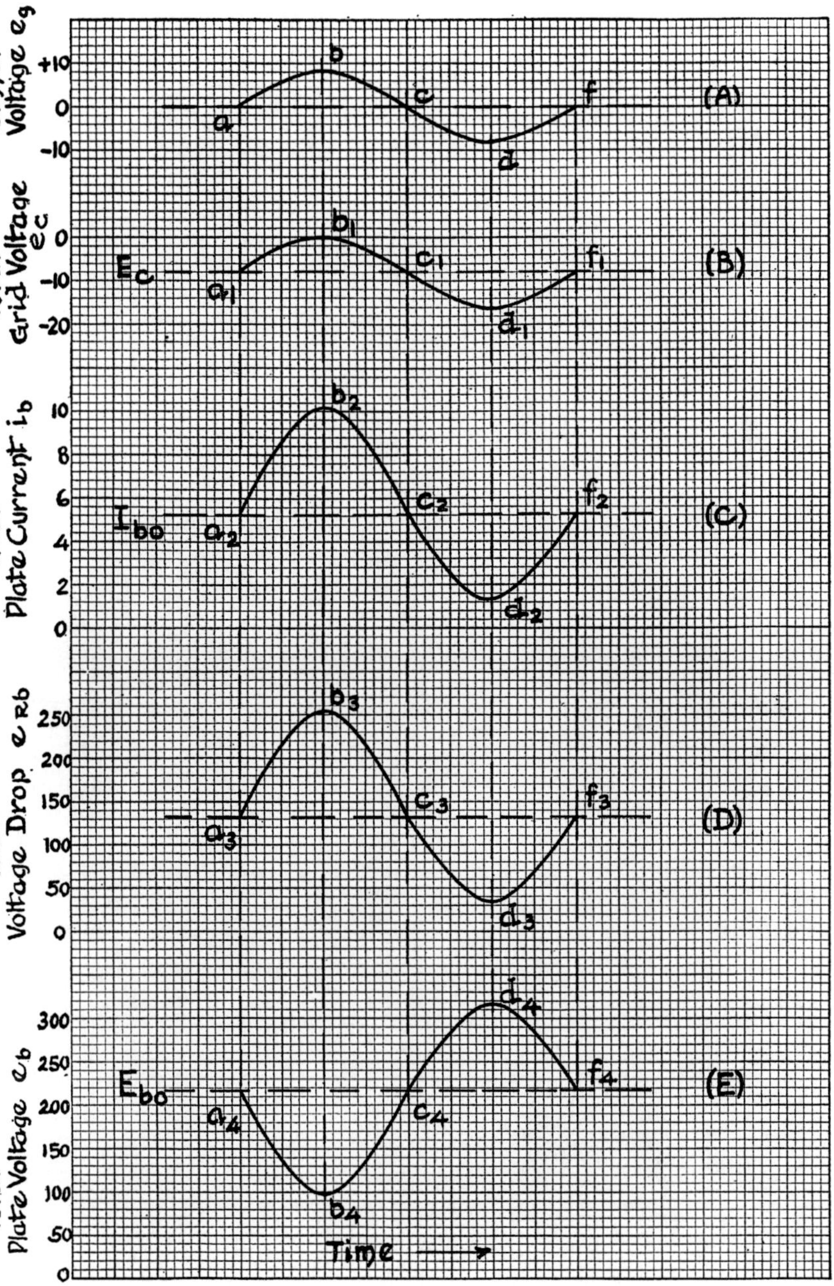

IG. 11-8. This series of waves shows that the input voltage e_g is in phase with the plate cur-ent i_b and that e_g is 180° out of phase with the plate voltage e_b.

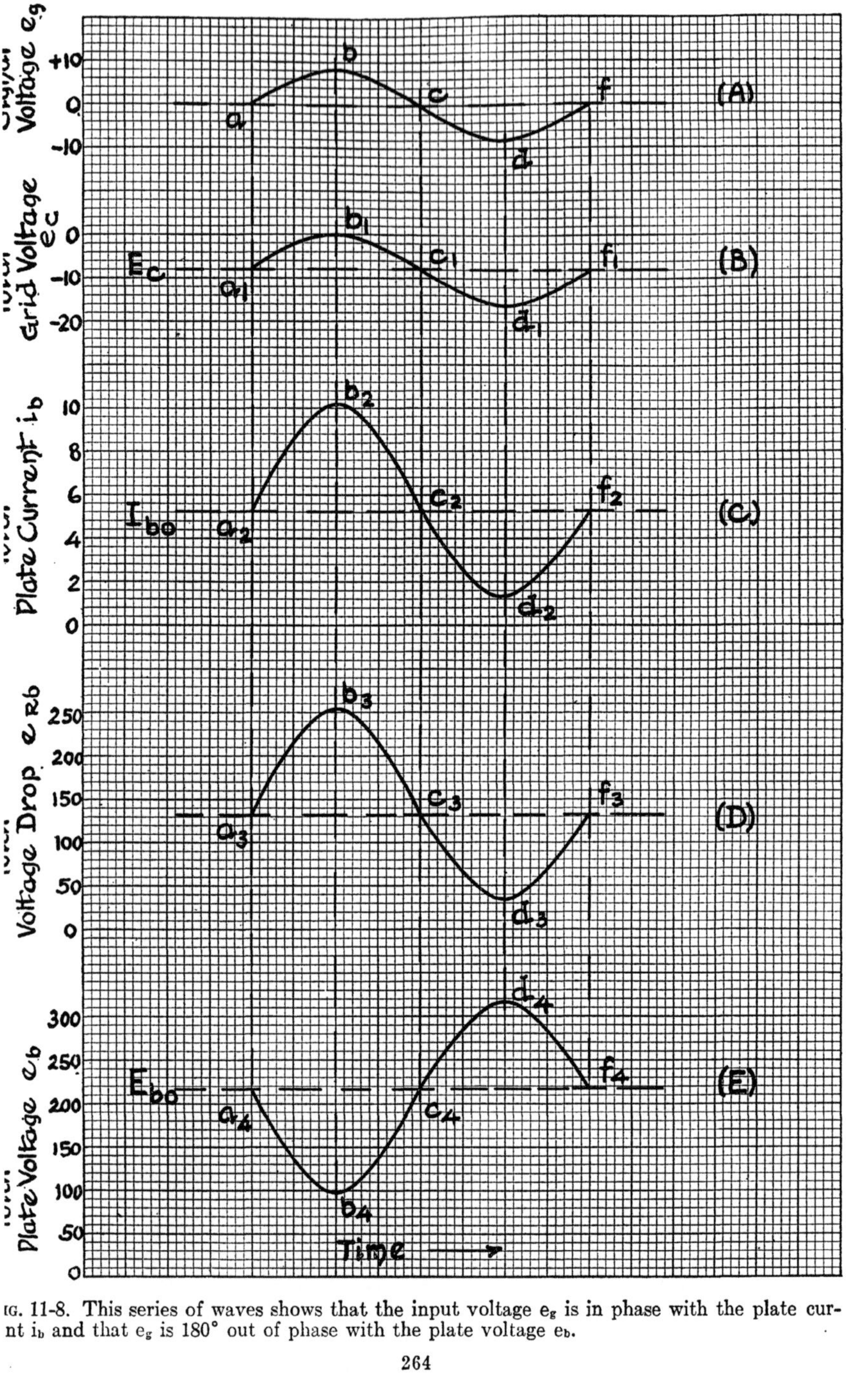

IG. 11-8. This series of waves shows that the input voltage e_g is in phase with the plate current i_b and that e_g is 180° out of phase with the plate voltage e_b.

and negative peaks at the same instants, they are in phase. Thus, we can state that *both the plate current and the voltage-drop curves are in phase with the input signal voltage,* as from the explanation given above you can understand that the input signal voltage e_g is really the determining factor in the increase or decrease of the total voltage e_c impressed on the grid. (Of course, this can also be seen from Fig. 10-6).

As you can see from the fifth wave E in Fig. 11-8, the conditions are entirely different as far as the phase relations are concerned. Here, the sine wave, representing the voltage e_b effective on the plate, varies first below and then above the quiescent value, E_{bo}—just the opposite of the other waves. We want to call your attention to the fact that the value about which this wave varies is the quiescent value E_{bo} and not E_b, the plate-supply voltage, which remains constant at 350 volts, whereas E_{bo} is 219 volts in the case we are considering.

You will remember that in vacuum-tube circuits the cathode is considered as the point to which voltages are referred; for instance, all voltages are positive or negative with respect to the cathode. When plate current i_b flows through the load resistor R_b in the plate circuit, a voltage drop $e_{Rb} = i_b R_b$ occurs, this being the product of the instantaneous total plate current i_b and the resistance of the load R_b. Now the instantaneous total plate voltage e_b, that is, the voltage actually existing at the plate at any particular instant, is equal to the difference between the plate-supply voltage E_b and the voltage drop e_{Rb} across the load resistor R_b. This we can express mathematically as

$$e_b = E_b - e_{Rb} = E_b - i_b R_b$$

From this you can see that the actual voltage effective at the plate depends on three factors: the plate-supply voltage E_b, the plate current i_b, and the load resistance R_b. From what has gone before, you know that the plate-supply voltage E_b and the value of the load resistance R_b are constant; therefore the variable factor with which we are concerned here is the plate current i_b. Also you know that the greater the plate current, the larger will be the voltage drop across the load, and consequently, the lower the plate voltage. Conversely, a small plate current will result in a small voltage drop across the load and therefore, the plate voltage e_b will be greater.

Returning now to Fig 11-8, you have seen that the magnitude of the plate current i_b is dependent upon the signal voltage e_g that is applied to the grid as e_c. In the preceding paragraph it was pointed out that the instantaneous plate voltage e_b is dependent on the plate current; therefore *the voltage e_b effective at the plate of a tube depends on the* *signal voltage applied to the grid.* Now when the signal voltage e_g is zero (point a on wave A), the plate current i_b is at its quiescent value $I_{bo} = 5.25$, point a_2, which produces a voltage drop $e_{Rb} = i_b R_b = 131$ volts (point a_3 on wave D) in the load $R_b = 25000$ ohms. This voltage drop $e_{Rb} = 131$ volts subtracted from the plate-supply voltage $E_b = 350$ volts, gives us a quiescent value E_{bo} for the plate voltage e_b of 219 volts, indicated at a_4 in Fig. 11-8 on the curve E. As this occurs at the same instant as a, a_1, a_2, and a_3, we can show this by continuing the vertical line connecting these points down to the bottom wave.

When the input signal e_g increases to its positive peak ($e_g = +8$), point b in wave A, which makes the grid more positive (less negative) as shown in wave B as point b_1, the value of the plate current increases, curve C, and the voltage drop across the load resistor increases, curve D. This voltage drop must be subtracted from the value of the plate-supply voltage and the instantaneous plate voltage is consequently reduced. Therefore, when the input signal is at its maximum in the positive direction and places the grid at its minimum negative potential (point b_1), maximum plate current flows, as indicated at point b_2. Accordingly, the voltage drop across the load is a maximum (point b_3) and since this must be subtracted from the plate-supply voltage, the instantaneous plate voltage e_b is a minimum, as indicated at point b_4 in wave E.

Now when the input signal decreases from its positive peak, as from b to c on wave A, the grid voltage curve B becomes more negative and this causes a decreasing plate current, as from b_2 to c_2. This results in a decreasing voltage drop across the load, as from b_3 to c_3, and this in turn means an increasing effective plate voltage shown from b_4 to c_4 in wave E. When the signal voltage reaches zero, point c, the quiescent value of the plate current is reached at the same instant, c_2, and this results in the quiescent value of plate voltage, point c_4 in wave E.

From c to d in wave A, the signal input is becoming more negative than the quiescent value and so the plate current decreases, as from c_2 to d_2 in wave C. This means a decrease also in the voltage drop across the load resistance, as from c_3 to d_3, and the plate voltage accordingly increases as from c_4 to d_4. When the signal input is at its maximum negative value, d on wave A, the plate current is a minimum, point d_2 on wave C, and this results in a minimum voltage drop across the load, point d_3. The plate voltage at this point d_4 is therefore at its maximum. Then when the input signal approaches its quiescent value f, the plate current also increases, the voltage drop across the load increases, and the plate voltage decreases, finishing the cycle at f_4 in wave E, corresponding to the quiescent value, E_{bo}.

We have gone into this amount of detail so that you will understand that when the signal input e_g to the grid is at its maximum positive value, the plate voltage e_b is at its minimum value; also that when the input voltage e_g is a minimum (maximum negative value), the plate voltage is at its maximum. In other words, we can state that *the plate current is in phase with the grid voltage, but the plate voltage is 180 degrees out of phase with the grid voltage.* This can be plainly seen in the waves of Fig. 11-8 and also in Fig. 11-9.

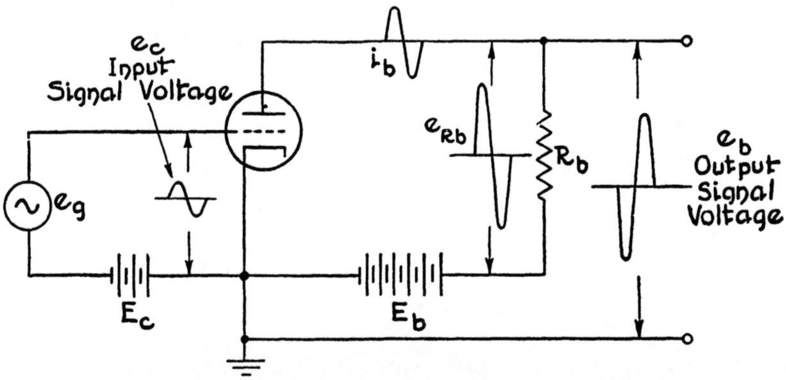

FIG. 11-9. The waveforms shown indicate the relative values of the voltages or currents, their phase relationships, and their locations in the triode circuit.

A graphical summary of the conditions created within and without the triode for different conditions of the grid circuit is shown in the following table and is to be used in conjunction with Fig. 11-8.

GRID BIAS E_c	SIGNAL VOLTAGE e_g	TOTAL GRID VOLTAGE e_c	PLATE SUPPLY VOLTAGE E_b	PLATE CURRENT i_b	VOLTAGE DROP ACROSS LOAD $e_{Rb} = i_b R_b$	TOTAL PLATE VOLTAGE e_b
■	↑ 1	↑ 1	■	↑	↑	↓
■	●	●	■	●	●	●
■	↓ 2	↓ 2	■	↓	↓	↑

● Values subject to change, but as shown indicates quiescent condition

■ Values always fixed ↑ Increase ↓ Decrease

1. An increase in grid voltage e_g is equivalent of a <u>less</u> negative condition at e_c

2. A decrease in grid voltage e_g is equivalent of a <u>more</u> negative condition at e_c

Optimum Value of Load Resistor

We have previously seen that the greater the value of the plate load resistor R_b, the less the resulting distortion of the output waveform would be. In order, then, to obtain an output voltage wave that is practically identical in wave-shape as the applied input signal it is necessary to employ a plate load resistor that is many times the value of the plate resistance r_p of the tube.

On the other hand, we have also seen that an increase in the plate load resistor beyond from 3 to 7 times the plate resistance did not result in any appreciable increase in the obtainable voltage amplification A_v.

From the preceding it might be thought that the best value of load resistor would be one the resistance of which is very many times that of the plate resistance in order to minimize distortion, and at the same time, not being particularly concerned with the fact that the voltage amplification is not much greater than would be the case for a smaller value of load resistance. Other factors are involved, however, which must be taken into consideration.

One of these factors is that when the total grid voltage e_c becomes highly negative so that no plate current flows (see graph of Fig. 11-1), there is no voltage drop across the load resistor R_b. Consequently the entire plate-supply voltage E_b will be effective on the plate of the tube, and this may result in a flashover inside the tube thereby destroying it. It is seen that the use of a very high value of load resistance, together with a high plate-supply voltage in order to obtain reasonable values of plate current, is not advisable. It is not good practice to operate a triode at very low values of plate current for if this is done, there is a tendency to noisiness.

Another factor, which is mentioned only in passing, is the loss of the high-frequency response due to the output (plate-to-cathode) capacitance of the tube. This high-frequency attenuation becomes greater as the value of the load resistor is increased. Therefore, too large a value of load resistance should not be used.

From other considerations we have observed that a load resistance having a magnitude of from 3 to 7 times that of the plate resistance is suitable. For the sake of illustration let us choose a plate load resistor of 25,000 ohms to be used in the plate circuit of the 6J5 tube, the plate resistance of which is 7700 ohms. In other words, the plate load resistance is 3.2 times as great as the plate resistance, a condition which is very close to the lower recommended condition. The load line, under these circumstances, is shown in Fig. 11-1.

As has been previously described, from the intersection of this load line with the curves in the plate family, we may obtain the dynamic transfer characteristic shown in Fig. 11-10. Now one of the factors that is involved in both the choice of a plate load resistor and of a quiescent operating point is to be able to apply a fairly large input signal to the grid without obtaining an excessively distorted output wave-shape. A further consideration is that we do not wish the total grid voltage e_c to go beyond zero and into the positive region, for reasons which will be discussed at the end of this chapter.

If we were to choose a quiescent (or bias) grid voltage E_c of -8 volts corresponding to point P and an input signal voltage e_g of 8 volts peak (or 16 volts peak-to-peak) as has been done in Fig. 11-10, we observe that the waveform of the resulting plate-current wave is only slightly distorted. This comes about because we were operating over the fairly linear portion S—R of the dynamic transfer characteristic, and excluding use of the curved S—T portion.

If we had employed a larger plate load resistance for the same conditions of operation, the output waveform would be even less distorted, because the dynamic transfer characteristic would be even more linear. We may therefore state that an optimum value of the plate load resistance of a voltage amplifier would be approximately 5 times the plate resistance.

Phase Relations Again

We have previously remarked that the output plate-current wave is in phase with the input grid-voltage wave, and this is shown in Fig. 11-10. We have also stated that the output plate-voltage wave was 180° out of phase with the input grid-voltage wave. Although we have already shown this relationship graphically, it is interesting to examine it in a somewhat different light.

Just as we were able to derive the dynamic transfer characteristic of Fig. 11-10 from the points of intersection of the load line with the grid-voltage curves of Fig. 11-1, in a similar manner we can correlate the various values of grid voltage with the corresponding values of plate voltage. Only this time, instead of projecting the load-line intersections to the left to obtain the appropriate values of plate current, we project the load-line intersections downward to the plate-volts axis in order to obtain the corresponding value of plate voltage e_b. When this is done we may plot the dynamic grid voltage (e_c)—plate voltage (e_b) characteristic, as shown in Fig. 11-11.

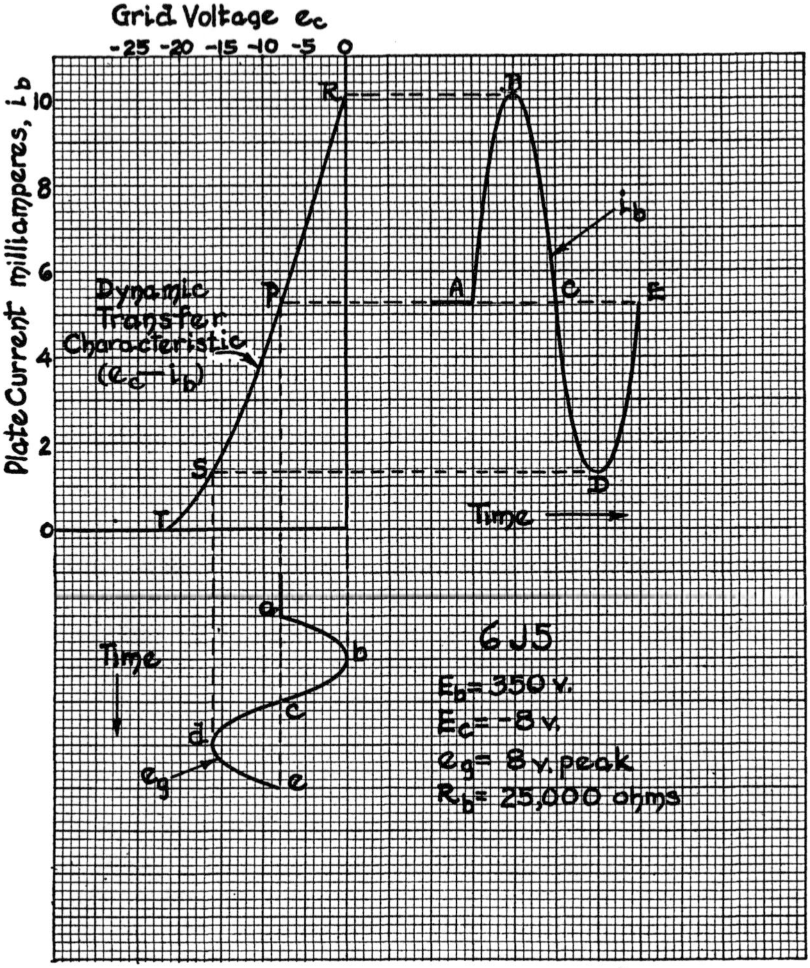

FIG. 11-10. This dynamic transfer characteristic of a 6J5 triode has an 8-volt peak signal input impressed on the circuit. The plate-current output wave is in phase with the input voltage.

Now applying the input signal-voltage e_g wave to this characteristic, we obtain the output plate-voltage e_b wave. Comparing the plate-current i_b wave of Fig. 11-10 with the plate-voltage e_b wave of Fig. 11-11, it is seen that when the plate-current wave goes upwards, the plate-voltage wave goes downwards, and vice versa. So, once again, we observe that the plate current i_b is in phase with the signal voltage

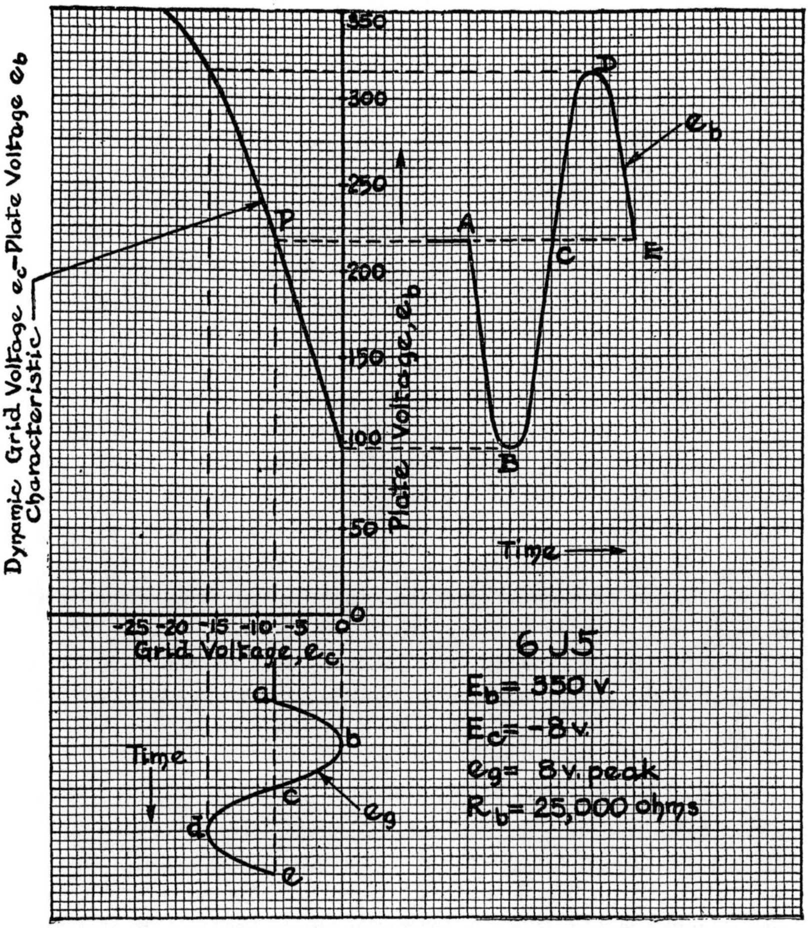

FIG. 11-11. The dynamic grid voltage—plate voltage characteristic is drawn according to the directions in the accompanying text. When the input voltage wave is projected onto this characteristic, the resulting waveform on the right shows that the plate voltage is 180° out of phase with the input.

e_g, while the plate voltage e_b is 180° out of phase with the signal voltage e_g.

When the Grid Swings Positive

Thus far in our considerations of the triode we have discussed the conditions when the grid was maintained at a negative potential with respect to the cathode by a fixed voltage source, E_c, and employment of an input signal e_g of such a value that the total voltage e_c on the grid never swung past zero to a positive value. Many occasions arise, deliberately or otherwise, when the potential on the grid does become positive with respect to the cathode and you should be acquainted with what occurs within the tube when this happens. Let us consider some of the effects of this operating condition in general.

You will recall that during our discussion of the grid-cathode and plate-cathode fields, it was stated that when the grid is positive with respect to the cathode, the grid-cathode field aids the plate-cathode field by exerting more force on the electrons in the space charge thereby increasing the number that leave that area and increasing the plate current. This is the case, of course, when the plate voltage is relatively high, but it is not the whole story. Remember that the grid is also positive with respect to the cathode and therefore some of the electrons will be attracted to it. Here, then, is another circuit similar to the plate circuit: an element of a tube that is positive with respect to the cathode, and which is connected to the positive terminal of a source of voltage, the negative terminal of which is connected to the cathode. Is it not logical to assume that the electrons will behave similarly in this circuit as they do in the plate circuit? Certainly it is, and they do follow the same procedure; those which are attracted to the grid wires flow along the external circuit to the grid voltage source, through this, and on to the cathode. This flow of electrons in the grid circuit is called *grid current.*

Returning once more to the matter of fields within a tube, the higher the difference in potential between two electrodes, the greater will be the force exerted upon an electron within the field set up by the potential existing across the electrodes. Conversely, the smaller the voltage difference, the less will be the force exerted on an electron in the field so set up. In the triode we have two fields that can be varied in strength so that the combined force exerted on an electron will be different under various circumstances, and it is these combinations that are of interest. Let us consider some of the variations in connection with the triode circuit shown in Fig. 11-12, which has a milliammeter in the plate circuit for reading the plate current, and a means for varying the voltage impressed on the plate. The battery E_c, in the grid circuit is shunted by a potentiometer whose slider is connected to the grid so that the grid can be made positive or negative with respect to the cathode.

If we start with the movable arm of the potentiometer next to the negative terminal of the battery E_c, the grid will be negative with respect to the cathode and some plate current i_b will be indicated on the milliammeter even if the plate voltage is low, whereas nothing will be indicated on the grid microammeter, for electrons will be repelled by the negative grid and they will therefore pass on to the plate. Keeping

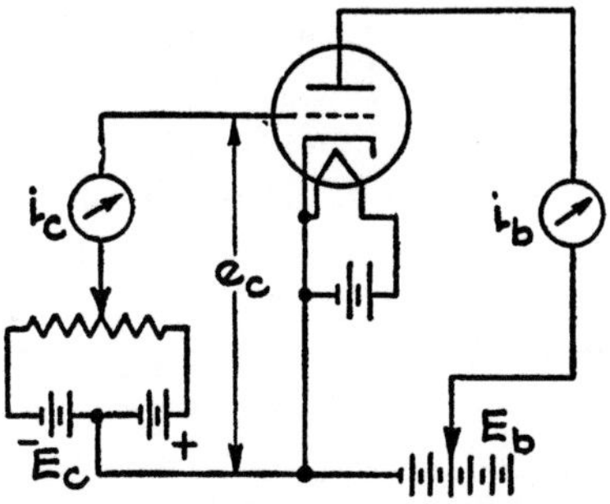

FIG. 11-12. The potentiometer in the grid circuit enables either a positive or negative bias to be impressed on the grid. The current meter i_c shows the amount of grid current flowing when the grid is positive with respect to the cathode.

the plate voltage at a low value, we slide the movable arm of the potentiometer towards the middle, making the grid voltage less and less negative. While this is being done, the plate milliammeter shows an increasing plate current, but still nothing is indicated on the grid microammeter. More electrons go to the plate (as indicated by the increasing plate current) because the plate-cathode field is less and less counteracted by the gradual reduction of the grid-cathode field.

Now let us assume that the movable arm of the potentiometer has been moved past the middle and that a positive potential is applied to the grid. The grid now aids the plate-cathode field in extracting electrons from the space charge. Also, since the grid is positive, quite a large number of electrons will be attracted by the grid and flow through its external circuit back to the cathode. This means that a fairly large grid current will flow under these conditions of a positive grid and a relatively low plate voltage.

If the grid voltage is further increased so that the grid becomes still more positive with respect to the cathode, a stronger grid-cathode field will result. This has the effect of partially neutralizing the space charge and so permitting freer movements of the electrons in it. More electrons will be attracted to the grid with a consequent rise in grid current and fewer will go to the plate. This is because some of the electrons that otherwise would go to the plate are stopped by the grid wires being in their paths and others, which are not obstructed by any part of the grid structure, will still be attracted to the grid by its positive influence pulling them to one side or the other from out of their normal paths.

Now let us increase the plate voltage so that it is substantially higher than the positive voltage applied to the grid. This increase materially strengthens the plate-cathode field and increases the force of the field at the space charge. Thus the electrons coming within the pull of this force will tend to travel towards the plate with a greater velocity than they did in the previously described case when the plate voltage was lower and the plate-cathode field weaker. Such faster traveling electrons are less liable to be pulled out of their paths by the positively charged wires comprising the grid and, as a consequence, a great many will speed through the openings of the grid wires without feeling its influence. This, of course, will increase the plate current. The paths of other electrons may be directly toward the wires forming the grid structure and these electrons will be attracted to the grid wires and will strike them. These electrons constitute a grid current that is less than in the former case, since with the higher voltage on the plate exerts a greater attraction for the electrons.

Now let us hold the plate voltage constant at some particular value. Starting with some negative voltage on the grid as we decrease this negative voltage we find that the plate current increases. We further find that when a positive voltage is applied to the grid, the plate current is still greater. Then, as we make the grid more and more positive with respect to the cathode, we observe that the plate current continues to increase. Does this plate current continue to increase indefinitely as the grid is made more positive? Over a very large range this effect does occur for modern receiving tubes since the electron emission from the cathode is more than ample to supply both the plate and the grid currents. There comes a point, however, when the impact of the electrons on the grid causes it to become overheated and it may buckle and make contact with another electrode, thereby destroying the usefulness of the tube. Alternatively, the impact of the electrons on the grid and plate causes them to become so hot as to release gas that has been trapped within the metal of the electrodes, thereby destroying the vacuum, and so the usefulness of the tube.

Influence of the Grid Resistor

Our remarks regarding operation of the grid in the positive region have concerned themselves with the circuit shown in Fig. 11-12 which you will recall is similar to that which was used to obtain the static characteristics of the tube. Now let us employ the circuit shown in Fig. 11-13 where a plate load resistor R_b of 100,000 ohms is incorporated in the plate circuit. Further provision has been made to have

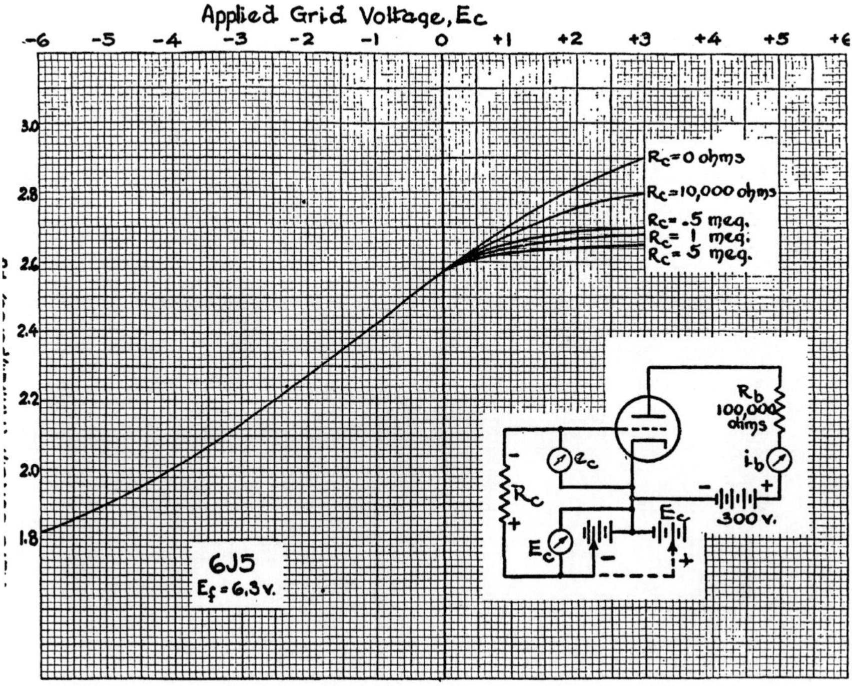

FIG. 11-13. When a resistance is in the grid circuit, its effect is to flatten the end of the dynamic characteristic as it extends over into the region where the grid becomes positive. The greater the grid resistance, the more the curve is flattened.

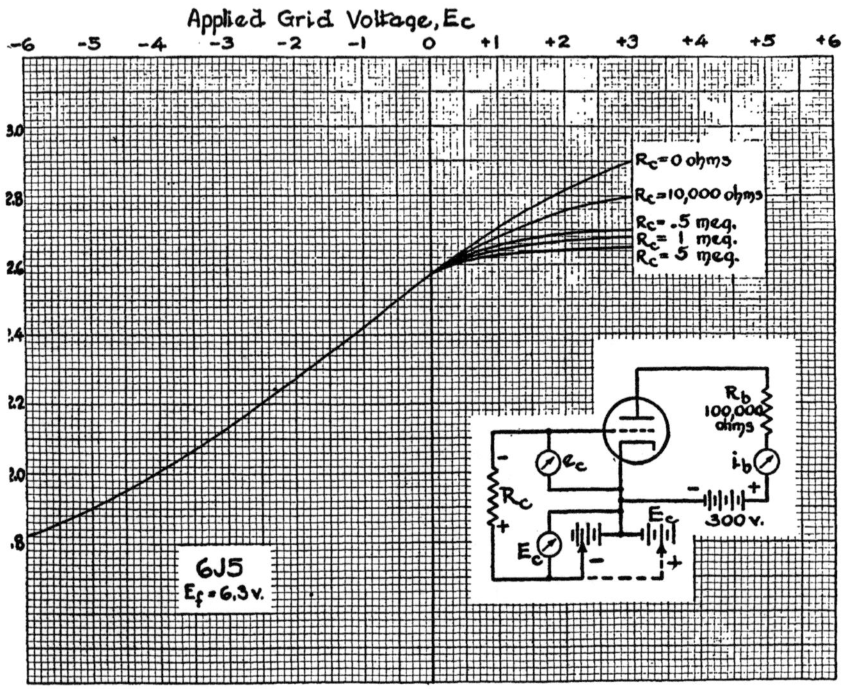

. 11-13. When a resistance is in the grid circuit, its effect is to flatten the end of the dy-nic characteristic as it extends over into the region where the grid becomes positive. The ater the grid resistance, the more the curve is flattened.

zero resistance in the grid circuit (similar to Fig. 11-12) or to have a definite amount of resistance in the grid circuit. The value of this grid load resistance R_c has a definite effect on the shape of the dynamic characteristic curve.

You will observe that all the curves are identical in the entire negative region of applied grid voltage, and that they differ from each other only in the positive grid-voltage region. You will further notice that all of the curves have a tendency to flatten out (become more horizontal), this effect being more pronounced the higher the value of the grid resistor R_c.

Let us first take the case when the grid load resistance is zero ($R_c = 0$). Several paragraphs back, in connection with the circuit of Fig. 11-12 which had no load resistor in either the plate or grid circuit, it was remarked that the plate current continued to increase indefinitely (until destruction of the tube occurred) as the grid was made more and more positive. When we examine the $R_c = 0$ curve, we find that despite the fact that the grid load resistor R_c is zero, that this curve in the positive region is not a continuation of the curve in the negative region, but instead has a slight tendency to flatten out. Now remember that there is zero resistance in the external grid circuit, so that the flow of grid current does not result in a voltage drop in the external grid circuit. Consequently, the total grid voltage e_c is equal to the supply voltage E_c. The cause of the slight flattening of the $R_c = 0$ curve is not to be accounted for by what happens in the grid circuit. We observe, however, that the plate circuit contains the 100,000-ohm plate-load resistor R_b. When the grid is driven positive a large plate current flows, and therefore a large voltage drop occurs across the load resistor. Since this voltage drop opposes the plate-supply voltage, only a small voltage is effective at the plate. As the voltage applied to the grid becomes more and more positive, then, due to a greater and greater plate current, the voltage drop across the plate load resistor becomes larger and larger, so that the voltage effective at the plate becomes less and less. So although the plate is quite effective in attracting electrons to it, its effectiveness is not as complete as would be the case if there were no plate load resistor in the plate circuit.

The plate resistance r_p of the tube and the plate load resistor R_b are in series with the plate-supply voltage E_b. The plate resistance becomes less and less as the grid becomes more and more positive. Since the plate resistance becomes very small in comparison with the plate load resistance, we may disregard the plate resistance and consider the plate load resistance as the factor which limits the total plate current.

Accordingly, by Ohm's Law, the plate current can never exceed the value.

$$i_b = \frac{E_b}{R_b}$$

For the circuit of Fig. 11-13, the plate supply voltage $E_b = 300$ and the plate load resistor $R_b = 100{,}000$. Consequently the plate current i_b can never exceed

$$i_b = \frac{E_b}{R_b} = \frac{300}{100{,}000} = 0.003 \text{ ampere} = 3.0 \text{ milliamperes}$$

In Fig. 11-13 the flattening of the $R_c = 0$ curve is due to the effect of the plate load resistor, and that this curve approaches, but never reaches, a plate-current value of 3 milliamperes.

Although the plate load resistor does have some effect in causing a flattening of the curves, its effect is but a minor one in comparison with that of the grid load resistor. Let us now consider the insertion of a grid load resistor R_c in the grid circuit. When the grid becomes positive with respect to the cathode, grid current i_c flows. In other words, inside the tube some of the electrons which leave the cathode go to the positively charged grid. In the external grid circuit, the electrons which were attracted to the grid, flow through the grid load resistor R_c and then back to the cathode. The electron flow i_c through the grid load resistor R_c results in a voltage drop across this resistor, and the polarity of which is such as to make the end of the resistor nearer the grid, negative with respect to the cathode side of the resistor, as illustrated in the diagram of Fig. 11-13. It is, therefore, seen that the voltage drop across the grid load resistor R_c bucks the grid supply voltage E_c, so that the total grid voltage e_c is

$$e_c = E_c - i_c R_c$$

The larger the magnitude of the grid current i_c or the larger the value of the grid load resistor R_c, the greater the value of the voltage drop across this resistor.

We therefore see that when the grid goes positive the total grid voltage e_c will be less than the applied grid voltage E_c. This is equivalent to the grid being less positive than it would be if the grid load resistor were not present, which, in turn, means that less plate current will flow, and so the tube characteristic curve flattens out. This flattening effect is particularly noticeable for large values of grid load resistance. For example, as shown in Fig. 11-13, the tube characteristic becomes practically horizontal when the grid load resistor R_c equals 5,000,000 ohms (5 megohms).

When the grid voltage—plate current curve flattens out in this manner even though a greater and greater positive voltage is applied to the grid, the condition is sometimes known as *saturation* or *plate-current saturation.* This terminology is unfortunate and this type of "saturation" should not be confused with genuine saturation wherein the plate current flattens off because the cathode can not supply sufficient electrons to the plate, but since this latter condition is not encountered in modern receiving type tubes, we need not concern ourselves with this topic.

Production of Square Waves

We have thus far emphasized the importance of obtaining as nearly faithful an amplified replica in the output circuit as was applied to the input circuit. Although it is true in the majority of cases that we

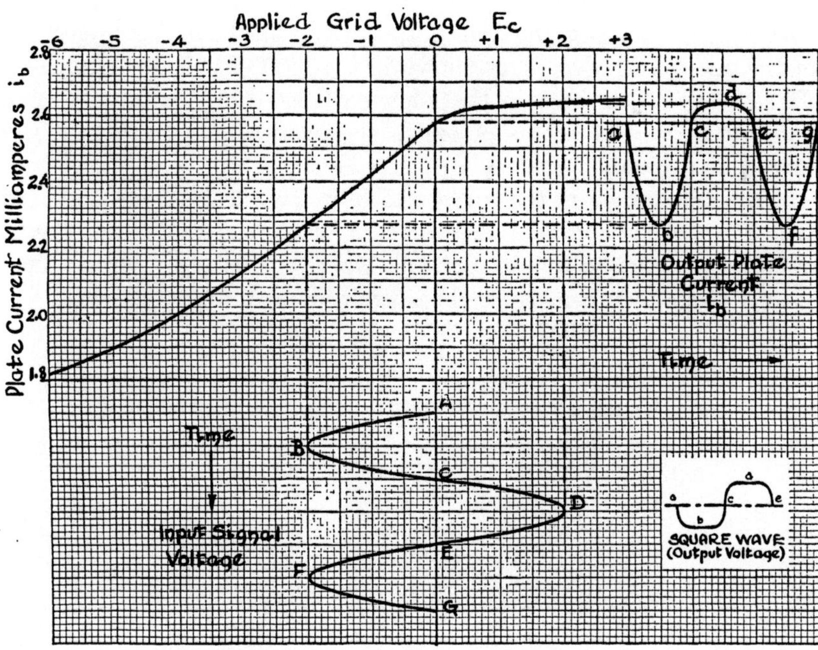

FIG. 11-14. By including a large resistance in the grid circuit, and driving the grid voltage into the positive region, the output waveform will have one side considerably flattened.

wish to retain the same wave-shape, this is not invariably the case. We have furthermore, limited the scope of our inquiry to waveforms of the sine-wave type. There are, however, other useful types of waveforms, such as square waves and triangular waves.

A case of interest occurs when we deliberately employ the vacuum tube to cause distortion, for example, to make a vacuum-tube circuit convert a sine wave into a square wave. Of the various curves shown in Fig. 11-13, let us choose the one for which the grid resistor $R_c = 5$ megohms, and redraw this curve on Fig. 11-14. For our operating point we shall choose a point in this characteristic where there is a decided curvature, namely, the zero bias point $E_c = 0$. The application of the sine-wave voltage *ABCDEFG* to the grid then results in an output plate-current wave *abcdefg*.

It will be seen that portions *abc* and *efg* of the plate-current wave are quite faithful replicas of the portions *ABC* and *EFG* of the grid voltage wave. Portion *cde* is entirely different in shape from portion *CDE*, as it has been considerably flattened and is almost rectangular in shape. If this output wave is applied to another tube which is also operating at zero grid bias, the portions *ABC* and *EFG* will also be flattened, so that now the final output wave will have the appearance of the one-cycle wave *abcde* shown in the insert at the lower right of Fig. 11-14. It can be seen that this wave has fairly squared corners and, although it is not as perfect a specimen as can be produced, is known as a *square wave*.

A re-examination of the curves shown in Fig. 11-13 shows that they become more horizontal as the grid resistor R_c is made larger. Consequently, in order to obtain a square wave with sharper corners, it is merely necessary to increase the magnitude of the grid resistor.

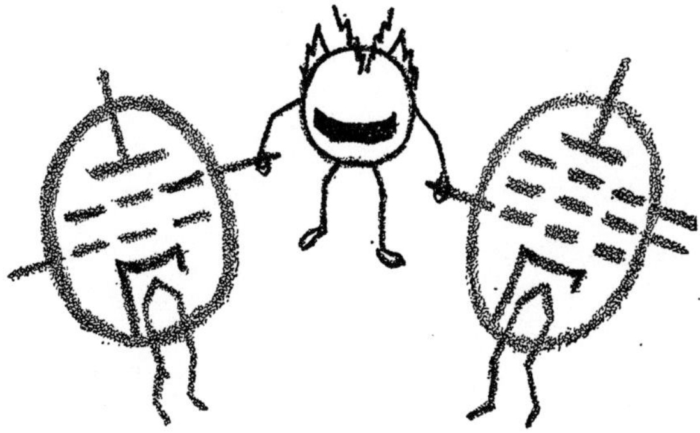

Chapter 12

TETRODES AND PENTODES

THUS FAR in our considerations of vacuum tubes, we have dealt with the fundamental operations and the principles of the diode and triode. During the early days of radio broadcasting these were the only available tubes and many of the receiver circuits were designed to compensate for their shortcomings; this was especially true in the case of the triode. It was not until late in the 1920's that tube manufacturers introduced the *multi-electrode* or *multi-element* vacuum tube that made possible the efficient receivers that are common today.

Before starting our discussion of *multi-electrode* tubes, a matter of nomenclature should be clarified that has often caused misunderstanding. In one group of tubes with four or more electrodes are those which embody certain electrodes that enable the tube to function better than a triode. Such tubes are the tetrode or pentode, having four and five electrodes respectively, and about which you will be told in this chapter. In the second group are those tubes which have more than three electrodes, but which perform *more than a single function in a circuit*. These are the so-called *multi-purpose* tubes, such as the duo-diode triode and the dual triode, and which will be discussed in a later chapter. The duo-diode-triode, for example, has a single cathode for the two diode plates and the grid and plate of the triode portion of the tube. The difference between these two general types of tubes—the multi-electrode and the multi-purpose—will be apparent if you will take into consideration whether the purpose of the tube in question is single or multiple.

THE TETRODE

First we will consider the *screen-grid* or *tetrode* tube, which is so named because it contains four elements: the usual three found in a triode and a second grid located between the control-grid and the plate. The physical arrangement of these four elements is illustrated in Fig. 12-1.

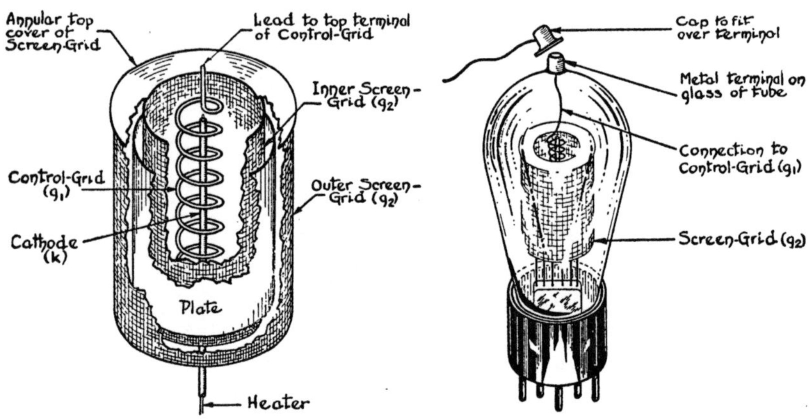

FIG. 12-1. The arrangement of the four elements of the tetrode is shown at the left. The control grid is connected to the metal cap at the top of the glass envelope, as shown on the right.

The cathode may be either of the filament or the indirect-heater type; the control-grid is a spiral wire with the turns fairly close together. The cylindrical plate is enclosed on its inside and outside by a cylindrical wire mesh which is capped with an annular metallic disc. This mesh constitutes the added element and is called the *screen-grid*. The lead to the control-grid is introduced into the glass envelope of the tetrode through a cylindrical metal terminal at the top onto which a metal cap fits which is in turn connected to the external grid input circuit. The plate, screen-grid, heater, and cathode connections are made through the conventional prongs at the bottom of the tube base.

The symbolic representations of the tetrode are shown in Fig. 12-2. The old-style symbol was used to show the enclosure of the plate by the screen-grid, but the modern symbol, which we will use hereafter,

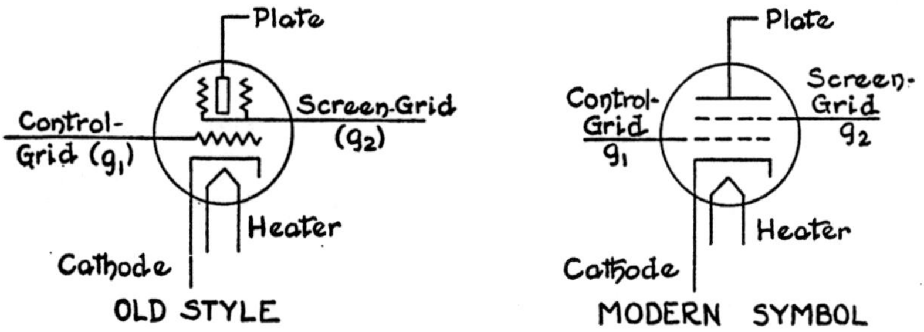

FIG. 12-2. The use of zigzag lines representing the grids in tube symbols has been replaced by dashed lines making for simpler representation of multi-element tubes.

supposes that the reader knows the physical make-up of the tube and shows the screen-grid as a dashed line similar to the control-grid. Incidentally, in this as well as other multi-electrode tubes, the grids are sometimes numbered as g_1, g_2, g_3, *etc.* The lowest subscript is given to that grid or element nearest the cathode, that is, in the tetrode g_1 is the control-grid and g_2 is the screen-grid.

First of all let us consider some of the difficulties which led to the development of the tetrode; afterwards we will deal with the operation of the tube. Before starting off on this, we want to bring some facts to your attention that have a bearing on the development of the tetrode.

It was previously shown that a field existed in a triode between the plate and the grid, that is to say, the field existed between two electrically charged bodies. It was unnecessary then to point out that this arrangement of the grid and plate with the electrostatic field between them, was the equivalent of an electric condenser, but now this fact becomes important. The capacity of this condenser, C_{gp}, formed by the grid and plate of a triode, as shown in Fig 12-3, is very small—only a few millionths of a microfarad—and at the low frequencies in the audio-frequency range (20 to 15,000 cycles) the capacity effect is small. That is to say, at low frequencies changes in grid voltage affect the plate current and the plate voltage, but this last-mentioned potential has no significant effect on the potential on the grid. At higher frequencies, however, the plate potential materially affects the grid potential to such an extent that it interferes with the proper operation of the triode as an amplifier.

The screen-grid provides the answer to this problem of the unwanted feedback of energy between the plate and the grid. As was mentioned above, the screen-grid encloses the plate more or less completely and so functions as a shield. This grid is maintained at a potential that is positive with respect to the cathode but not generally at as high a

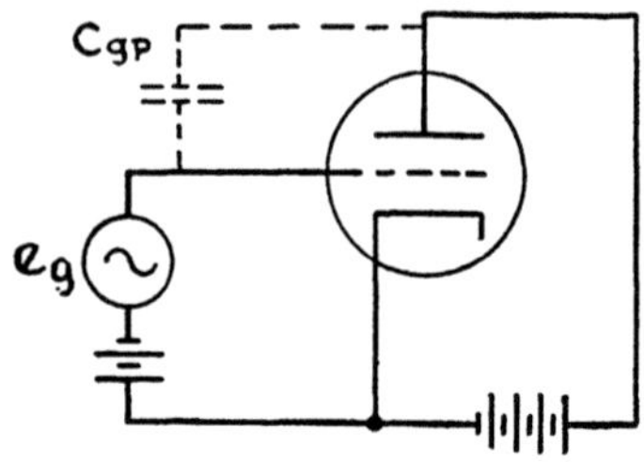

FIG. 12-3. The capacity C_{gp} existing between the grid and the plate in a triode can be disregarded at low frequencies, but at high frequencies (above 15,000 cycles) this capacity effect interferes with proper operation of the tube. The screen-grid of the tetrode greatly reduces this condition.

voltage as that at which the plate is maintained. Thus a capacitive effect is built up between the plate and the screen-grid and any detrimental effects at the high frequencies are practically limited to those two elements of the tube and so do not interfere with its overall operation. By this means the grid-to-plate inter-electrode capacity of the tetrode is only about 0.01 $\mu\mu$f.

Now there is another point of view which we wish to bring to your attention and that is the dual function of the plate of a tube. First, in combination with the grid, the plate sets up an electrostatic field which determines the number of electrons that pass from the space charge surrounding the cathode to the plate; second, the plate provides a means for collecting the electrons at the end of their inter-electrode travel and starting them on their trip to the cathode through the external plate circuit. Perhaps you did not separate these two functions of the plate in your mind when you read about them, but they are certainly separate and distinct. For example, if the grid is positive with respect to the cathode and the voltage on the plate is very low, the grid becomes the main electron collector rather than the plate, but at the same time the combined fields set up by the grid and plate determine the amount of electron flow from the space charge. You can see from this that it is quite logical to separate the functions performed by the plate, although both are closely associated in the triode.

In the tetrode the screen-grid separates the dual functioning of the plate just described. The screen-grid, which is maintained positive with respect to the cathode as indicated in the circuit of Fig. 12-4, and the control-grid set up the electrostatic fields that control the amount of the electron flow. The plate in this tube therefore performs just one of the two functions it does in the triode: it collects the electrons and starts them off on their way to the cathode. Of course, the cathode

FIG. 12-4. The voltage applied to the screen-grid is called E_{c2} and is generally less than the plate supply voltage E_b. The current flowing in the screen-grid circuit is called i_{c2}. The total current that returns to the cathode is the sum of i_b, the plate current, and i_{c2}.

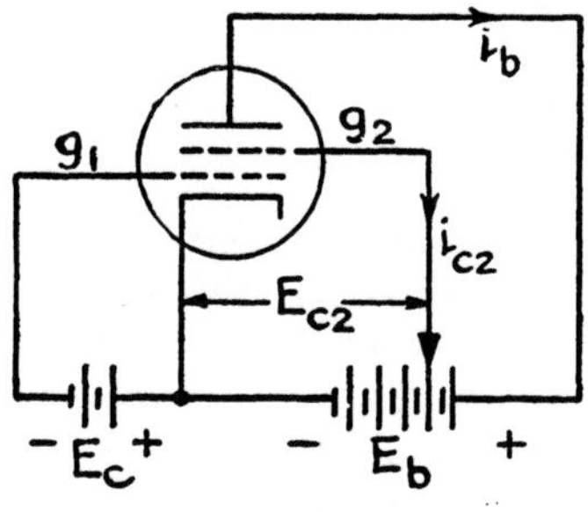

and the control-grid perform in the tetrode the same as they do in the triode.

The plate of the tetrode is usually kept at a higher positive potential than the screen-grid, which is also positive with respect to the cathode. This means that in this type of tube we have two collectors of electrons and two external circuits over which current flows back to the cathode. In other words, when the plate voltage is at a certain value electrons will be attracted from the space charge and travel towards the plate

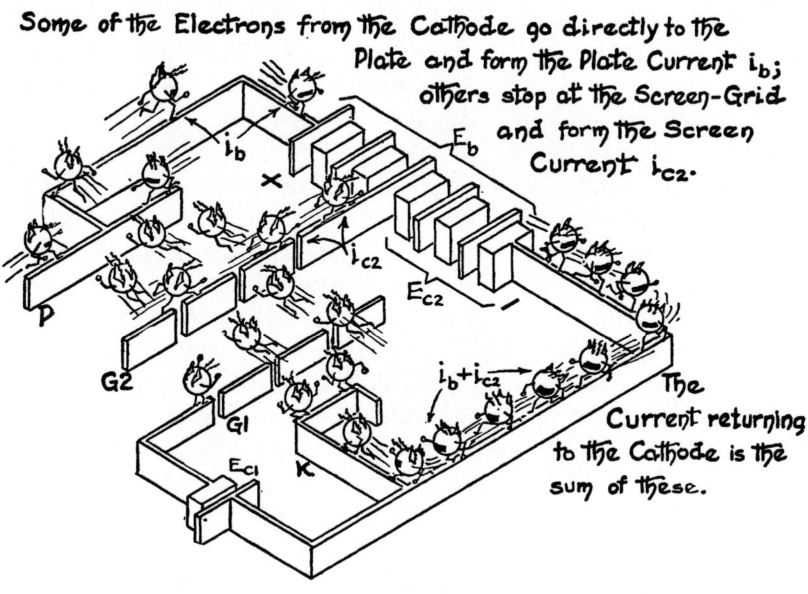

with a certain velocity. Some of these electrons will make direct hits on the wires of the screen-grid and flow out into its external circuit; this electron flow is the *screen current* i_{c2}. The rest of the electrons will travel on to the plate and form the plate current i_b. Most of the electrons pass through the openings of the screen-grid mesh and continue on to the plate. So by and large, the plate is a collector of electrons in the tetrode as it is in the triode.

It was stated above that the screen-grid performs one of the two functions of the plate in a triode. Do not forget that the screen-grid is so constructed around the plate that it acts as a shield and therefore *any variation in potential on the plate is not felt back at the space charge;* the screen-grid effectively prevents that happening to any great extent. You will see later on that the plate voltage—total space current characteristic curves are practically flat horizontally, indicating that even with a decided increase in plate voltage (after a critical value) the total current in the tetrode is practically constant. This total current is, of course, the sum of the screen current and the plate current and has just a bit larger value than the plate current alone, for, as it was stated above, almost all the electrons pass through the screen-grid and go on to the plate within the operating voltage range of the tube.

Now let us see how this affects the amplifying properties of the tetrode. Do you recall in Chapter 10 the voltage amplification of a triode was defined as the ratio of the change in plate-voltage to the change in grid-voltage? Furthermore, the voltage effective at the plate of a triode is dependent on the voltage drop across the load resistor in the plate circuit and this means that the plate voltage e_b is always less than the source of the plate voltage E_b. As the plate current depends directly on the input signal voltage, the voltage gain or amplification is limited by this voltage effective at the plate.

In the case of a tetrode this condition is changed. It has already been stated that the plate voltage does not affect the strength of the electrostatic field at the space charge, as the plate is more or less shielded by the screen-grid. As the screen takes the place of the plate in establishing the electrostatic field, the voltage applied to the screen must be considered in determining the obtainable voltage amplification. Because of the absence of any appreciable resistance in the screen circuit, as shown in Fig. 12-4, the voltage E_{c2}, effective at the screen, will remain constant. Hence in the tetrode we have a constant voltage unaffected by current variations, and as a consequence the gain may be several hundred instead of around 10 or 15, as in the case of a triode. This will be considered in more detail later on.

Tetrode Characteristic Curves

The behavior of a tetrode under various conditions can be analyzed from its plate voltage—plate current characteristic curves, just as was done in the case of the triode. The schematic diagram for obtaining the curves is shown in Fig. 12-5 and the points found are plotted on

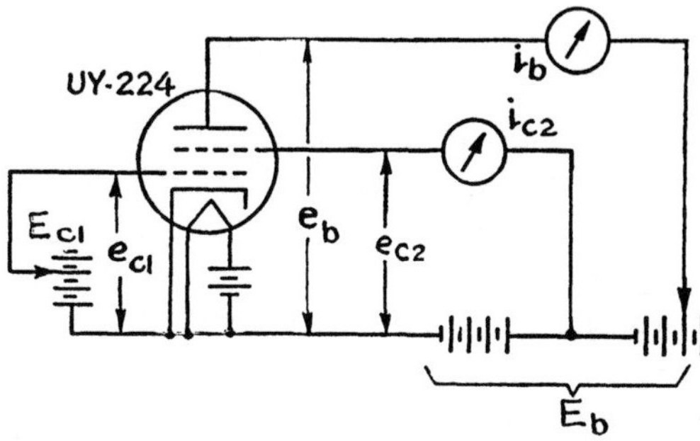

FIG. 12-5. The tetrode circuit used for determining the plate family of characteristic curves, one of which is shown in Fig. 12-6.

the graph of Fig. 12-6. The tetrode used in this test was the now obsolete filament-type UY-224, because it exhibits certain properties of fundamental interest which later types of tetrodes do not show. Curves for just one control-grid and screen-grid voltage setting are shown in Fig. 12-6 for illustrative purposes: the control-grid voltage e_{c1} is equal to -1.5 volts and the screen-grid voltage e_{c2} is equal to $+75$ volts.

It was stated that changes in plate voltage e_b do not affect the total current through a screen-grid tube to any great extent. This is quite true as evidenced by the flatness of the curve marked $i_b + i_{c2}$ in Fig. 12-6, this being the sum of the other two curves, one of which, marked i_b, shows the changes in plate current for changes in plate voltage and the other, marked i_{c2}, shows the changes in screen current for similar changes in plate voltage. You can also see that great changes in these two curves occur below a certain value of plate voltage, above which the curves are practically flat. It is in the region below this certain value of plate voltage that great interest lies, although above this voltage is the operating range of the two when used as an amplifier.

If it is true that the plate voltage has no appreciable effect on the field controlling the electrons emitted from the cathode, how does it happen that such a great variation occurs in the plate current when the plate voltage is varied from 0 to 90 volts, as shown in Fig. 12-6? Remember that in this instance we are discussing, the screen-grid has $+75$ volts applied to it constantly, and that the control-grid voltage does not change. The answer is the variation in the electrostatic field existing between the plate and the screen-grid and the different force exerted on the electrons in this region when the plate voltage is varied. You will see how this functions as you read farther.

Before we apply any voltage to the plate of the tetrode, a field exists between the screen-grid and the cathode. This field exerts a force on the emitted electrons and pulls them past the control-grid just as a positive potential on the plate of a triode causes the electrons to go to it. Hence we have a reading of 4 milliamperes on the meter in the screen-grid circuit, shown in Fig. 12-5. Even though the plate has 0 voltage some few of the electrons going to the screen will pass through the wires of the screen and continue on to the plate. This is evidenced by the fact that at zero plate voltage the plate-current curve shows a reading of 0.5 milliampere. Naturally with the screen at +75 volts with respect to the cathode, most of the electrons will go to the screen, giving a current reading of 4 milliamperes, as shown at the 0-volt line at the left in Fig. 12-6.

Let us now raise the voltage on the plate from 0 to +2 volts. This is indicated by the point A at the bottom of Fig. 12-6. This means that the plate is +2 volts with respect to the filament (or cathode) and the screen-grid is +73 volts (75 − 2) with respect to the plate. Some of the electrons leaving the space charge due to the force exerted by the field of the screen-grid and the small field due to the plate voltage, attain a certain velocity that is sufficient to take them all the way to the plate. Many of these electrons, however, are deflected from their paths while passing through the positively charged wires of the screen-grid so that they never reach the plate. Yet because some of them do arrive at the plate, the plate current is greater than it was at 0 volts, while the screen-grid current i_{c2} is less. These conditions are indicated by the upward slope of the plate-current curve and the downward slope of the screen-current curve.

The next step is to raise the plate voltage e_b by 3 more volts, making the total +5 volts with respect to the cathode, and it is indicated by the point B at the botton of Fig. 12-6. With this increase in plate voltage, the velocity of more electrons will be increased and so more will reach the plate with each plate-voltage increase. Thus you would expect that the plate-current curve would continue to go upwards, but at +5 volts, the point B, you can see that just the opposite occurs: the curve turns and starts downwards.

In order to explain this we must discuss the phenomenon known as *secondary emission*. When an electron has acquired a certain amount of kinetic energy through an increase in its velocity and then collides

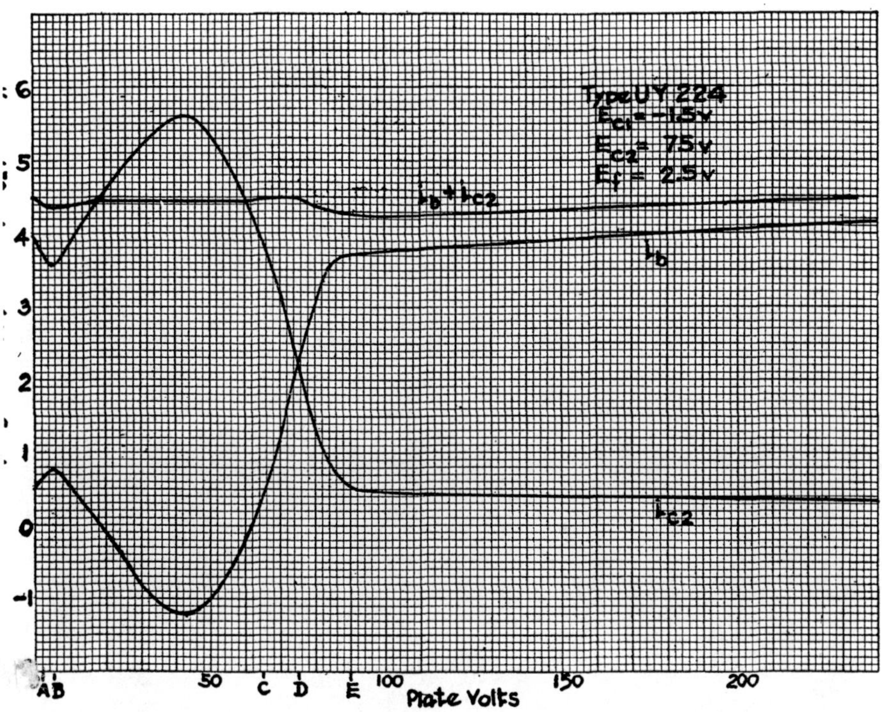

g. 12-6. The plate voltage—plate current characteristic curves showing the variations of cur- nt flow in the plate and screen-grid circuits, (curves marked i_b and i_{c2} respectively) and the al current flow, which is the sum of these two currents.

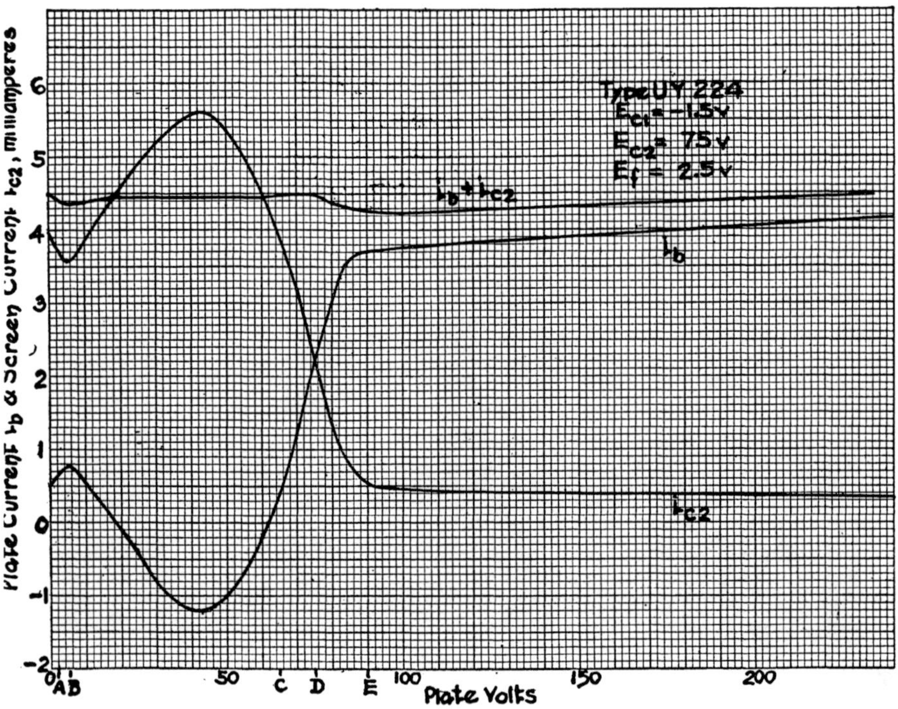

FIG. 12-6. The plate voltage—plate current characteristic curves showing the variations of current flow in the plate and screen-grid circuits, (curves marked i_b and i_{c2} respectively) and the total current flow, which is the sum of these two currents.

with another electron which is part of an atom, some of the kinetic energy from the high-velocity electron is imparted to the second electron and this second electron flies off from its parent atom. This is what occurs in this instance in the tetrode: the electrons that arrive at the plate have sufficient kinetic energy so that when they collide with electrons in the atoms of the plate structure, these plate electrons are freed and go off into space. Inasmuch as the nearby screen-grid is 70 volts positive (75 − 5 volts) with respect to the plate, many of these secondary electrons go to it and supplement the screen-grid current i_{c2}.

Going into this increase in screen-grid current in more detail, with a further increase in plate voltage more electrons will be attracted to the plate, and the more that arrive, the more secondary electrons will be liberated. Hence more electrons eventually go to the screen-grid than are ordinarily attracted to the screen on their way to the plate. This means, of course, an increase in screen current and a corresponding decrease in plate current. In other words, the net plate current is the difference between the number of primary electrons arriving at the plate and the number of secondary electrons that go to the screen-grid.

When the plate voltage is raised beyond the point B in Fig. 12-6, the electron velocity is also increased; therefore the kinetic energy of the incoming electrons to the plate is greater the more the plate voltage is raised. After such a fast-moving electron frees an electron from an atom, it may go on to collide with an electron of a second or even a third atom and so free several others besides the one originally freed. Of course, this will cause more electrons than ever to go to the screen, which is still positive with respect to the plate, and will mean a greater screen current for each volt increase on the plate. You can see from this that we can have more secondary electrons leaving the plate to go to the screen than there are primary electrons arriving at the plate, so the result is really a backward or negative plate current. This accounts for the downward trend of the plate-current curve.

With a further increase in plate voltage, say up in the neighborhood of 43 volts, more and more secondary electrons are freed from the plate, but inasmuch as the plate voltage is approaching the voltage value of the screen grid, i.e. 75 volts, some of the secondary electrons are attracted back again into the plate instead of going on to the screen. Of course, many others which have sufficient kinetic energy will go to the screen, but nevertheless, the quantity returning to the plate is sufficient to cause the plate-current curve to turn towards an increase. Naturally the screen current decreases accordingly, as is indi-

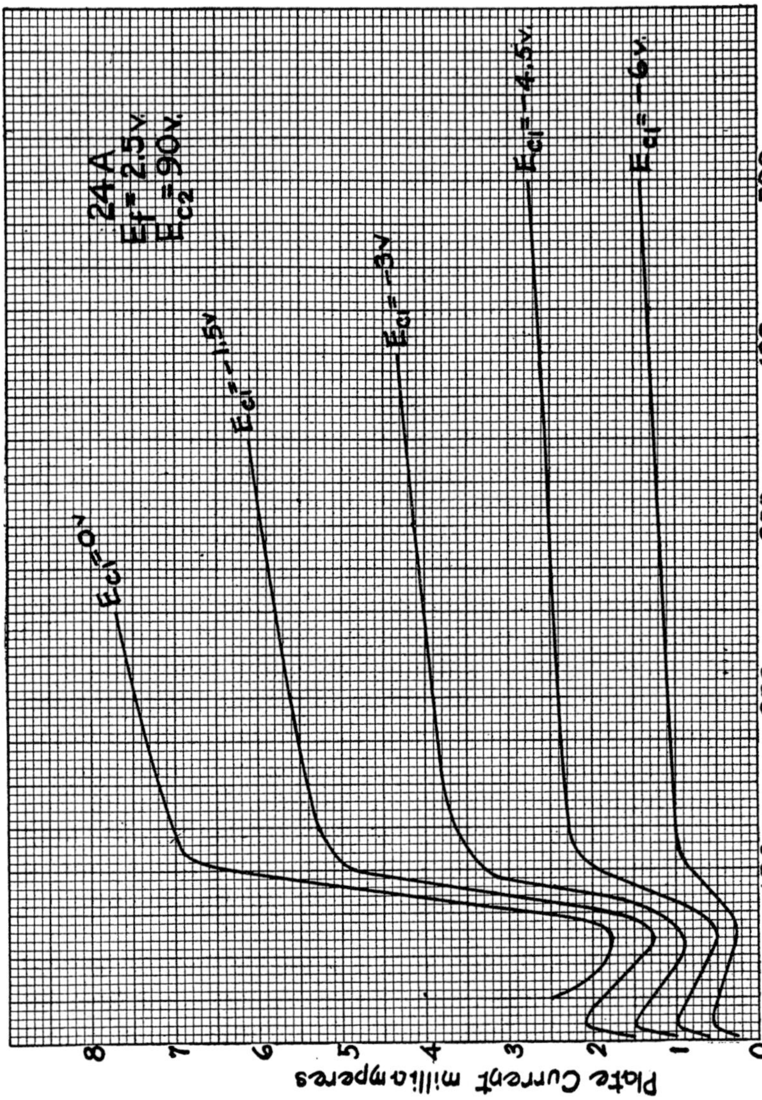

FIG. 12-7. The plate family for the 24A tetrode for a constant screen-grid voltage of 90 and different voltages applied to the control grid.

cated in Fig. 12-6. When the plate voltage has been increased up to within a few volts of the value of the screen-grid voltage, as at point C, more of the secondary electrons return to the plate and these along with the electrons from the space charge that get to the plate anyway, cause the plate current to increase at a fairly rapid rate per volt-increase of plate voltage.

When the plate voltage has been made the same as the screen voltage —in our example 75 volts shown at point D in Fig. 12-6—secondary electrons are freed from the plate as before, but because of the higher voltage now applied to the plate, the more powerful plate field prevents the secondary electrons from leaving the vicinity of the plate, because they are nearer the plate than they are to the screen. This same condition applies, of course, the higher the plate voltage becomes and the increase in plate current per volt-increase on the plate is quite rapid, as evidenced by the steep rise in the plate-current curve. Conversely, the decrease in screen-grid current is almost as rapid, as may be seen by the steepness of the screen-current curve i_{c2} in Fig. 12-6, up to about 90 volts on the plate. The two curves are not exactly opposite or mirror like, for secondary electrons are produced at the screen just as they are at the plate and as these secondary electrons appear on the side of the screen away from the plate, most of them are attracted by the screen and so only a very few get on to the plate. From this point on, the increase in plate current per volt-increase in plate voltage is very slow, as shown in Fig. 12-6 by the flatness of the curve.

The normal operating range of this type of screen-grid tube is above 100 volts and the normal quiescent value of the plate voltage is between 150 and 175 volts. It seems needless to mention that the portion of the curve below 100 volts is unused due to its different slopes and the distortion that would be caused if the plate voltage were this low. The curve indicating the total of the plate and screen currents, $i_b + i_{c2}$, in Fig. 12-6, is practically flat throughout its length, as may be expected, because when the plate current increases in value, the screen current decreases accordingly for like values of plate voltage. Moreover the uniformity of the total-current curve bears out the statement made earlier in this chapter that the total current flow through the tetrode is independent of the value of the voltage applied to the plate, of course, within certain limits.

It has mentioned that the UY-224 tube which was used for the curves in Fig. 12-6 is now obsolete; a more modern tetrode, the 24-A, was used for obtaining the more complete plate family of characteristics shown in Fig. 12-7. In this family of curves the control-grid voltage has been varied from 0 to −6 volts, but the screen voltage has been held constant at 90 volts throughout. In order that you can compare the plate family of a tetrode with the plate family of a triode, typical plate families of these two tubes have been included on the same page in the Appendix.

You will notice that none of these plate-current curves of the 24-A go below the zero current line as occurred for the UY-224 tube. This is due to the change made in the plate material. In the UY-224 the plate material permitted one or more secondary electrons to be emitted for each primary electron colliding with it; in the 24-A type of tube, the surface of the plate is treated so that fewer secondary electrons are liberated by the primary electrons. This, of course, reduces the number of secondary electrons that can be attracted to the screen-grid when its voltage is greater than that of the plate, and so the plate current never goes below zero under ordinary operating conditions, as shown in Fig. 12-7.

Whether we employ the old or the new type of tetrode we must always be careful not to allow the plate voltage ever to go below a value which would cause the tube to be operated in the region of erratic curvature, because in this region the resulting distortion becomes very great. On the other hand, if the operation of the tetrode is limited to the linear horizontal portion of the characteristics, then the alternating output voltage of the tube is greatly reduced, as compared to a pentode. It is mainly for these reasons that tetrodes are rarely used today and have been almost completely displaced by pentodes, which is a type of tube we shall discuss immediately. Furthermore, with the exception of these phenomena which depend upon secondary emission, practically everything that is stated about a pentode is applicable to a tetrode.

THE PENTODE

The Suppressor-Grid

If a fifth element, known as a *suppressor-grid* is now added to the tube, we obtain a *five-element tube* or *pentode* The action of the suppressor-grid, which is located between the screen-grid and the plate, serves, as its name indicates, to suppress the secondary emission from the plate and prevent it from reaching the screen-grid. This is accomplished in the following manner. Primary electrons emitted by the cathode successively pass through the control-grid g_1, the screen-grid g_2, and the suppressor-grid g_3, and then strike the plate. The impact of the primary electrons upon the plate results in the liberation of

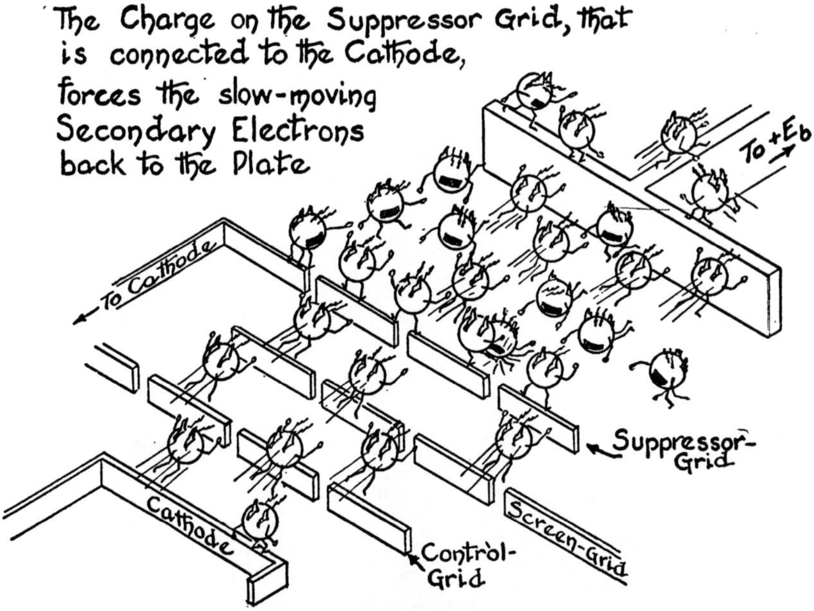

secondary electrons. In the absence of the suppressor-grid these secondary electrons would, because of the positive potential on the screen-grid, flow to the screen-grid. The suppressor is, however, at cathode (or ground) potential and therefore exerts a retarding effect upon these secondary electrons so that they are forced to return to the plate.

The screen-grid is maintained at a positive potential with respect to the cathode and, although, in Fig. 12-8, the screen is shown to be at a lower voltage than the plate this is not necessarily the case. If de-

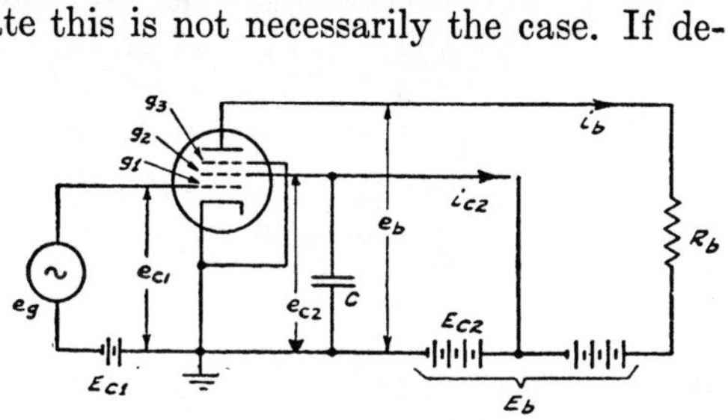

FIG. 12-8. Schematic showing the voltage and current relationships for a pentode. Note that the suppressor-grid is connected to the cathode, thus making these two elements have the same potential.

sired, the screen may be at the same potential as the quiescent plate voltage, since the harmful effects of secondary emission have been eliminated as a result of suppressor-grid action. With regard to signal voltages, however, the screen-grid is effectively at ground potential in virtue of the bypassing effect of condenser C.

Pentode Plate Family

The effect of the suppressor-grid upon the characteristics of the tube can be seen by an examination of Fig. 12-9. You will observe that the erratic kinks that appeared in the characteristic curves of the tetrode have entirely disappeared and that these have been replaced by gradually rounding knees and then a sudden downward drop. (Compare the three plate families in the Appendix.) This means, as we shall see later, that we are no longer so restricted in our linear operating region as was the case with the tetrode, and consequently our output is very greatly increased.

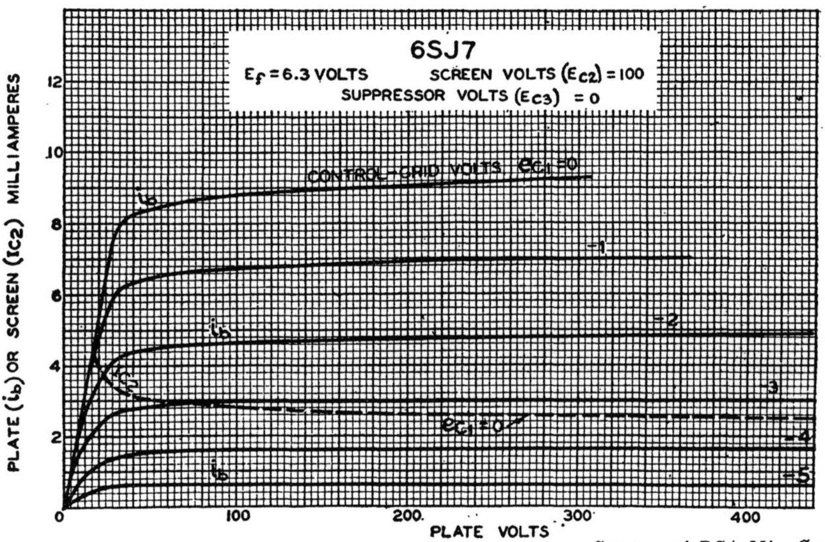

Courtesy of RCA Mfg. Co.

FIG. 12-9. The plate family of the 6SJ7 pentode. The solid curves show plate current variations and the dashed line shows the screen-grid current variations with changes in plate voltage.

You will observe that the spacing between the curves corresponding to equal changes in grid voltages are unequal. Thus, in Fig 12-9, we see that the distance between the $e_{c1} = 0$ and $e_{c1} = -1.0$ is considerably greater than the distance between the lines from $e_{c1} = -4.0$ to $e_{c1} = -5.0$. This indicates that equal changes in grid voltage, such as might be produced by the positive and negative alternations of the input signal, will result in unequal changes in plate current and this,

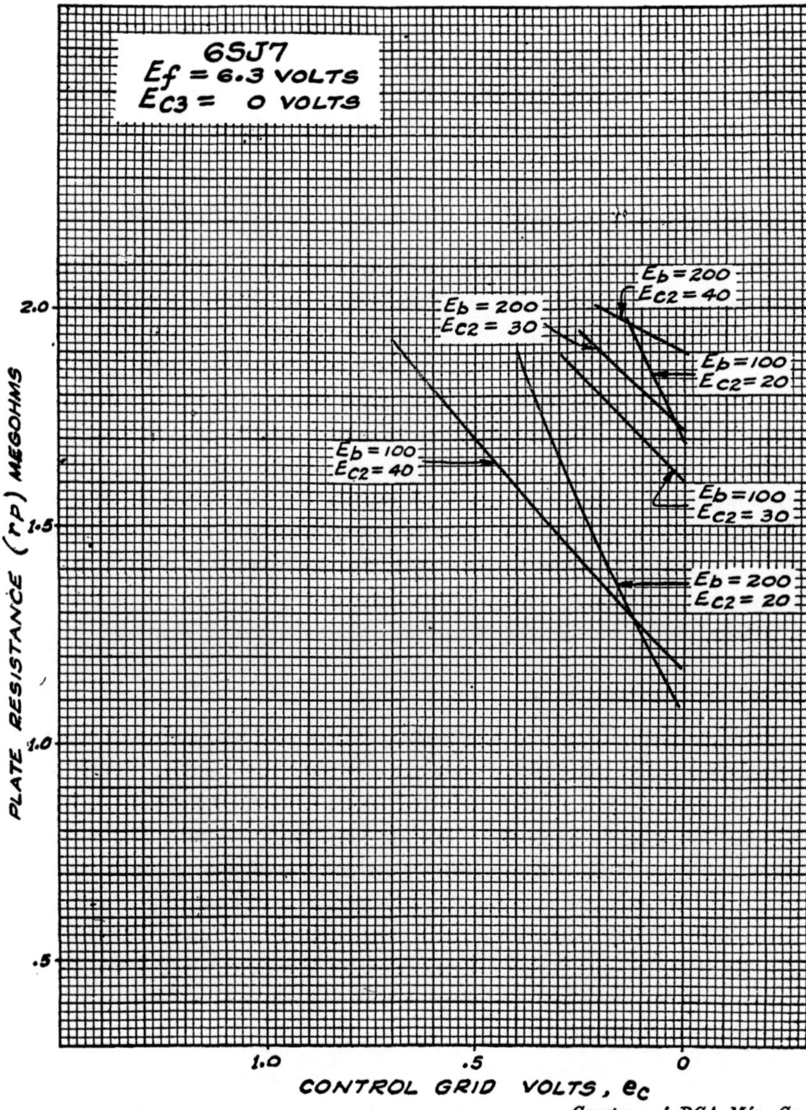

Courtesy of RCA Mfg. Co.

FIG. 12-10. These curves show the variation of the very high plate resistance of the 6SJ7 pentode as the control-grid voltage is varied. These curves are discussed on the next page.

as we have learned, signifies that the signal will be distorted. This consequent distortion which accompanies amplification is much greater for pentodes than for triodes, and it is for this reason when using a pentode as a voltage amplifier that we must limit our input signal to very small values. For example, the input signal to a 6SJ7 tube is usually about 0.1 volt or less.

Amplification Constant

The fact that for plate voltages above about 50 volts the characteristic curves are almost horizontal indicates that even very large changes in plate voltage produce but very small changes in plate current, as is seen in Fig. 12-9. In other words, the plate current is practically independent of the plate voltage. The plate current is, on the other hand, greatly influenced by the control-grid and screen-grid voltages.

Since it requires a very large change in plate voltage to produce the same change in plate current that results from a small change in control-grid voltage, it is seen that the amplification factor of a pentode is very great. For example, the amplification factor of a 6SJ7 is greater than 1500, in other words, approximately 100 times as great as for a triode.

Plate Resistance

Further significance of the nearly horizontal character of the characteristics is that the plate resistance is extremely high. For the 6SJ7, for example, this is of the order of from 1.0 to 2.0 megohms, which is 100 to 200 times as great as that for a typical triode. This is better shown in Fig. 12-10, which illustrates the variation of plate resistance of a 6SJ7 for plate potentials of 100 and 200 volts and for screen-grid potentials of 20, 30, and 40 volts, respectively. You will observe that the plate resistance is also dependent on the control-grid voltage.

Transconductance

Although the amplification factor and the plate resistance of a pentode are much greater than would be the case for a triode, the transconductance of pentodes and triodes are approximately of the same order. The 6SJ7 pentode, for example, has a transconductance of about 1650 micromhos, while the transconductance of a 6J5 triode is about 2600 micromhos. The variation of the transconductance of a 6SJ7 for different values of screen-grid voltage and control-grid voltage is illustrated in Fig. 12-11.

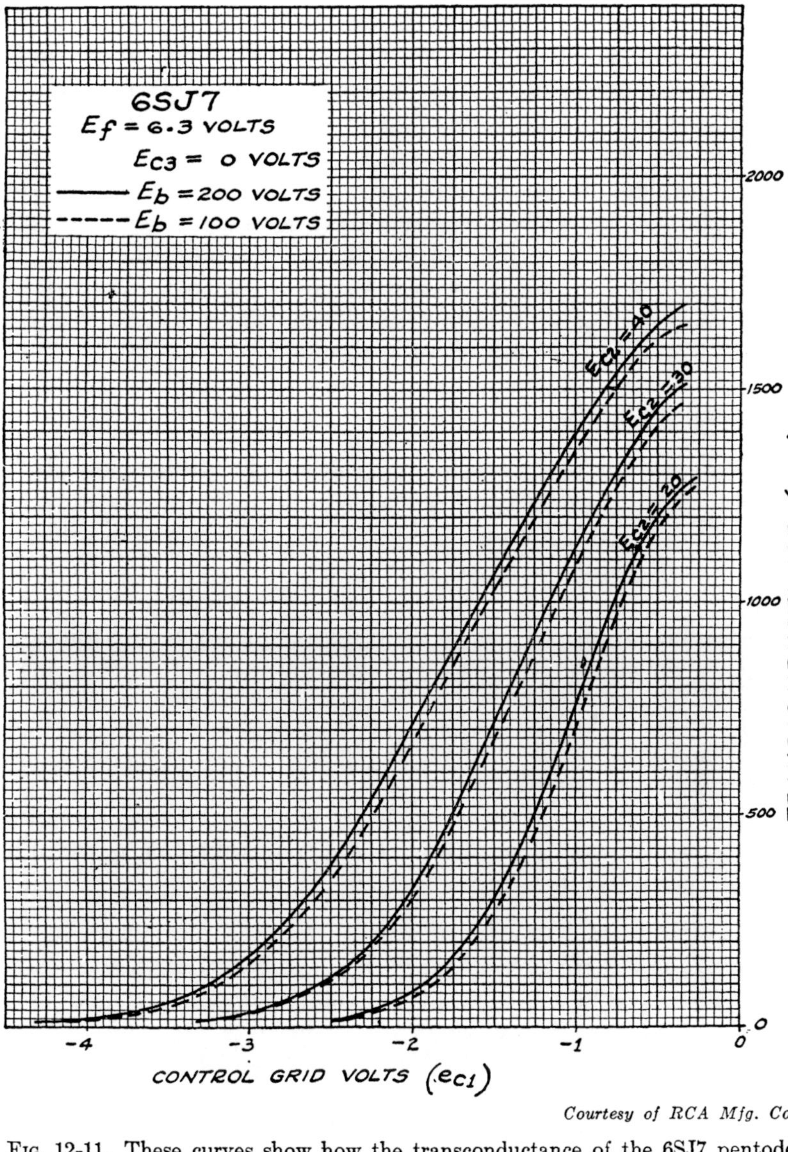

Courtesy of RCA Mfg. Co.

FIG. 12-11. These curves show how the transconductance of the 6SJ7 pentode varies for changes in control-grid and suppressor-grid voltages.

Load Resistor

Due to the extremely high value of plate resistance encountered in pentodes used for voltage amplification the load resistor is ordinarily but a fraction of that of the plate resistance. This condition is necessitated among other things, by virtue of the fact that a load resistor approximating the plate resistance would have so large a voltage drop across it that very little voltage would appear at the plate of the tube. The attempt to employ very large values of load resistor in order to obtain maximum voltage amplification would require an excessively large plate-supply voltage and this in turn would increase the possibility of the occurrence of flashover within the tube, and its possible destruction. A compromise must therefore be effected which results in a reasonable value of voltage amplification together with a safe and economical source of plate supply voltage.

Voltage Amplification

Because of the relative magnitudes of the plate resistance and load resistance, the formula which we have previously obtained to express the voltage amplification for triodes is usually employed in a modified form, the derivation of which follows.

We have already found that the voltage amplification of a vacuum tube having a resistive load in the plate circuit is given by the expression

$$A_v = \frac{R_b}{r_p + R_b} \mu$$

where

A_v = voltage amplification, R_b = resistance of load resistor,
r_p = plate resistance of tube, and μ = amplification factor of tube

But we have also seen that

$$\mu = g_m \; r_p$$

where g_m is the transconductance of the tube. If we now substitute this value of μ in our equation for voltage amplification, we obtain,

$$A_v = \frac{g_m \; R_b \; r_p}{r_p + R_b} = g_m \; R_b \left(\frac{r_p}{r_p + R_b} \right)$$

When the plate resistance r_p is very much greater than the load resistance R_b the expression

$\frac{r_p}{r_p + R_b}$ is approximately equal to 1 and we can accordingly obtain

$$A_v = g_m \; R_b$$

You should keep in mind that our explanation of pentodes is related

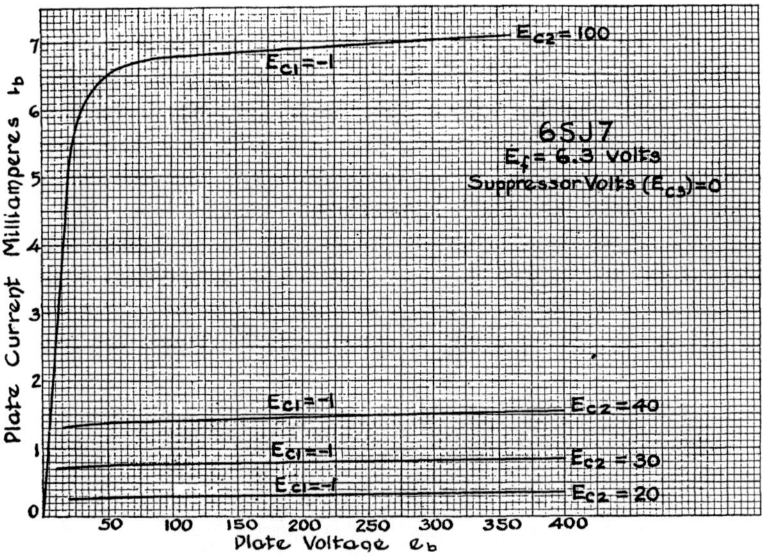

FIG. 12-12. These characteristics were made with the same control-grid voltage (—1 volt) and different values of screen-grid voltage. Note that the plate current increases with an increase in E_{c2}.

to their use principally as voltage amplifiers. When employed as power amplifiers the properties of pentodes, while following the same general principles which have been given, are, however, somewhat modified.

Screen-grid Voltage

One of the most important factors in the operation of a pentode is the voltage applied to the screen-grid. We have already seen that very large changes in plate voltage cause but slight changes in plate current. In contrast to this, any alteration of the screen-grid voltage is immediately indicated by an appreciable change in the plate current. This effect is shown in Fig. 12-12, which shows the curves resulting from screen-grid voltages of 20, 30, 40, and 100 volts respectively, while the control-grid voltage is held constant at —1 volt. The curves shown are of particular interest in view of the fact that when used as an audio-frequency voltage amplifier the screen-grid of a pentode is generally operated at approximately the lower voltages indicated, namely, from about 20 to 40 volts. The plate family characteristic curves generally published by tube manufacturers, on the other hand, show these characteristics when the screen-grid voltage is 100 volts, which value is suitable when the pentode is to be employed as a *radio-frequency* amplifier. We see, therefore, that when the pentode is to be used as an audio-frequency voltage amplifier the screen-grid voltage is low, with the result that the tube is operated at very small values of plate current. In fact, the plate current will ordinarily be even smaller than is indicated in Fig. 12-12, for the control-grid bias will customarily be more negative than the −1 volt shown.

By picking off the appropriate points from the four plate voltage—plate current curves of Fig. 12-12, and the given constant screen-grid voltages $E_{c2} = 20$, 30, 40 and 100, we may obtain the curves shown in Fig. 12-13. This latter group of curves clearly reveals the important

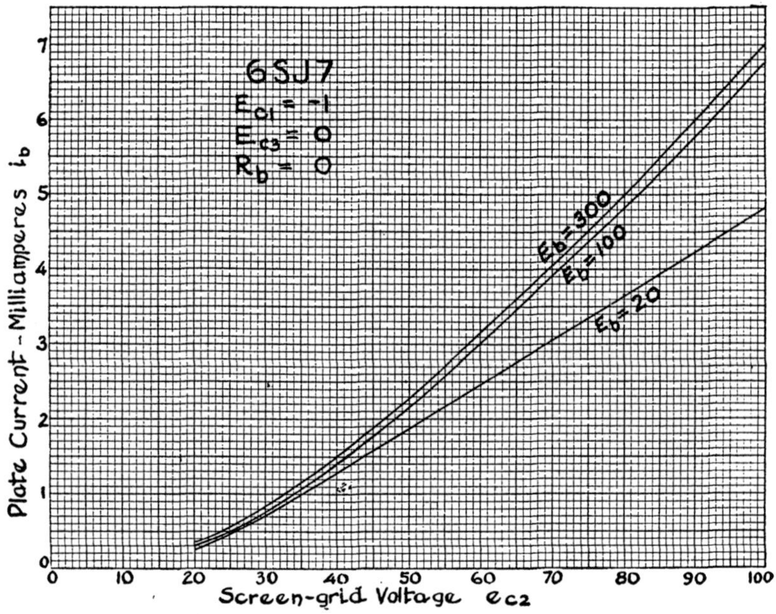

FIG. 12-13. The screen-grid voltage—plate current curves were derived from the curves of Fig. 12-12. These show that the plate current increases rapidly with an increase of screen-grid voltage.

influence that the screen-grid voltage exerts upon the magnitude of the plate current. Even when the plate voltage is very low, as, for example, $E_b = 20$ volts, we see that an appreciable plate current will flow when the screen-grid voltage is suitably adjusted.

The effect of the screen-grid voltage becomes even more striking when we plot curves for plate current i_b versus control-grid voltage e_{c1}

for various values of screen-grid voltage, as is done in Fig. 12-14. Although these curves represent static operating conditions, since the plate load resistor is zero ohms, they nevertheless are indicative of

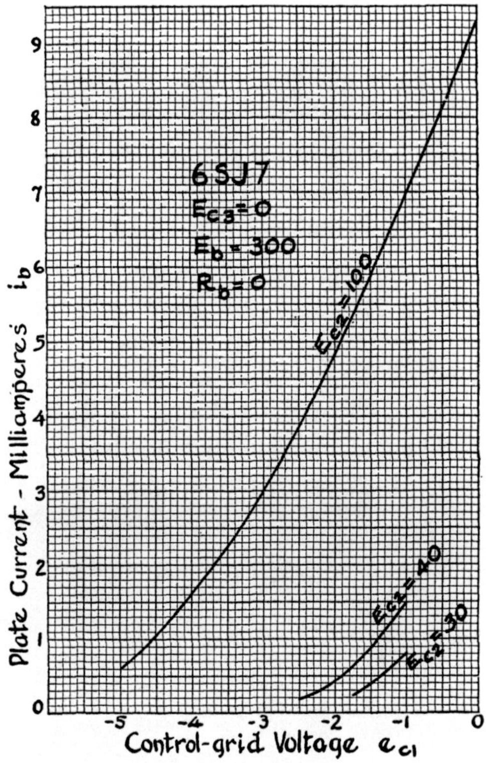

FIG. 12-14. This family of control-grid voltage—plate-current curves illustrate that only a very small value of input voltage swing is necessary; also note what a great difference a change in the screen-grid voltage E_{c2} makes in the plate current. This group of curves and those in Figs. 12-12 and 12-13 were made with no load in the plate circuit.

actual working conditions, insofar as they represent more nearly ideal operation than would be the case with a resistive load of high magnitude. Since, as we have already indicated, audio-frequency voltage amplifiers are generally operated at low screen-grid voltages, we observe how limited the permissible control-grid swing becomes. This partly accounts for the rule that the input signal to a voltage amplifier pentode should never exceed several tenths of a volt.

Now let us examine the effect of the screen-grid voltage when various values of resistance are inserted in the plate circuit of the tube. Accordingly, in Fig. 12-15 we have drawn several load lines representing plate loads of 0, 100,000, and 200,000 ohms, the control-grid being maintained constant at -1 volt and the plate-supply voltage being fixed at 300 volts. By picking off the various points of intersection of these load lines with the curves of constant screen-grid voltage, we may obtain the curves shown in Fig. 12-16. We must remember that the voltage drop e_{Rb} across the plate load resistor is subtracted from the available

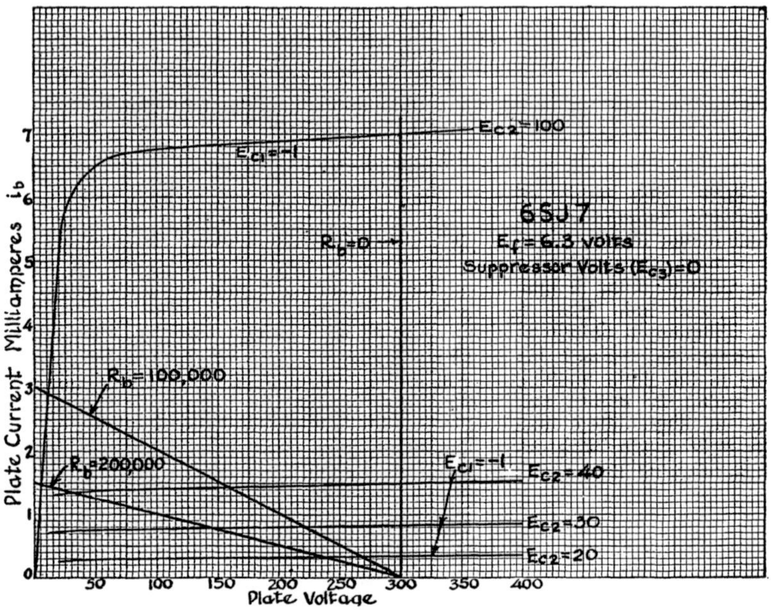

Fig. 12-15. Three load lines have been drawn upon the characteristics of the 6SJ7 that was shown in Fig. 12-12, these being for $R_b = 0$, 100,000 and 200,000 ohms as indicated.

plate-supply voltage E_b, the difference being that available for the plate voltage e_b. Since the greater the current through the load resistor the larger becomes the voltage drop across it and consequently the lower the available plate voltage, we might gain the impression, from what has previously been stated regarding the influence of screen-grid voltage upon plate current, that the use of large values of screen-grid voltage would reduce the available plate voltage to a very low value. That such a view is untenable when plate loads of the magnitude commonly employed are located in the plate circuit is quickly disclosed by an examination of Fig. 12-16. For, considering the curve illustrating the action when a load resistor of 200,000 ohms is employed, we see that any increase in the screen-grid voltage above about 45 volts results in a negligible increase in plate current. Consequently, the plate voltage, insofar as it is influenced by the screen-grid voltage, will remain unchanged for values of screen-grid voltage in excess of 45 volts.

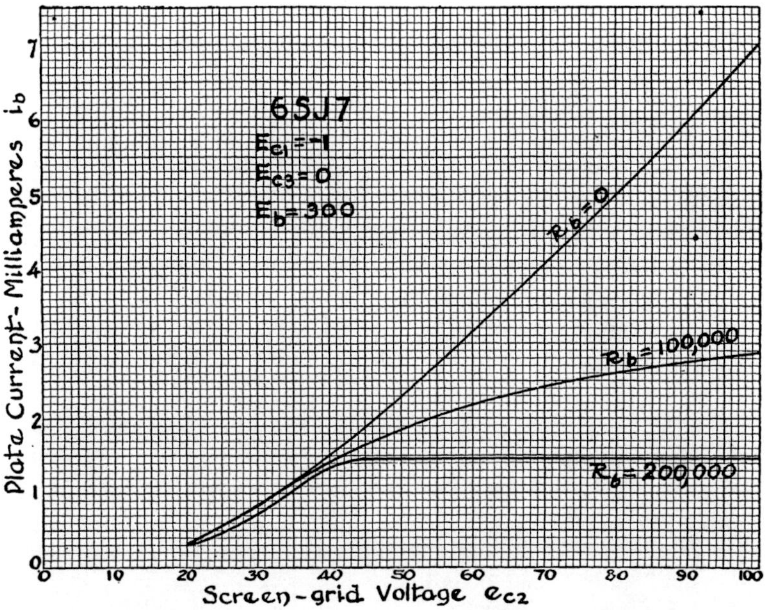

Fig. 12-16. These curves were derived from the load lines of Fig. 12-15 on the opposite page. In order to keep the d-c voltage drop across the load resistor at a reasonably low value, so that a fairly high plate voltage will be available, the screen-grid voltage must be maintained at a low value.

In contrast to the foregoing effect, further examination of Fig. 12-16 graphically shows one reason for the employment of low values of screen-grid voltage in actual voltage amplifiers employing pentodes. For, as we shall presently see, the operation of a pentode with too low a value of plate voltage results in an excessive curvature of the dynamic transfer characteristic, and this in turn causes excessive distortion. In order to operate the tube at a reasonably high plate potential it is necessary to limit the plate current flow in order that too great a d-c voltage drop is not produced across the load resistor. In the case of the 200,000-ohm load, as indicated in Fig. 12-16, this can be accomplished by decreasing the screen-grid voltage to values which are less than 45 volts. In this manner, the smaller plate current does not cause an excessive d-c voltage drop across the load resistor, in consequence of which a greater portion of the plate-supply voltage is made available to act as plate voltage. This partly accounts for the low values of screen-grid voltage which are encountered in actual audio-frequency pentode voltage amplifiers.

Plate Load Resistor

We have already seen that the larger the magnitude of the plate load resistor the greater is the voltage amplification obtainable from the tube. Several offsetting factors, however, prevent the full utilization of all of the voltage gain of which the tube is capable of producing by employing very large values of plate load resistor.

Let us examine the plate family of the 6SJ7 for a screen-grid voltage of 100 volts, as shown in Fig. 12-17. As we have indicated, this value

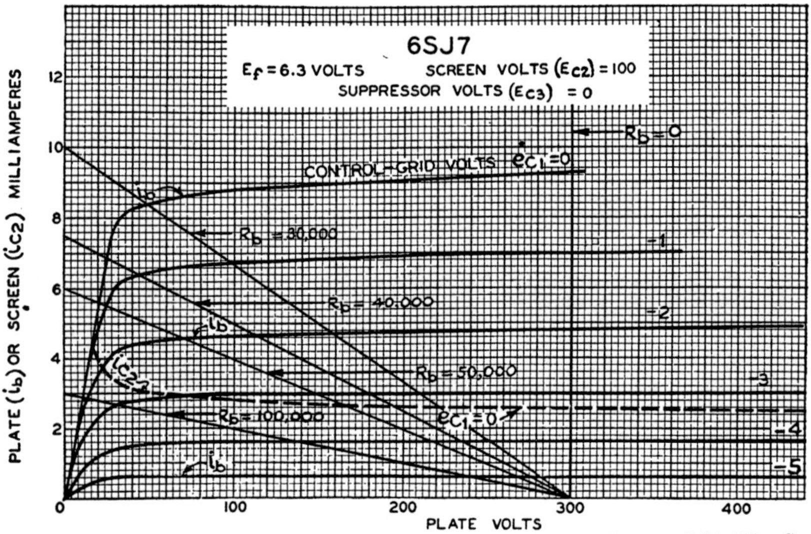

Courtesy RCA Mfg. Co.

FIG. 12-17. Load lines of 0, 30,000, 40,000, 50,000, and 100,000 ohms are shown upon the plate family of the 6SJ7 pentode.

of screen-grid voltage is much larger than that utilized in an audio-frequency pentode voltage-amplifier stage. Nevertheless, an examination of the pentode for this condition will enable us to obtain some general information which is applicable to the tube under actual operating conditions.

Load lines for loads of 0, 30,000, 40,000, 50,000, and 100,000 ohms

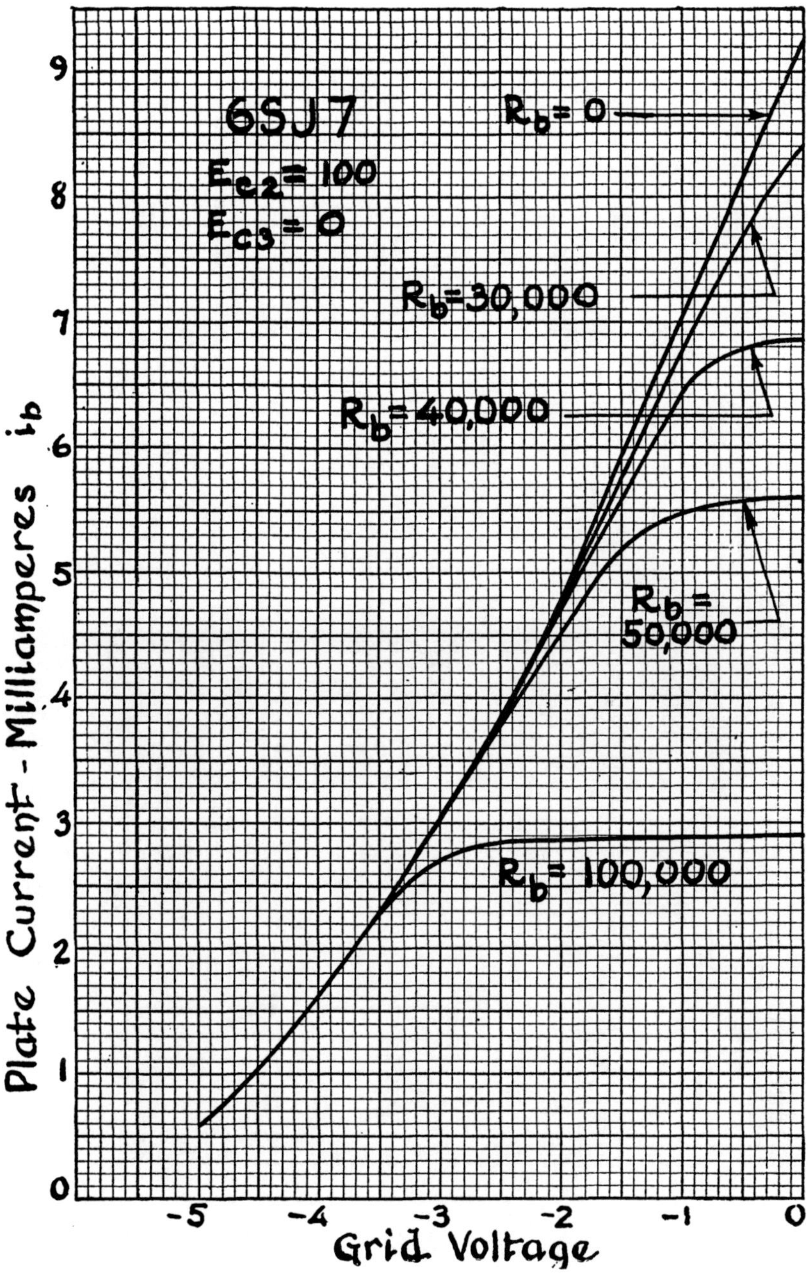

FIG. 12-18. The dynamic transfer characteristic curves of the 6SJ7 pentode derived from the load lines on Fig. 12-17.

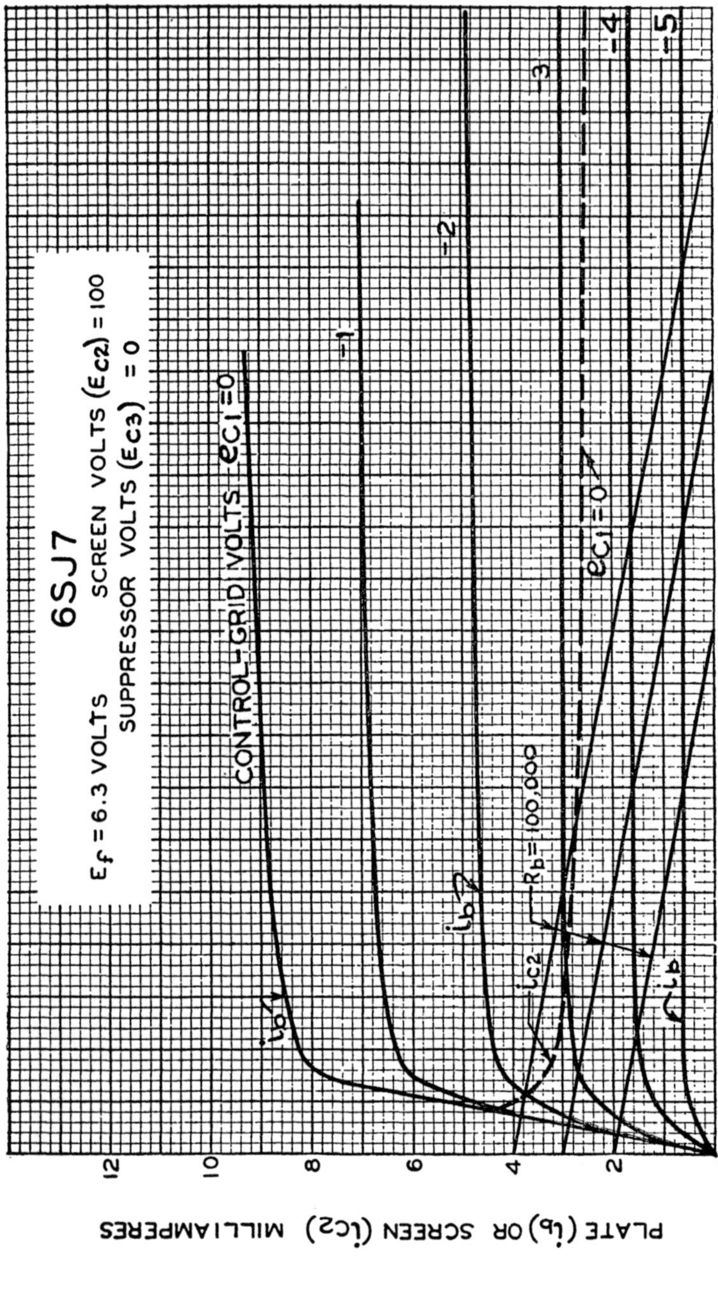

Courtesy RCA Mfg. Co.

FIG. 12-19. The three load lines, for the same 100,000-ohm load, are drawn for three different values of plate-supply voltage, 200, 300, and 400 volts.

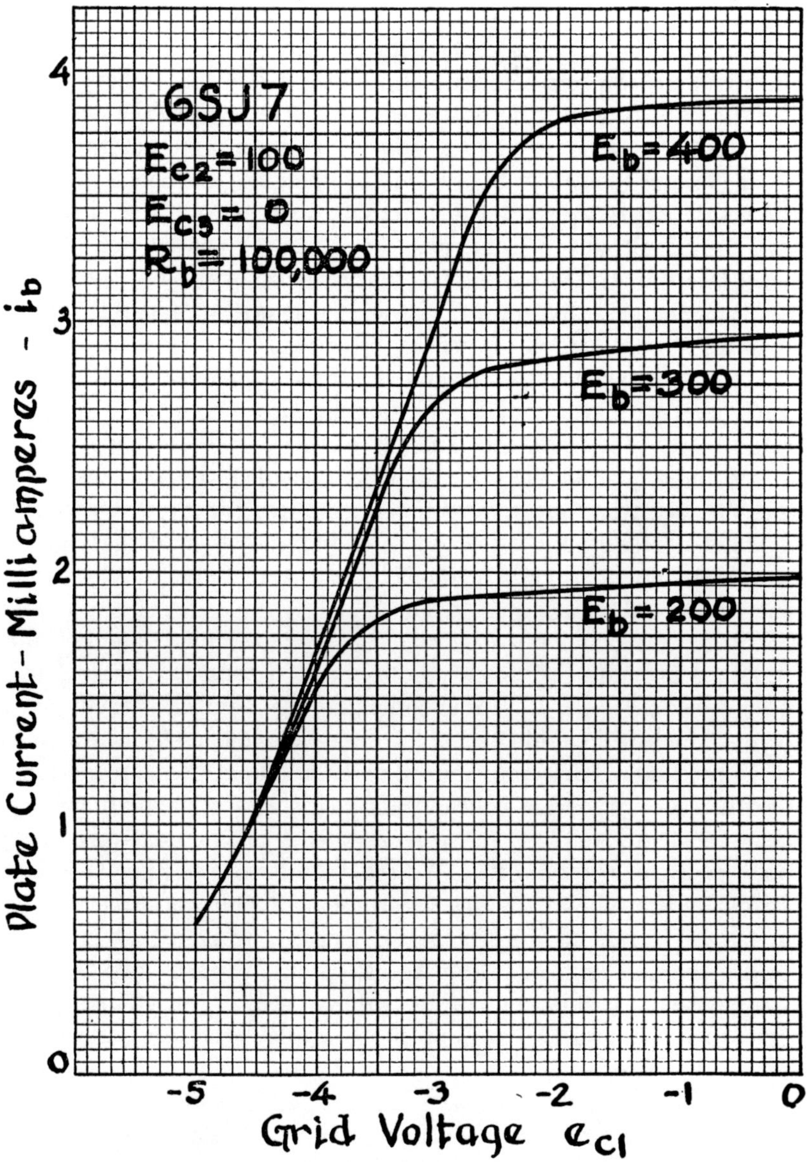

FIG. 12-20. These three dynamic transfer characteristics are derived from the load lines of Fig. 12-19 on the opposite page.

are drawn on Fig. 12-17, the plate-supply voltage being maintained constant at 300 volts. Taking the various points of intersection of the load lines with the pentode plate-family curves we are enabled to plot the dynamic transfer characteristic of the tube, as has been done in Fig. 12-18. From this figure we see how very different the action of a triode and pentode are. In the case of a triode the dynamic transfer characteristic (shown in Fig. 9-11) became more and more straight as the magnitude of the load was increased, provided that the control-grid was not permitted to become positive. With the pentode, on the other hand (see Fig. 12-18) we observe that increased values of load resistor result in increased curvature of the characteristic, and this, in turn, signifies the production of greater distortion.

Thus we become impaled upon the horns of a dilemma, for we are confronted with two undesirable choices: for a pentode the larger the load resistor the greater the gain (since $A_v = g_m R_b$) and the greater the distortion, while, on the other hand, the smaller the load resistor the less the distortion and unfortunately, the less the gain. Fortunately, it is not necessary for us to choose one or the other of these alternatives, for there are several methods by which a satisfactory compromise can be effected.

Let us first examine the effect of changing the plate-supply voltage while maintaining the magnitude of the load resistor constant, as has been done in Fig. 12-19. From this figure we may select the necessary points which will enable us to plot the dynamic transfer characteristic of Fig. 12-20. It becomes immediately apparent that an increase in the plate supply voltage, which means an increase in the plate voltage of the tube for a given load resistor, lengthens the comparatively straight portion of the characteristic. We see, therefore, the importance of not operating the tube at too low values of plate voltage. Referring back to our comments on the screen-grid voltage, we note that in order to secure the maximum plate voltage from any given fixed plate-supply source, it is desirable to use a low value of screen-grid voltage in order not to lose too great an amount of the available voltage supply across the load resistor.

In addition, we shall choose a value of load resistor which is neither excessively large or small. This compromise will enable us to retain a satisfactory voltage gain and at the same time the straight portion of our characteristic will be adequate, particularly when we apply only small input signals to the grid of the tube. Furthermore, this inability to utilize the maximum amplification possible is not too serious a matter when we consider that 10% of the maximum possible amplification available from a pentode is still considerably in excess of that obtainable from a triode which is operated to give 95% of its maximum amplification.

Typical Pentode Characteristics

We shall now consider a pentode voltage amplifier as it is customarily used as an audio-frequency voltage amplifier, namely, with a screen-grid voltage of about 40 volts. The plate family of a 6SJ7 for this condition is illustrated in Fig. 12-21. We have also drawn load lines for plate loads ranging from 100,000 to 500,000 ohms since this is the range of commonly used load resistors. In addition, we have drawn the load line for a plate load of zero ohms in order to serve as a reference line. Loads of this magnitude are of the order of from 1/10 to 1/3 the plate resistance of the tube, unlike a triode where proper operation required that loads which were several times the plate resistance be employed. From Fig. 12-21 we note that the larger the load resistor employed, the more negative must the control-grid bias be made. This latter point is even better seen on the dynamic transfer characteristic of Fig. 12-22, which was derived from Fig. 12-21.

In Fig. 12-22, portions of the dynamic transfer characteristics for the plate load $R_b = 250{,}000$ and $R_b = 500{,}000$ ohms are shown as dashed lines. These sections of the characteristic were not derived from the plate family of Fig. 12-21, as this information was not obtainable from the plate family for the scale of the drawing did not permit this. Instead, this information of the curves behavior was computed in a manner similar to that already described in Chapter 11 when we discussed what happened when the grid became positive. We there learned that the maximum possible plate current i_b depended upon the plate supply voltage E_b and the magnitude of the plate load resistor R_b, namely,

$$i_b = \frac{E_b}{R_b}$$

For example, when $E_b = 300$ volts and $R_b = 500{,}000$ ohms, we have

$$i_b = \frac{E_b}{R_b} = \frac{300}{500{,}000} = 0.0006 \text{ ampere} = 0.6 \text{ milliampere.}$$

In Fig. 12-22, we therefore know that for the $R_b = 500{,}000$ ohm curve, the plate current can never exceed 0.6 milliampere. Although it is not shown in Fig. 12-22, the dynamic transfer characteristics for loads of

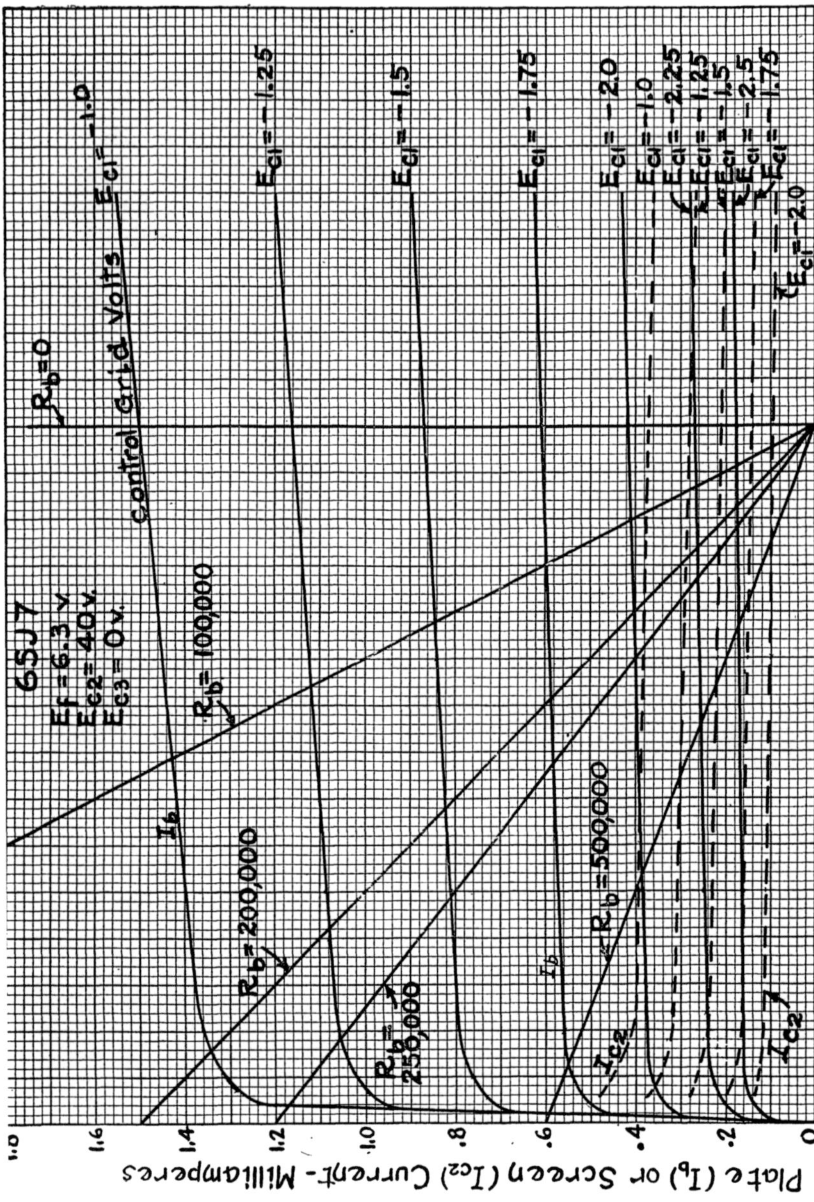

FIG. 12-21. The plate family for a constant screen-grid voltage of 40 volts. Load lines for various values of load resistance are shown, the plate-supply voltage in each case being 300 volts.

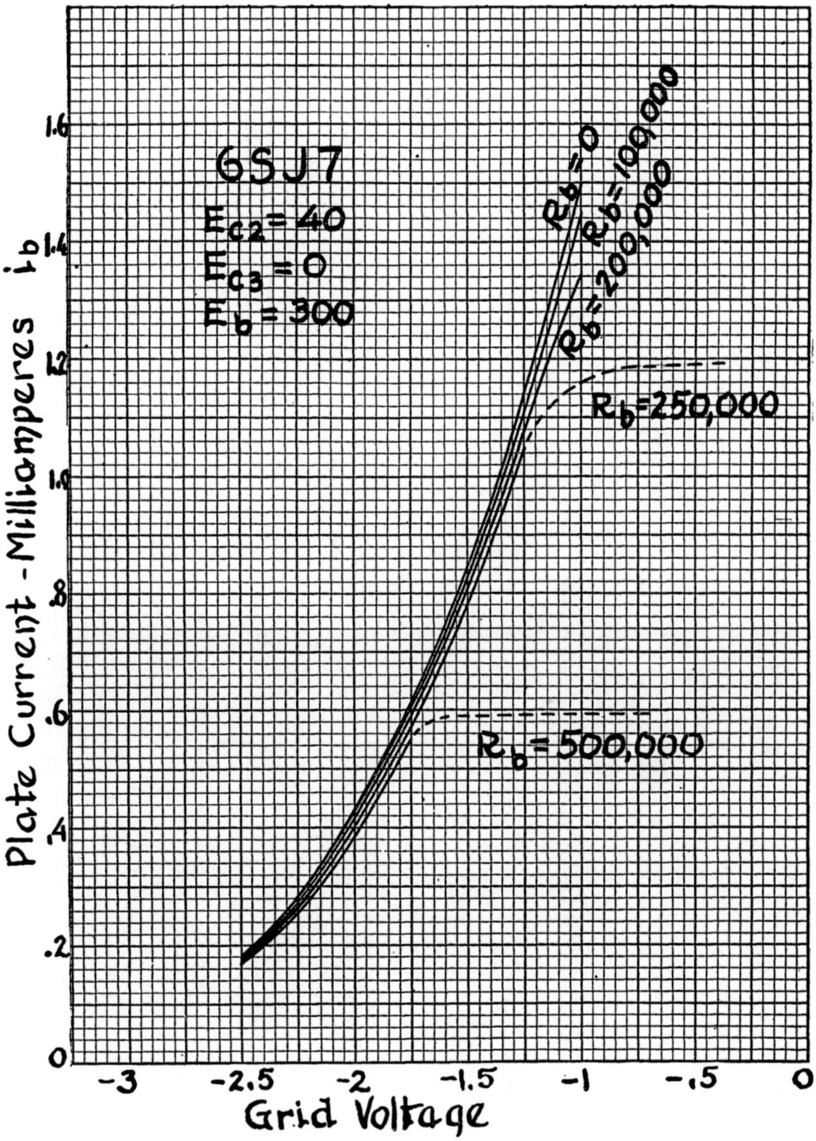

FIG. 12-22. The dynamic transfer characteristics were derived from the load lines on the opposite page.

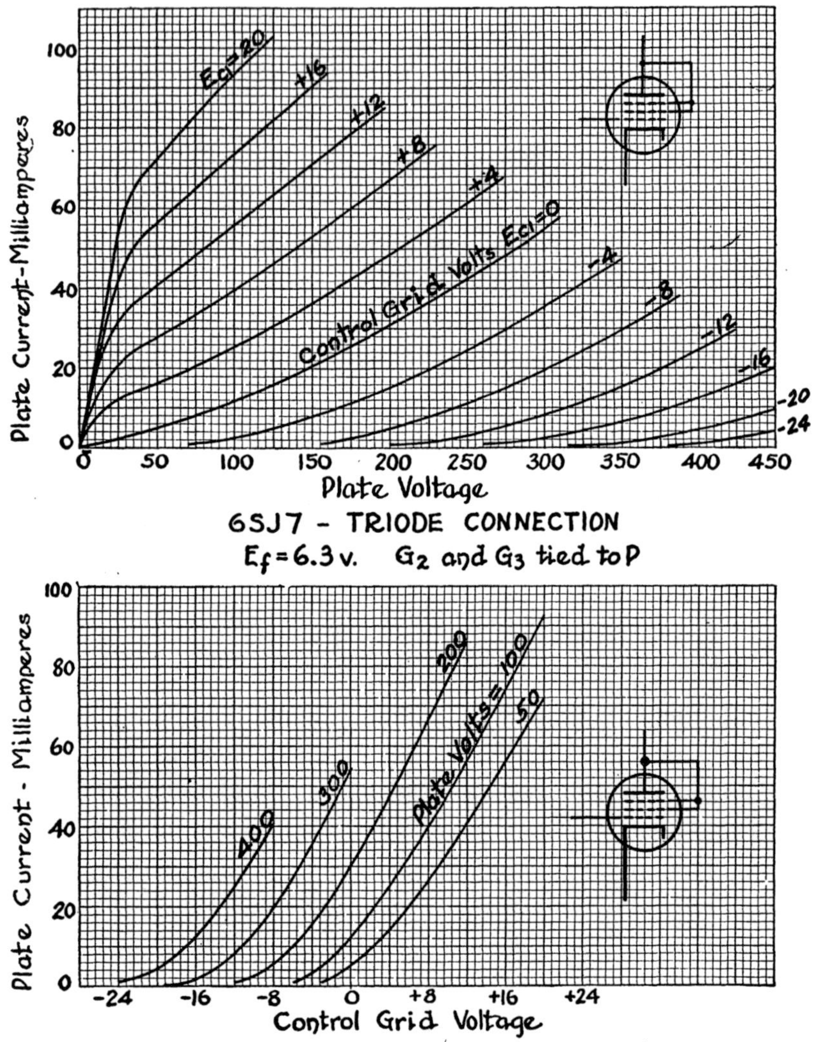

FIGS. 12-23 (above), 12-24 (below). The plate and grid families of a 6SJ7 connected as a triode with the screen and suppressor grids connected.

100,000 and 200,000 will also level off, but not as rapidly as was the case for $R_b = 500{,}000$.

From the shape of the curve representing $R_b = 500{,}000$ it can be seen that a square wave can be produced even though the grid is maintained negative. This differs from the triode of Fig. 11-13 and 11-14 where

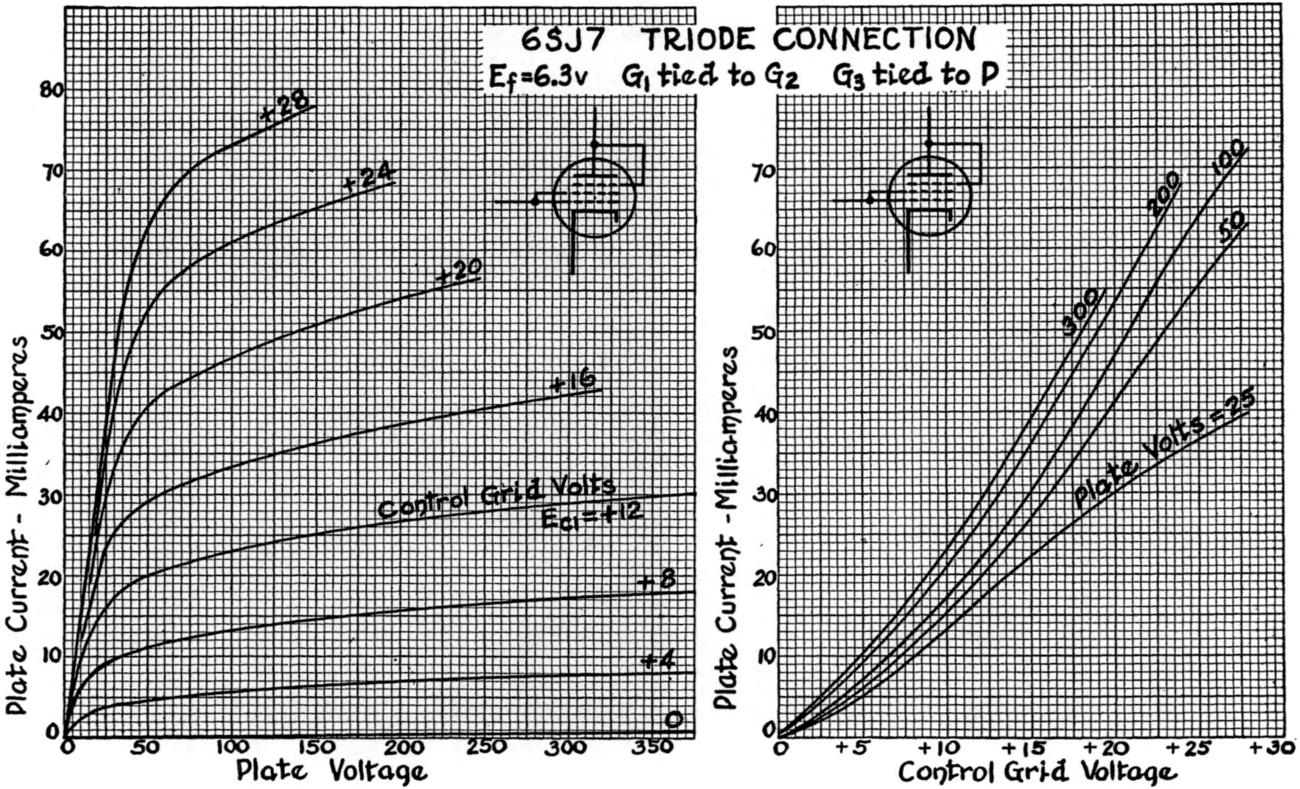

FIGS. 12-25 (left), 12-26 (right). The plate and grid families of a 6SJ7 connected as a triode when the control-grid is connected to the screen-grid and the suppressor-grid is connected to the plate.

it was necessary to drive the grid positive in order to produce a square wave.

Both Figs. 12-17 and 12-19 show a dashed curve for the screen-grid current i_{c2} when the screen-grid voltage $E_{c2} = 100$ volts and the control-grid voltage $e_{c1} = 0$ volts. It will be observed in Fig. 12-17 that the screen-grid current is approximately one-quarter that of the plate current, for the same value of control-grid voltage ($e_{c1} = 0$ in the case illustrated).

Pentodes Connected as Triodes

In view of the fact that the various grids of a pentode are available at the socket it becomes obvious that a pentode need not necessarily be connected in its usual manner. For, by tying the grids in various ways to the other elements, it is also possible to utilize the pentode as either a triode or as a tetrode. As tetrodes are but infrequently used in this country, we shall not concern ourselves with this type of connection.

For certain special applications it is sometimes found desirable to utilize a pentode as a triode. An example of this is illustrated in Fig. 12-23 and 12-24 where Fig. 12-23 represents the plate family and Fig. 12-24 the static grid-family characteristics of a 6SJ7 when grids 2 and 3 are tied to the plate. When used with this connection, the tube becomes a medium-mu triode and has a mu of about 18. Although the plate resistance of the tube is then approximately 6,000 ohms, a load resistor upwards of 40,000 ohms should be employed in order to limit the plate current to a reasonable value.

The type of connection just mentioned should not be confused with that shown in Figs. 12-25 and 12-26. Here the 6SJ7 is again utilized as a triode, but in this case grid 3 is connected to the plate while grid 2 is connected to grid 1. Only positive values of grid voltage are shown, since for negative values the plate current becomes very small. Except for very special applications this manner of connection is to be avoided, not only because of the danger to the tube due to the large plate current flow, but also because positive values of control-grid voltage will result in an excessive grid current flow which will result in distortion and excessive loading of the preceding circuit.

Chapter 13

THE CATHODE CIRCUIT

THUS FAR we have examined the grid circuit and the plate circuit of vacuum tubes in some detail. The cathode had always been considered, in our previous discussion, to be at ground potential, and a bias voltage for the grid had been obtained by means of a C-battery. We now want to consider the action of the cathode circuit in the vacuum tube.

It has already been mentioned that the grid circuit was comprised of all components that are connected between the grid and the cathode; similarly, the plate circuit is comprised of all components connected between the plate and the cathode. According to the definition just given, it is seen in Fig. 13-1 that the grid circuit consists of a source of signal voltage e_g and the cathode resistor R_k. Likewise the plate circuit consists of the load resistor R_b, the plate-supply voltage E_b, and the cathode resistor R_k. Since the cathode resistor R_k appears as a component of both the grid circuit and the plate circuit, we can say that the cathode resistor is common to both the grid and the plate circuits.

We have hitherto considered the grid circuit and the plate circuit as distinct from each other, and the only item that interrelated them was the vacuum tube. You can see from Fig. 13-1 that the cathode resistor R_k being common to both circuits, must have a decided effect on the operation of the tube and that this effect will prove of considerable interest.

Self-Bias

First let us consider the cathode resistor in relation to the plate circuit. As can be seen in Fig. 13-1, electrons are emitted by the cathode, go to the plate, flow through the load resistor, and into the positive

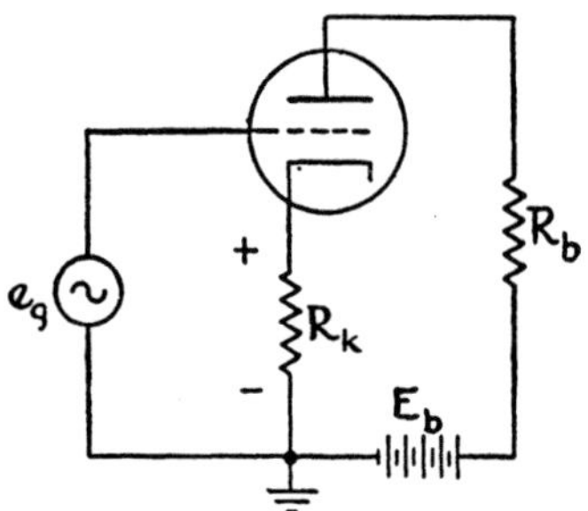

FIG. 13-1. The plate current returning to the cathode passes through the cathode resistor R_k and causes a voltage drop equal to the product of the current and the value of the resistance.

terminal of the plate-supply battery E_b. A corresponding number of electrons leave the negative terminal of the plate-supply battery, flow through the cathode resistor R_k, and return to the cathode. In consequence of this flow of plate current a voltage e_k is developed across this cathode resistor, the magnitude of this voltage depending upon the magnitude of the plate current i_b and the value of the cathode resistor R_k. More specifically, this is $e_k = i_b R_k$. For example, if the plate current is 4.0 milliamperes and the cathode resistor is 2000 ohms, the voltage drop across the cathode resistor is

$$e_k = 0.004 \times 2000 = 8.0 \text{ volts.}$$

With regard to the polarity of the voltage across the cathode resistor, the direction of the plate current through this resistor is such as to make the cathode end of the resistor positive with respect to the end furthest from the cathode (the grounded end). Or, otherwise expressed, the grounded end of the cathode resistor is negative with respect to the cathode end. Since the grid return is connected to the grounded end of the cathode resistor, it is seen that the grid is negative in potential with respect to the cathode. In other words, the grid has a negative bias as a result of the voltage drop across the cathode resistor. Since this grid bias is obtained through the plate current of the tube itself, it is known as *self-bias*.

As long as the plate current is constant in magnitude, the voltage drop across the cathode resistor will remain constant. However, since the cathode resistor is common to both the grid and plate circuits, the application of a signal voltage to the grid of the tube will result in the appearance of a signal voltage not only across the load resistor R_b but also across the cathode resistor R_k. In certain applications, which will be discussed later, the signal voltage appearing across the cathode resistor serves a definite function. However, when the cathode resistor is employed in order to obtain self-bias operation, these voltage fluctuations are undesirable, since the grid bias should remain constant in value. This self-bias voltage can be maintained constant by means of a bypass condenser across the cathode resistor, but before describing how this is accomplished, let us compare the previously illustrated, battery obtained, fixed-bias and self-bias operation when no bypass condenser is employed.

Effect of Unbypassed Cathode Resistor

In the left column of Fig. 13-2 there is shown the fixed-bias operation of a triode, while the other column illustrates self-bias operation. We shall start with Fig. 13-2(C), which shows a fixed-bias of 5 volts and a condition of zero volts input signal. The voltage that is effective on the grid is then −5 volts. Fig. 13-2(D) also shows a zero voltage input signal. Here, however, the magnitude of the plate current and cathode resistor is such as to result in a 5-volt drop across the cathode resistor. (It should be remarked that the voltages developed across the cathode resistor as here given are chosen for ease of understanding the operation, whereas the method for determining the exact values is described later.) Again the voltage that is effective on the grid is −5 volts. Thus far, with regard to the voltage effective on the grid, there is no difference between fixed-bias and self-bias operation.

Now let us apply a 2-volt peak input signal to the fixed-bias circuit of Fig. 13-2(A) and having a polarity which makes the grid less negative. The 2-volt input signal and the 5-volt fixed bias buck each other, so that the voltage that is effective at the grid is −3 volts. This same 2-volt peak input signal is also applied to the self-bias circuit, as shown in Fig. 13-2(B). Since the polarity of this input signal is such as to make the grid less negative, more plate current voltage is developed across the cathode resistor. This 6 volts across the cathode resistor

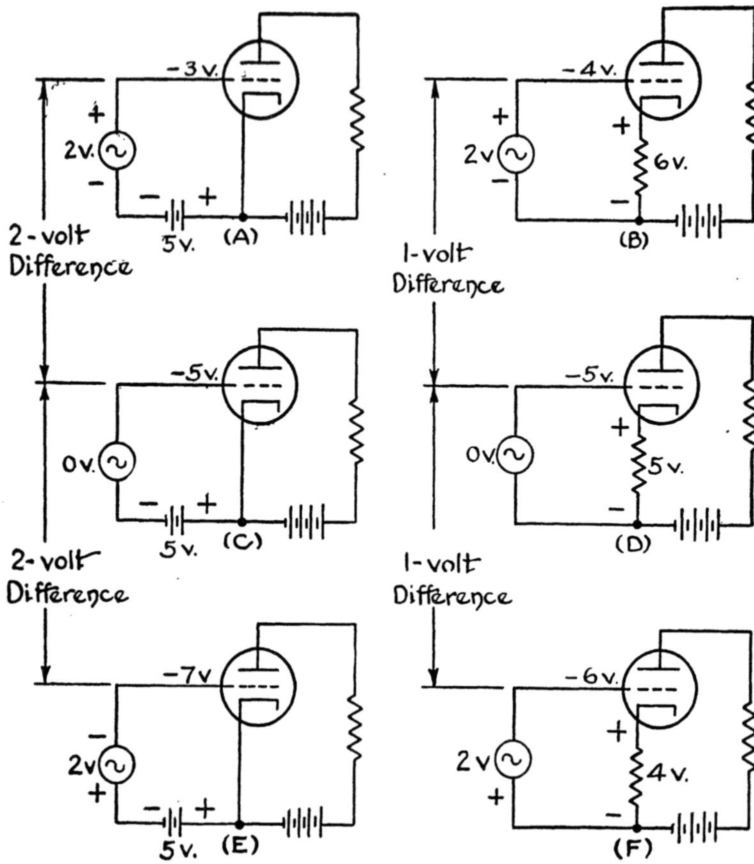

Fig. 13-2. The three diagrams on the left show different conditions of fixed-bias operation and those on the right illustrate corresponding conditions of self-bias operation.

bucks the 2-volt input signal, so that the voltage effective on the grid is −4 volts.

It is therefore seen that starting from the quiescent bias voltage of −5 volts, shown in Fig. 13-2(C), the application of a 2-volt signal voltage Fig. 13-2(A) to the fixed-bias circuit results in −3 volts that is effective on the grid. In other words, the application of a 2-volt signal voltage results in a 2-volt change in grid voltage. On the other hand, starting with the quiescent value of −5 volts in Fig. 13-2(D), the application of a 2-volt signal Fig. 13-2(B) results in only a 1-volt change in grid voltage.

Let us start again with the quiescent conditions of Figs. 13-2(C) and 13-2(D). A 2-volt peak signal is applied, as shown in Fig. 13-2(E), so as to make the grid more negative. The fixed-bias of 5 volts and the signal voltage of 2 volts are series aiding, so that −7 volts are effective at the grid. When this same 2-volt signal is applied to the self-bias circuit, as illustrated in Fig. 13-2(F), and since the grid is made more negative, the plate current becomes less, in consequence of which the voltage drop across the cathode resistor becomes, less, namely, 4 volts. The 4-volt drop across the cathode resistor and the 2-volt signal voltage are series aiding, so that the voltage effective at the grid is −6 volts. Again it is observed that the application of a 2-volt input signal voltage applied to the fixed-bias circuit results in a 2-volt change at the grid, whereas a 2-volt input signal voltage applied to the self-bias circuit results in a 1-volt change of voltage at the grid.

We are now able to draw some important conclusions. For a circuit employing fixed-bias, the change in grid voltage *is equal to* the change in the input signal voltage. On the other hand, for a self-bias circuit employing an unbypassed cathode resistor, the change in voltage effective at the grid is always *less than* the applied input signal voltage. Since the entire signal voltage is not effective in producing grid voltage variations, this is equivalent to a smaller amount of amplification from the tube, which is usually undesirable. In passing, it may be mentioned that this reduced amplification, because of certain compensation features, is not always a disadvantage.

Cathode Resistor Bypass Condenser

Throughout this book we have deliberately omitted any mention of circuital components, such as condensers, coils, transformers, etc. Our intent has been to limit our discussion to certain fundamental properties of the vacuum tube itself. At this point, however, we shall make an exception in order to discuss the action that occurs when a condenser is connected across the cathode resistor. This discussion will be very brief, so if some of the terms employed are not entirely clear, remember that immediately following this description we shall again return to the vacuum tube itself divorced of such circuital components.

Let us suppose that a condenser c_k is connected across the cathode resistor R_k, as shown in Fig. 13-3. It will be further assumed that the reactance of this condenser is low in comparison with the resistance of the cathode resistor. The alternating component of plate current will then flow through the condenser, and only the d-c quiescent value of plate current will flow through the resistor. This condenser is called a *bypass condenser* because it causes the a-c component to be detoured away from the cathode resistor.

Since only the d-c portion of the plate current flows through the resistor, the cathode-to-ground voltage remains constant. Accordingly, the cathode resistor and bypass condenser combination, insofar as the grid bias is concerned, is equivalent to the battery as employed in fixed bias.

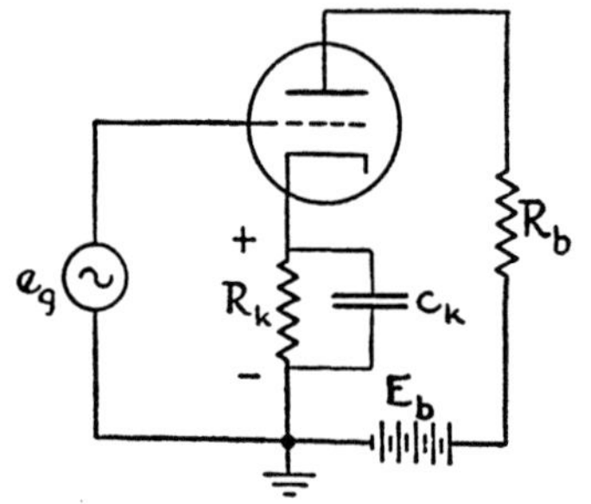

FIG. 13-3. The bypass condenser C_k in parallel with the cathode resistor bypasses the a-c component of the plate current and so only the d-c component flows through the resistor R_k. Thus the cathode-to-ground voltage remains constant.

Self-bias, as obtained from a cathode resistor and bypass condenser combination, is much more frequently used in actual practice than is fixed bias employing a battery. In our discussion of vacuum tube operation, we shall continue to show a fixed bias battery, where this is needed, since it offers the advantage of an easier visualization of phenomena.

Determining Value of Cathode Resistor

Perhaps the simplest method of determining the correct value of the cathode resistor is a graphical one which employs the plate family of the tube. Fig. 13-4 shows the plate family of a 6J5 triode with the two plate load lines $R_b = 25{,}000$ ohms and $R_b = 50{,}000$ ohms when a plate-supply voltage $E_b = 350$ volts is employed. Several other lines which spread out from the origin, are also drawn on the plate family, and it is the construction of these latter lines which is of interest to us.

We have already seen that the voltage drop e_k developed across the cathode resistor R_k when a plate current i_b flows, is given by the expression

$$e_k = i_b \; R_k$$

Solving this for the plate current, we obtain

$$i_b = \frac{e_k}{R_k}$$

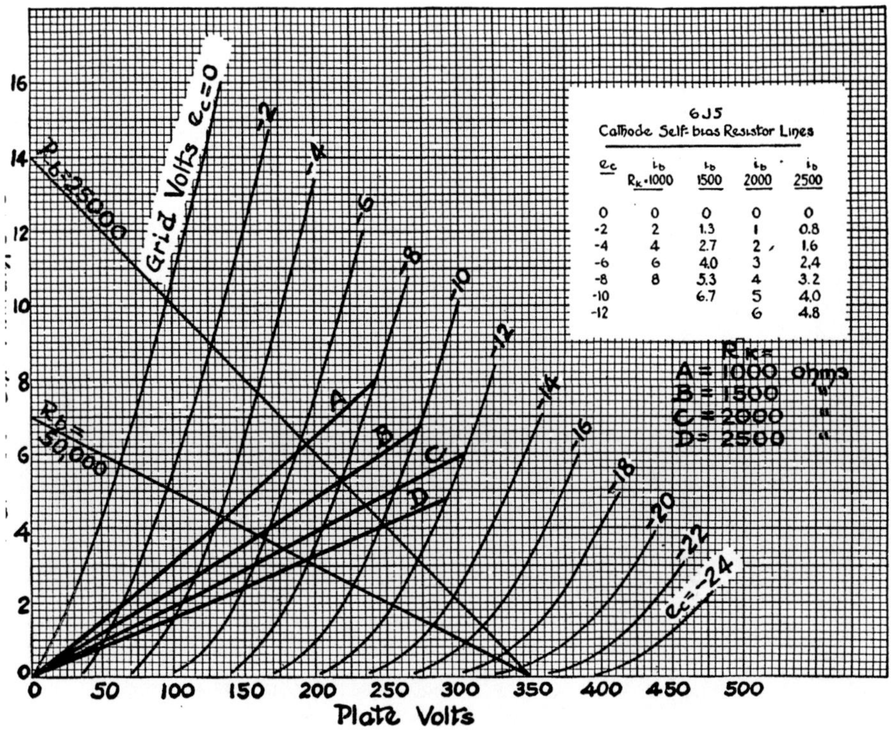

G. 13-4. The lines starting at the origin of this plate family of characteristic curves of the 5 are called the cathode self-bias resistor lines and their construction is described in the companying text.

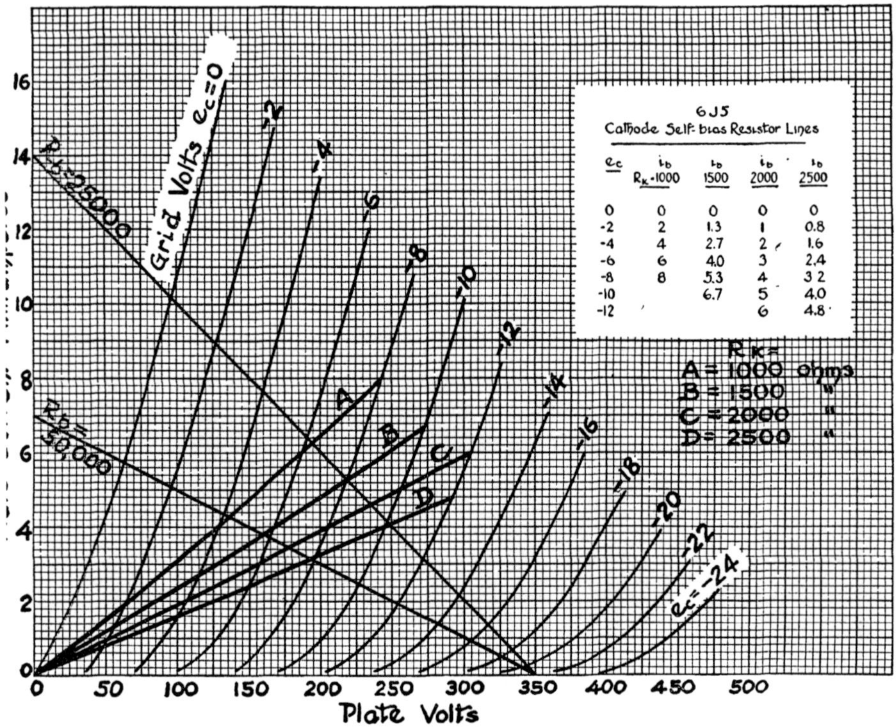

IG. 13-4. The lines starting at the origin of this plate family of characteristic curves of the 5 are called the cathode self-bias resistor lines and their construction is described in the companying text.

For any fixed value of R_k, each time e_k is changed, there is obtained a different value of i_b. Now it is easy enough to read the value of plate current i_b from Fig. 13-4, but we need some definite corresponding value for e_k. This can be secured by choosing values of e_k which correspond to various e_c curves on the figure; that is, we let $e_k = e_c$. By substituting in our last equation, we then obtain

$$i_b = \frac{e_c}{R_k}$$

For any fixed value of the cathode resistor R_k, we can now compute the magnitude of the plate current i_b for any grid voltage e_c. For example, suppose that we choose a cathode resistor of 1500 ohms. Then for $e_c = 0$ by Ohm's Law, we obtain

$$i_b = \frac{e_c}{R_k} = \frac{0}{1500} = 0$$

The intersection of the $e_c = 0$ curve of Fig. 13-4 and the zero-current axis is then one point of the self-bias resistor curve, and this point corresponds to the origin.

Still retaining the cathode resistor $R_k = 1500$ ohms, we now compute the plate current for $e_c = -2$, disregarding the minus sign in our computation. We then obtain

$$i_b = \frac{e_c}{R_k} = \frac{2}{1500} = 0.0013 \text{ ampere} = 1.3 \text{ milliamperes}$$

The intersection of the $e_c = -2$ curve of the plate family with the 1.3-milliampere horizontal line then determines another point on the desired self-bias curve.

Similarly, and still using the cathode resistor of 1500 ohms, for $e_c = -4$ we get

$$i_b = \frac{e_c}{R_k} = \frac{4}{1500} = 0.0027 \text{ ampere} = 2.7 \text{ milliamperes}$$

The intersection of the $e_c = -4$ curve with the 2.7-milliampere line results in still another point in the curve.

Other points are obtained in a similar manner for the $R_k = 1500$ ohms value of cathode resistor. These results, as well as those for other values of the cathode resistor, are tabulated on Fig. 13-4 for your convenience. A smooth curve is then drawn connecting the various points thus obtained for any fixed value of cathode resistor. It will be observed that each of the curves thus obtained, which shall be called a *self-bias resistor curve*, is almost, but not quite, a straight line.

We are now in a position to utilize the information that we have gained. *For any given fixed resistance, the intersection of its corresponding self-bias resistance curve with a load line (corresponding to the plate-load resistor) determines the quiescent operating point of the tube.* For example, the $R_k = 1500$-ohm self-bias resistance curve intersects the $R_b = 25{,}000$-ohm load line very close to the $e_c = -8$-volt curve of the plate family, so that the quiescent grid voltage (grid bias) is -8 volts.

On the other hand, if we had chosen a cathode resistor of 2000 ohms and still used a plate load of 25,000 ohms, the intersection of the $R_k = 2000$-ohm self-bias resistance curve with the $R_b = 25{,}000$-ohm load line falls approximately midway between the $e_c = -8$ and $e_c = -10$-volt curves of the plate family. Consequently, the grid bias of the tube would be -9 volts.

The reason for drawing a number of self-bias resistor curves is to enable us to choose a value of cathode resistor which will not only satisfy our requirements for the quiescent operating condition of the tube, but which will also enable us to choose a value of resistance that is commercially available. In addition by drawing a series of such self-bias resistor curves in conjunction with several load lines, the variation in the value of the cathode self-bias resistance can be determined for each value of the plate load resistor. For example, we have already determined that if we wanted to operate the tube with a grid bias of -8 volts when employing a 25,000-ohm plate load and a plate supply voltage of 350 volts, then we would require a cathode self-bias resistance of 1500 ohms. Now suppose that we wished to retain the grid bias at -8 volts and the plate-supply voltage of 350 volts, but wished to employ a plate load resistor of 50,000 ohms. Examination of Fig. 13-4 reveals that the 2500-ohm self-bias resistance curve intersects the 50,000-ohm load line very close to the $e_c = -8$ curve of the plate family. Consequently a 2500-ohm cathode self-bias resistor would be employed.

The preceding analysis gives a very close approximation to the exact value of cathode self-bias resistance that should be used. The fact that the bypassed self-bias resistor does have a definite value of impedance, and that the actual plate voltage differs from the value assumed in this discussion by the amount developed across the self-bias resistance, will not ordinarily affect the correctness of our results. This point is brought out to emphasize once again that it is usually desirable to simplify the complexities of natural phenomena by making certain approximations in order to obtain practical results.

Plate Family For Tube With Unbypassed Cathode Resistor

We have already seen in a general way that an unbypassed cathode resistor affects the characteristics of a tube. We shall now determine more accurately just what happens to the characteristics of the tube.

Fig. **13-5** shows the circuit of a triode incorporating an unbypassed cathode resistor. No load resistor is shown in the plate circuit, for our aim is to derive the *static* plate family for this new condition. The signal voltage applied between grid and ground is designated e'_c in order to indicate that it is equivalent in nature to the actual instantaneous grid voltage e_c, but that e'_c is the instantaneous *grid-to-ground* voltage, whereas e_c represents the instantaneous grid-to-cathode voltage. Similarly, e'_b signifies the instantaneous *plate-to-ground* voltage, as compared with e_b which is the instantaneous plate-to-cathode voltage.

FIG. 13-5. A triode circuit having an unbypassed cathode resistor, but no load resistor in the plate circuit. Note the various voltage designations.

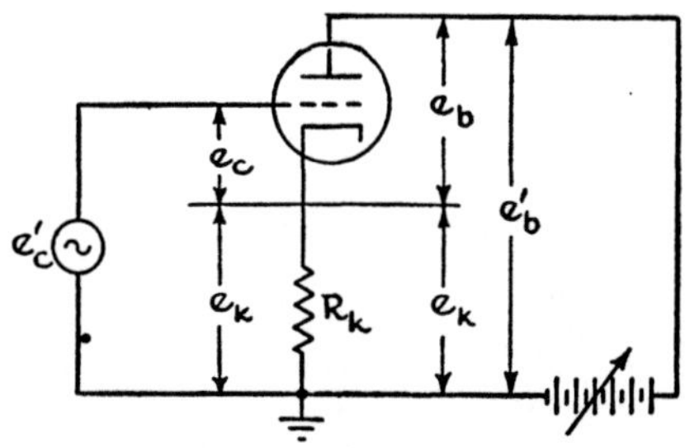

From Fig. 13-5 we observe that the grid-to-ground voltage e'_c is equal to the sum of the grid-to-cathode voltage e_c and the voltage drop e_k across the cathode resistor, namely,

$$e'_c = e_c + e_k$$

Transposing terms, we obtain

$$e_c = e'_c - e_k$$

Also

$$e_k = e'_c - e_c$$

We also know that the voltage drop e_k across the cathode resistor is equal to the product of the plate current i_b and the cathode resistance R_k, namely,

$$e_k = i_b \, R_k$$

Consequently,

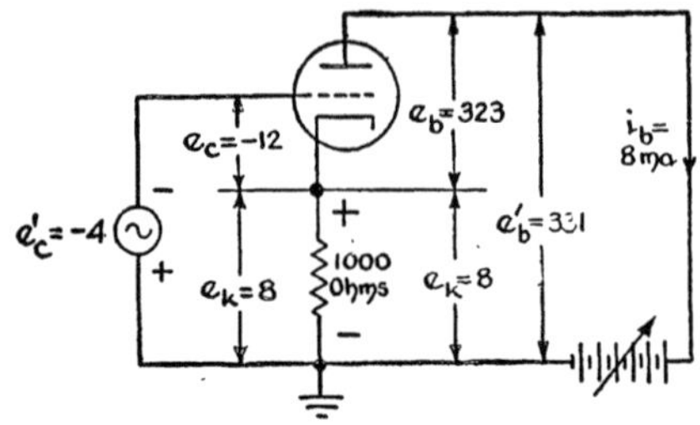

FIG. 13-6. Actual voltage and current values are indicated at the various points in this diagram that establish one point on the e'$_c$=—4 (dashed) curve in Fig. 13-7.

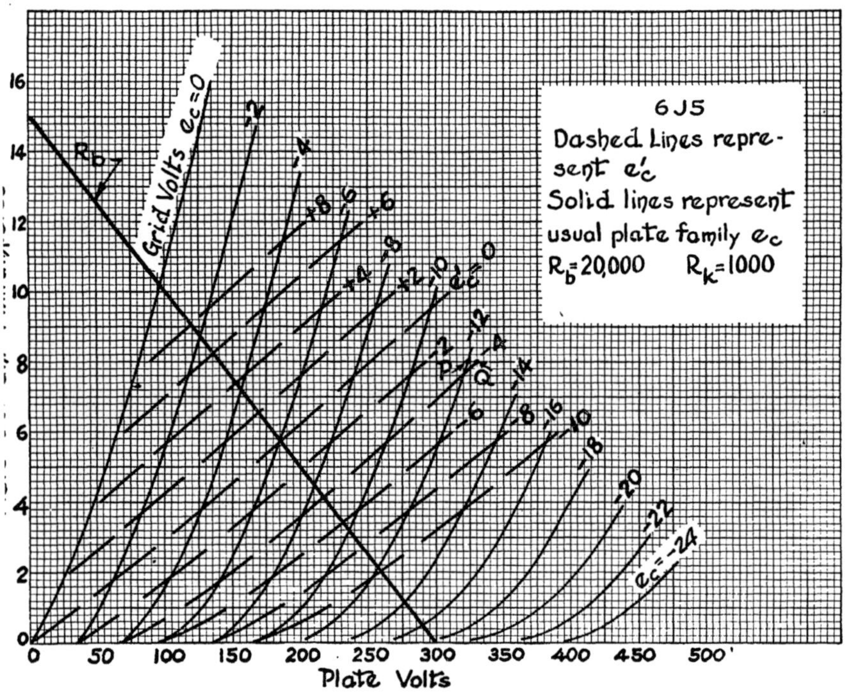

G. 13-7. The new plate family (dashed lines) when a cathode resistor of 1000 ohms is used, drawn upon the usual plate family (solid lines) of a 6J5 triode. The load line is for a plate ıd resistance of 20,000 ohms. Refer to Table A (page 331) for values of new plate family.

$$i_b = \frac{e_k}{R_k}$$

From Fig. 13-5 we further see that the plate-to-ground voltage e'_b is equal to the sum of the plate-to-cathode voltage e_b and the voltage drop e_k across the cathode resistor, namely

$$e'_b = e_b + e_k$$

With this information, we are now in a position to determine the new plate family.

Let us now refer to Figs. 13-6 and 13-7. Actual values for obtaining one particular point of the $e'_c = -4$ curve are shown on Fig. 13-6, while Fig. 13-7 shows the new plate family (dashed lines) superimposed on the usual plate family (solid lines). We shall now explain the procedure in a series of three major divisions, the third of these being further divided into additional steps.

1. A definite value of cathode resistor R_k is chosen. This value remains constant in obtaining *all* of the curves for the new plate family. (For example, we have chosen $R_k = 1000$ ohms.)

2. Select a value of e'_c for which it is desired to determine one curve of the new plate family. (For example, let $e'_c = -4$ volts, as indicated on Fig. 13-6). This value of e'_c remains constant until enough points have been obtained to draw this new member of the plate family.

3. Three variable factors are shown on the plate family: plate current i_b, plate voltage e_b, and grid voltage e_c. Anticipating our end result in advance, we find that the plate voltage will be a factor that will have to be determined from more data than is presently available. Accordingly, we have only the plate current i_b and the grid voltage e_c left, and it is from these that we shall determine our auxiliary points.

3a. First we choose some value of e_c that corresponds to one of the curves of the plate family. (In our illustration of Fig. 13-6 and Fig. 13-7, we chose $e_c = -12$.)

3b. Now we have already seen that

$$e_k = e'_c - e_c$$

We have previously chosen e'_c in Step 2 (for example, $e'_c = -4$) and we have chosen e_c in Step 3a (for example, $e_c = -12$). Consequently, the value of e_k is determined. (For example,

$$e_k = e'_c - e_c = -4 - (-12) = +8)$$

3c. We have also learned previously that

$$e_c = e'_c - e_k$$

and

$$i_b = \frac{e_k}{R_k}$$

We can therefore select a value of plate current i_b such that the instantaneous grid voltage e_c corresponds to a curve of the plate family. (In our example,

$$e_c = e'_c - e_k = -4 - 8 = -12$$

and therefore since R_k is fixed,

$$i_b = \frac{e_k}{R_k} = \frac{8}{1000} = 0.008 \text{ ampere} = 8.0 \text{ milliamperes.)}$$

3d. In Step 3a we selected a value of e_c (for instance, $e_c = -12$) and in Step 3c we obtained a corresponding value for i_b (for example, 8.0 milliamperes). These two values determine a point on the original plate family (solid lines). (For example, the intersection of the $i_b = 8.0$-milliampere line with the $e_c = -12$ curve gives us point P in Fig. 13-7).

3e. Find the corresponding value of plate voltage e_b for point P from the characteristics of Fig. 13-7. (In our example, where $i_b = 8.0$ milliamperes and $e_c = -12$ volts, then $e_b = 323$ volts.) This is the plate-to-cathode voltage required to produce the required voltage drop across the cathode resistance.

3f. The plate-to-ground voltage e'_b differs from the plate-to-cathode voltage e_b by the cathode-to-ground voltage e_k, as we previously saw. That is

$$e'_b = e_b + e_k$$

This plate-to-ground voltage e'_b and the plate current i_b previously obtained determine one point on the e'_c curve. (In our example, $e'_b = e_b + e_k = 323 + 8 = 331$ volts. Consequently, the intersection of the 331-volt and 8.0-milliampere lines determine point Q on Fig. 13-7.)

4. The various sections of Step 3 are repeated until sufficient points are obtained to draw one particular e'_c line of the new plate family. (In our example, the dashed line $e'_c = -4$ of Fig. 13-7.)

5. A new value of e'_c is chosen as indicated in Step 2, and the various sections of Step 3 are repeated until another line of the new plate family is secured. This process is repeated until as many lines of the new plate family are obtained as desired. For your convenience, Table A is given which shows all of the points that are necessary in obtaining the new plate family (dashed lines) of Fig. 13-7.

TABLE A

6J5 Tube $R_k = 1000$

e'_c	e_c	e_k	$i_b = \frac{e_k}{R_k}$	e_b	e'_b
0	0	0	0	0	0
	—2	2	2	65	67
	—4	4	4	129	133
	—6	6	6	187	193
	—8	8	8	243	251
	—10	10	10	300	310
—2	—2	0	0	34	34
	—4	2	2	107	109
	—6	4	4	168	172
	—8	6	6	225	231
	—10	8	8	284	292
—4	—4	0	0	67	67
	—6	2	2	148	150
	—8	4	4	206	210
	—10	6	6	266	272
	—12	8	8	323	331
—6	—6	0	0	95	95
	—8	2	2	182	184
	—10	4	4	246	250
	—12	6	6	305	311
—8	—8	0	0	135	135
	—10	2	2	220	222
	—12	4	4	283	287
	—14	6	6	347	353
—10	—10	0	0	165	165
	—12	2	2	256	258
	—14	4	4	323	327
	—16	6	6	384	390
+2	0	2	2	27	29
	—2	4	4	86	90
	—4	6	6	147	153
	—6	8	8	204	212
	—8	10	10	259	269
+4	0	4	4	48	52
	—2	6	6	104	110
	—4	8	8	162	170
	—6	10	10	220	230
+6	0	6	6	65	71
	—2	8	8	121	129
	—4	10	10	177	187
	—6	12	12	234	246
+8	0	8	8	81	89
	—2	10	10	136	146
	—4	12	12	192	204

From Fig. 13-7 it is immediately apparent that the plate resistance of a tube having an unbypassed cathode resistor is higher than the plate resistance of the tube when this unbypassed cathode resistor is absent. This is evident from the fact that the steepness of the new plate family is less than that of the original plate family. If your memory needs refreshing on this point, refer to Fig. 9-3.

It will be noted that the new plate family shows grid-to-ground curves for positive values as well as the more customary negative values. This is because a tube having an unbypassed cathode resistor is sometimes employed in this manner, since the self-adjusting voltage developed across the cathode resistor enables positive input signals to be applied to the tube, without the occurrence of undesirable grid current.

A load line may be drawn on the new plate family and a dynamic transfer characteristic obtained. This has been done on Fig. 13-7 for a load resistance $R_b = 20{,}000$ ohms and a plate-supply voltage $E_b = 300$ volts. In order to obtain a more graphic picture of the effect of the presence or absence of an unbypassed cathode resistance, points along the load line have been tabulated (see tables on Fig. 13-8) for both the original and the new plate family. From this tabulation, the dynamic transfer characteristics of Fig. 13-8 have been drawn.

When examining the dynamic transfer characteristics of Fig. 13-8, it should be remembered that e_c represents the instantaneous total grid-to-cathode voltage, while e'_c represents the instantaneous total grid-to-ground input voltage.

In comparing the two characteristics, it is interesting to note that the dynamic transfer characteristic for the tube employing an unbypassed cathode resistor ($R_k = 1000$ ohms) is much more linear than is the case when this resistor is not employed ($R_k = 0$ ohms). Consequently, distortion will be less when the unbypassed cathode resistor is employed as compared to the case when it is absent.

The Cathode Follower

One of the more interesting applications of a tube employing an unbypassed cathode resistor is when the output signal from the tube is taken from this resistor. A circuit, as is shown in Fig. 13-9, wherein the input signal voltage e'_c is applied between the grid and ground and the output signal voltage is obtained between cathode and ground, is called a *cathode follower*. This circuit derives its name from the fact that when the input signal is in a positive-going direction, the output

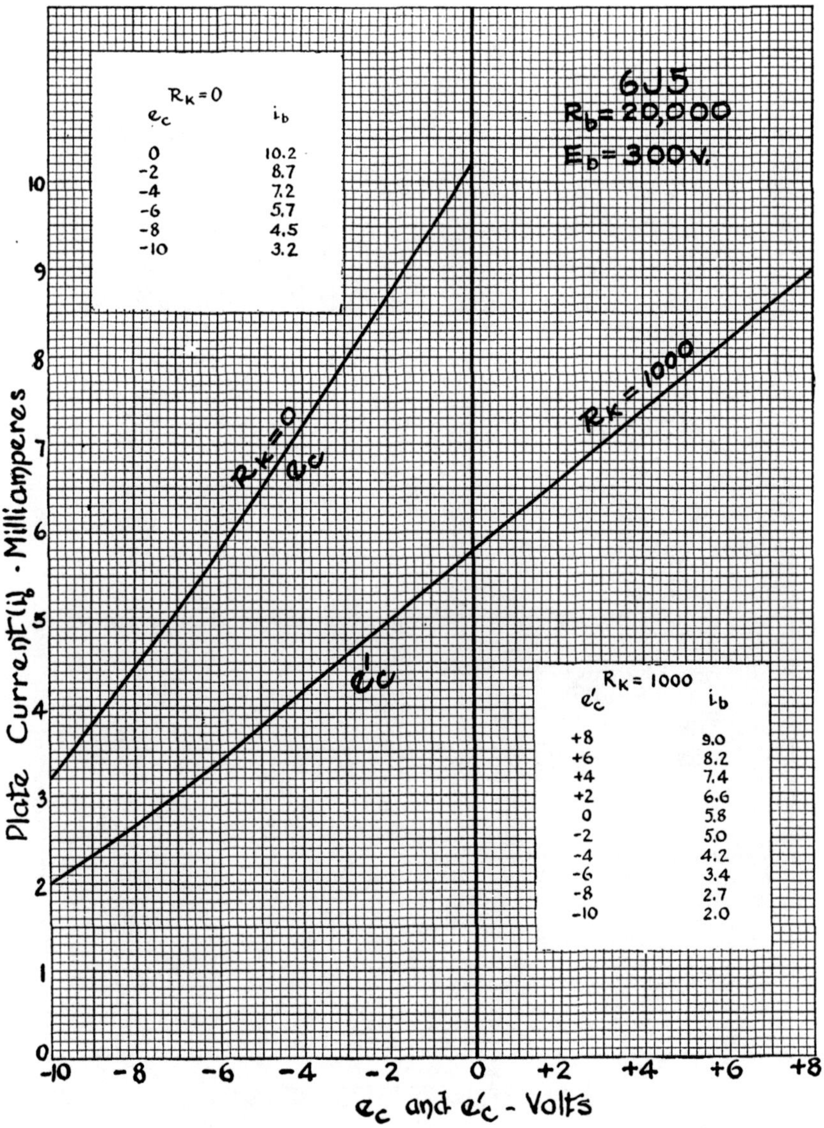

FIG. 13-8. The difference that a cathode resistor makes in the dynamic transfer characteristic of a triode is illustrated by these two curves. The curve marked e_c was derived when no cathode resistor was included, and the e'_c curve was derived when 1000 ohms was in the cathode circuit. In both instances a plate load of 20,000 ohms was used.

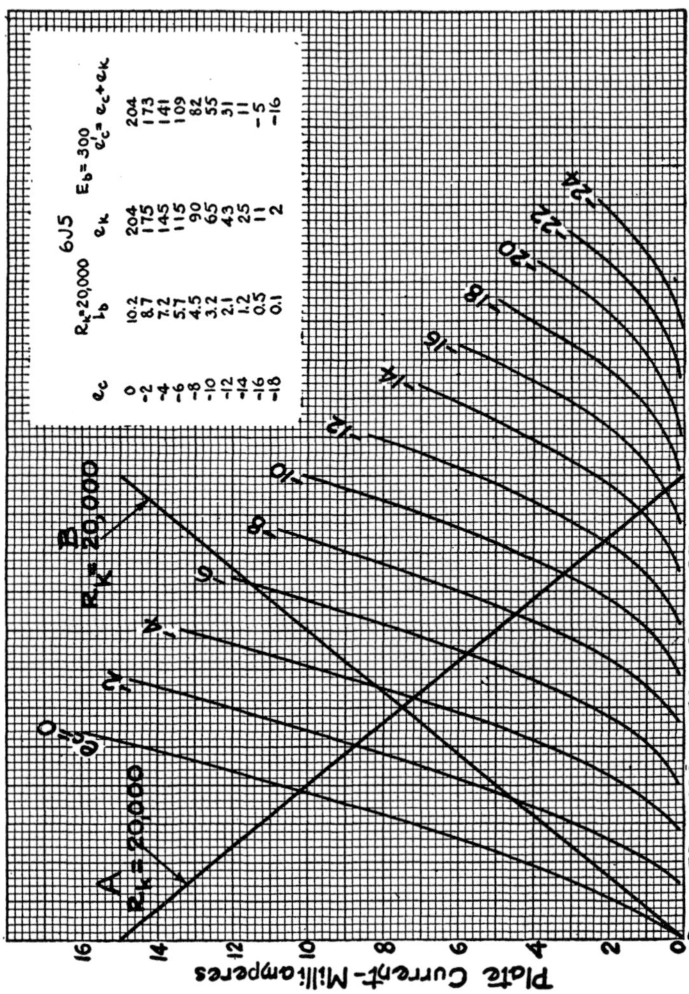

e_c	i_b	e_k	$e'_c = e_c + e_k$
0	10.2	204	204
-2	8.7	175	173
-4	7.2	145	141
-6	5.7	115	109
-8	4.5	90	82
-10	3.2	65	55
-12	2.1	43	31
-14	1.2	25	11
-16	0.5	11	-5
-18	0.1	2	-16

FIG. 13-10. Graphical method of determining the performance of a cathode follower.

signal is in a positive-going direction and when the input signal is negative-going, the output signal is negative-going. In other words, the output signal *follows* the input signal. This is in contrast to the case when the load resistance is exclusively in the plate circuit, the output signal then being 180 degrees out of phase with the input signal.

FIG. 13-9. When the output signal is taken off the cathode resistor, the output voltage e_k is in phase with the input. This circuit is called a cathode follower.

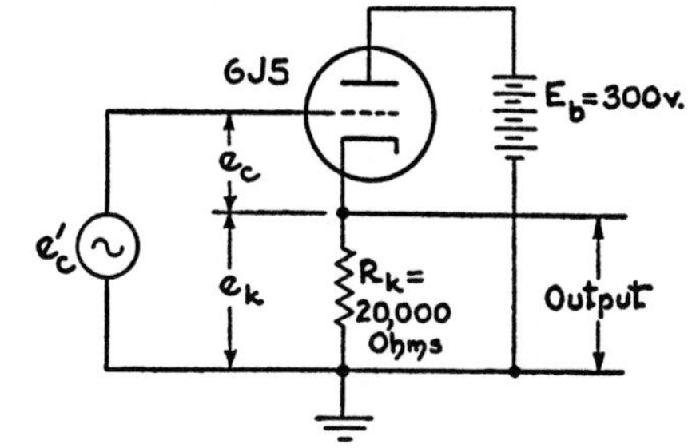

In order to find out the characteristics of a cathode follower circuit, let us examine the circuit of Fig. 13-9 more carefully. In particular, a 6J5 triode is shown, a cathode resistor R_k of 20,000 ohms, and a plate-supply voltage E_b of 300 volts. If a certain signal voltage e'_c is applied to the input, what will the output signal voltage e_k be?

We can most easily determine the answer to this question by employing the plate family for the 6J5 tube of Fig. 13-10. Since the cathode resistor is common to both the input circuit and the plate circuit, we shall first consider it as a resistance in the plate circuit. We can therefore draw a load line, just as if the cathode resistor R_k were the plate load resistor. This is the line A in Fig. 13-10. We shall not attempt to obtain a dynamic transfer characteristic, however, but shall utilize the information secured from this load line in a manner about to be described.

Let us choose a particular point where the load line intersects one of the curves of the plate family, say the $e_c = -8$-volt curve. This corresponds to a plate current i_b of 4.5 milliamperes. Since the 4.5-milliampere plate current flows through the cathode resistor R_k of 20,000 ohms, the voltage e_k across the cathode resistor is

$$e_k = i_b\ R_k = 0.0045 \times 20{,}000 = 90 \text{ volts.}$$

Now we are in a position to determine the value of the applied input signal e'_c. As we have already learned, and as is also evident from Fig. 13-9,

$$e'_c = e_c + e_k$$

For our example we chose $e_c = -8$, and by means of the plate family and a computation obtained $e_k = 90$ volts. Consequently,

$$e'_c = e_c + e_k = -8 + 90 = 82 \text{ volts.}$$

We are interested in obtaining a graph showing the relationship between the input signal e'_c and the output signal e_k. We have just determined one point on such a graph, namely, $e'_c = 82$ volts and $e_k = 90$ volts, and now require additional points. In order to lessen the effort required, we shall show a graphical method for evaluating e_k.

A line representing a resistance equal to the cathode resistance is drawn on the plate family, Fig. 13-10, in this case the resistance line B corresponding to 20,000 ohms. One end of this line starts at the origin, which corresponds to the point of zero plate voltage and zero plate current. The other end of this line may terminate at the point corresponding to 300 volts and 15 milliamperes. The voltage drop e_k across the cathode resistor R_k is then given by the reading on the plate volts axis corresponding to the point determined by the intersection of a horizontal plate-current line with resistance line B. Thus, if 10 milliamperes flow, then the voltage drop e_k is 200 volts.

We now have the interesting relation that (1) the intersection of load line A with a curve of the plate family determines a specific plate current, and (2) this same value of plate current in conjunction with resistance line B determines the voltage e_k. Let us return to our previous example. The intersection of load line A with the $e_c = -8$-volt curve of the plate family gives a plate current of 4.5 milliamperes. This current of 4.5 milliamperes, as read on resistance line B, determines a voltage $e_k = 90$ volts. The value of the input signal voltage e'_c is computed as previously described.

A complete tabulation of the results thus obtained is shown in the upper right-hand corner of Fig. 13-10. In order to clarify further how this tabulation was obtained, let us choose a few more examples. The intersection of load line A with the $e_c = 0$-volt curve of the plate family results in a plate current of 10.2 milliamperes and this value of plate current in conjunction with resistance line B gives a voltage $e_k = 204$ volts. Then, the input signal voltage $e'_c = e_c + e_k = 0 + 204 = 204$ volts.

In looking at the tabulation shown on Fig. 13-10, it will be seen that most of the values shown in the e'_c column are positive, but the last two are negative. Let us consider how one of these negative values is obtained. The intersection of load line A with the $e_c = -16$-volt curve of the plate family indicates a plate current of 0.5 milliampere, and this current in conjunction with resistance line B results in $e_k = 11$ volts. Then the required input-signal voltage e'_c is $e'_c = e_c + e_k = -16 + 11 = -5$ volts.

We have gone through the effort of making the tabulation shown on Fig. 13-10 in order to obtain the last two columns, for our interest lies in the relationship between the input signal voltage e'_c and the output signal voltage e_k. Upon plotting these values, and extending to the base line, as shown in Fig. 13-11, we observe, with the exception of a very slight curvature at the lower end, that a linear relation exists between the input and output. This signifies that a cathode follower introduces negligible distortion, so that the output voltage will be an almost perfect replica of the input voltage.

Since the cathode follower is practically distortionless, you may be curious why it is not used as an amplifier. The answer is that the output voltage swing is less than the input voltage swing. This can be seen from Fig. 13-11, where you can see that the output voltage e_k

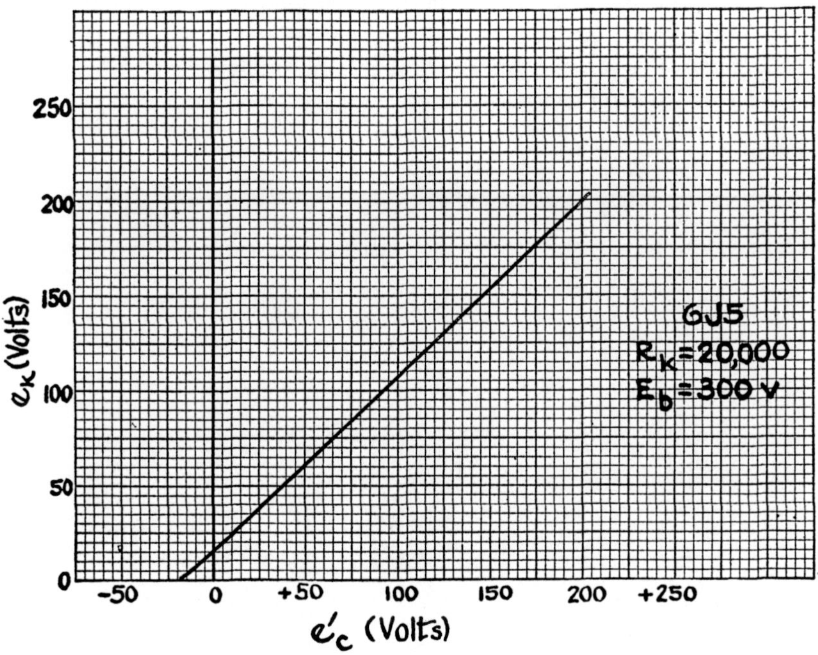

FIG. 13-11. The characteristic curve of the 6J5 as a cathode follower is practically a straight line, so the output signal is almost distortionless.

covers a range from zero to 204 volts in the positive region, while the input voltage e'_c goes from 0 to 204 in the positive region and from 0 to -18 in the negative region, or a total range of 222 volts. Now we may define the amplification A_v as the output voltage divided by the input voltage

$$A_v = \frac{e_k}{e'_c} = \frac{204}{222} = 0.92$$

We see therefore, that the amplification is less than unity, and this is true of all types of cathode followers.

You may then wonder why a cathode follower is used if it brings about a loss, rather than a gain, of signal strength. The answer to that question is outside the province of this book, but suffice it to say that cathode followers have some very important applications.

Before leaving the subject of the cathode follower, let us examine Fig. 13-11 again. You will notice that although the input signal can go all the way to 204 volts in the positive direction, it can only go to 18 volts in the negative direction. Any input signal more negative than 18 volts will be clipped off. There are other types of cathode followers; one type handles a large negative input signal but clips off the positive signal. Another type handles both positive and negative input signals. The particular type that will be employed depends upon the task to be accomplished.

Chapter 14

POWER AMPLIFIERS

THE VACUUM-TUBE CIRCUITS for triodes and pentodes with which we have been concerned in previous chapters, were discussed from the viewpoint of obtaining maximum voltage amplification with a minimum amount of distortion. There are occasions, however, when the delivery of *power* from a vacuum tube to its load is of primary interest. A *power amplifier* is then required. A book such as this would not be complete if this type of amplification were not considered, although the extent to which we can delve into the subject must be limited in order to conform with the general field of application of this volume.

Voltage amplification and power amplification have many things in common, differing primarily in the operating conditions established in the plate or output circuit of the tube. Accordingly, most of the discussion to follow will revolve around the plate circuit, with such references to the grid circuit as must be made in order to explain the subject properly.

The explanation to follow may at times appear like a discussion of the design of a power amplifier stage or system. Such a presentation is only a means to an end and is not in conflict with previous chapters; the most logical manner of conveying some of the fundamental ideas and requirements of power amplifiers to you is by imagining a problem. So many variable factors are involved in the design of power amplifiers that it becomes necessary to make a number of assumptions and approximations in order to simplify the analysis. Such simplification, however, is not a sacrifice of understanding of the subject, but rather the elimination of those more advanced thoughts which are not required at the level of this book.

If a complete description of a power amplifier were to be given it would become necessary to distinguish between two different load lines, a static load line (of the type with which we have previously been concerned) and a dynamic load line. It would further become necessary to distinguish between the quiescent operating point when the power stage is not excited by an input voltage and the operating point when the power amplifier is under excitation. In line with our aims, we shall disregard the foregoing factors, as well as others, and shall confine our attention only to the broader aspects of power amplifiers. It must therefore be realized that the following description of power-amplifier operation is a highly simplified one, although it does illustrate the paramount factors that must be considered.

Definition of Power

Power is defined as the *rate of doing work.* The unit used for measuring power in electrical work is the *watt,* and this is the rate at which work is done when one ampere of current flows against one volt. Or, more generally,

$$P = E\ I$$

where

$$\begin{aligned} P &= \text{power (in watts)} \\ E &= \text{electromotive force (in volts)} \\ I &= \text{current (in amperes)} \end{aligned}$$

For example, in a d-c circuit, the power expended by an electric light bulb which is connected to a 115-volt line and through which 2 amperes are flowing is $P = E\ I = 115 \times 2 = 230$ watts.

From Ohm's law $E = I\ R$, and substituting this value of E in the expression $P = E\ I$, we obtain $P = E\ I = (I\ R)\ I = I^2\ R$. Again from Ohm's law $I = E/R$, and substituting this value of I in the expression $P = E\ I$, there is obtained $P = E\ I = E\ \dfrac{(E)}{(R)} = \dfrac{E^2}{R}$. We therefore have the alternative ways of expressing power

$$P = E\,I = I^2 R = \frac{E^2}{R}$$

In order to get a physical picture, let us consider Fig. 14-1. A 5-ohm and a 20-ohm resistor are connected in series and connected to a 100-volt battery Since the resistance of the entire circuit is 25 ohms, a current $I = \frac{E}{R} = \frac{100}{25} = 4$ amperes will flow, and a voltage of $\frac{5}{25} \times 100 = 20$ volts will appear across the 5-ohm resistor, while $\frac{20}{25} \times 100 = 80$ volts will appear across the 20-ohm resistor. Fig. 14-1 illustrates these conditions, which were drawn in accordance with the data pertaining to the lines of Fig. 9-5.

Now consider the power dissipated in the 5-ohm resistor. This is $P = E\,I = 20 \times 4 = 80$ watts, that is, it is represented by the shaded area *OABC*. Similarly, the power dissipated in the 20-ohm resistor, (since $E = 100 - 20 = 80$ volts) is $P = E\,I = 80 \times 4 = 320$ watts, and is represented by the shaded area *CBDF*. The total power of 400 watts supplied by the battery is represented by the area *OADF*.

Plate Dissipation

The kinetic energy acquired by the electrons in traveling from the cathode to the plate inside a vacuum tube bombards the plate and heats the plate. If the energy thus dissipated becomes sufficiently great, the plate of the tube may become so hot that it releases absorbed gasses and thus partially destroys the vacuum in the tube, thereby rendering the tube inoperative. This factor is usually of no importance in the case of a tube used as a voltage amplifier, but is of extreme importance for a power-amplifier tube.

Manufacturers of power tubes generally state the maximum safe plate dissipation, and the first consideration in the design of a power amplifier is that this maximum plate dissipation should not be exceeded. Let us see how this works for a typical triode power-amplifier tube such as the 2A3. The maximum safe plate dissipation for this tube is 15 watts.

Consider the plate family for a 2A3 as shown on Fig. 14-2. From the relation $P = E\,I$, we can find corresponding values of plate current for each plate voltage such that the plate dissipation is always 15 watts.

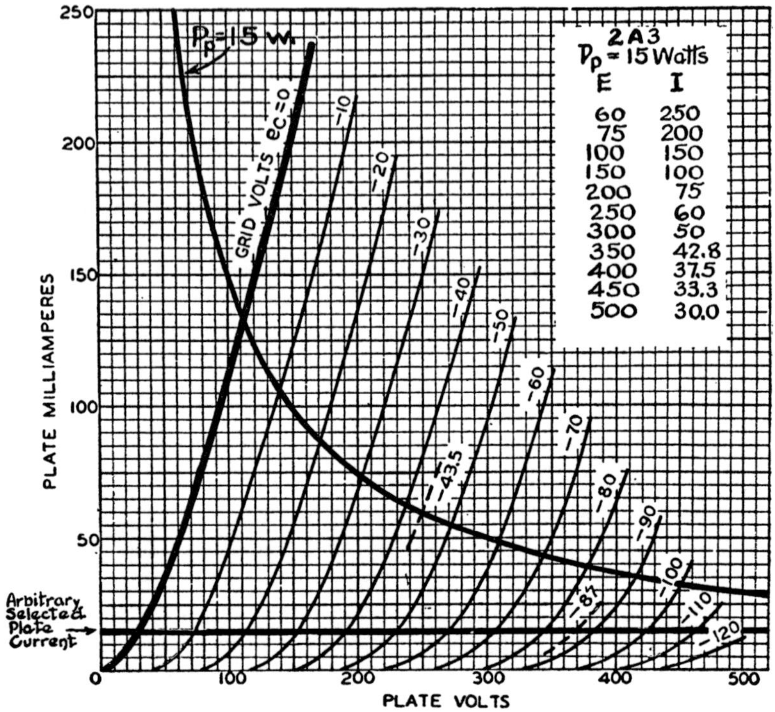

FIG. 14-2. The three boundaries for correct operation of a power tube are the plate-dissipation curve (P_p), the $e_c = 0$ curve of the plate family, and an arbitrarily selected plate-current line, in this case 15 milliamperes.

For example, for a plate voltage of 100, we obtain $I = \frac{P}{E} = \frac{15}{100} = 0.15$ ampere = 150 milliamperes. Consequently, the intersection of the 100-volt and 150-milliampere lines determine a point on the plate dissipation curve. Other points may be similarly computed as shown by the tabular insert of Fig. 14-2. These points are plotted on the plate family and a smooth curve drawn through them, the resulting curve having the shape of a hyperbola.

We now have a limiting factor: any quiescent operating point must lie *below* or *on* the plate-dissipation curve, since only then will the tube be functioning under safe conditions. Consequently, any load line that is drawn on the plate family must lie below (or just touch at one point) the plate-dissipation curve. This item will be considered in more detail shortly.

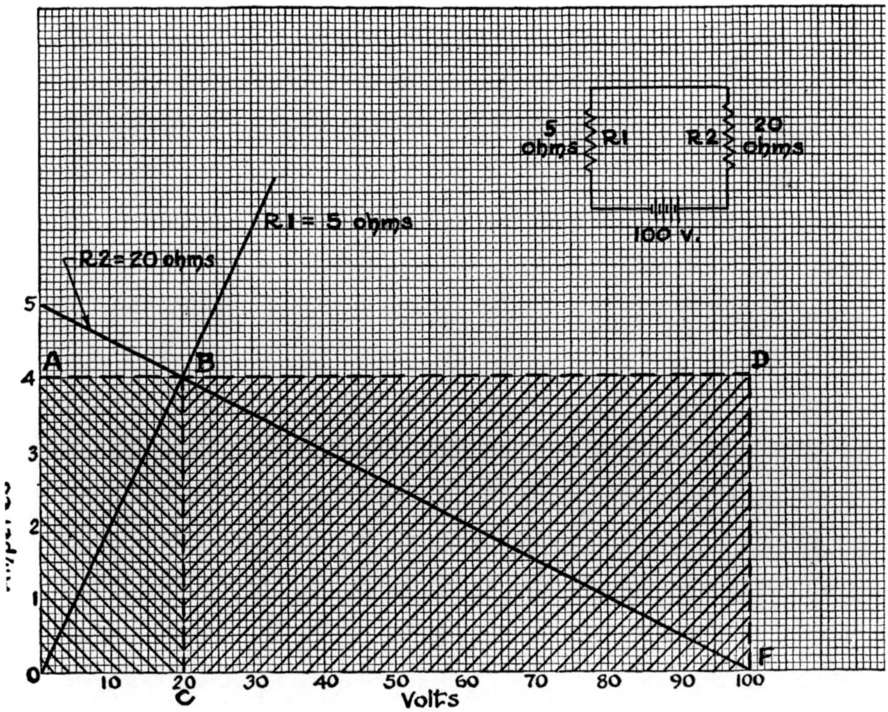

IG. 14-1. The area OABC represents the power dissipated in the 5-ohm resistor and are BDF represents the power dissipated in the 20-ohm resistor. The power supplied by th attery is represented by the area OADF.

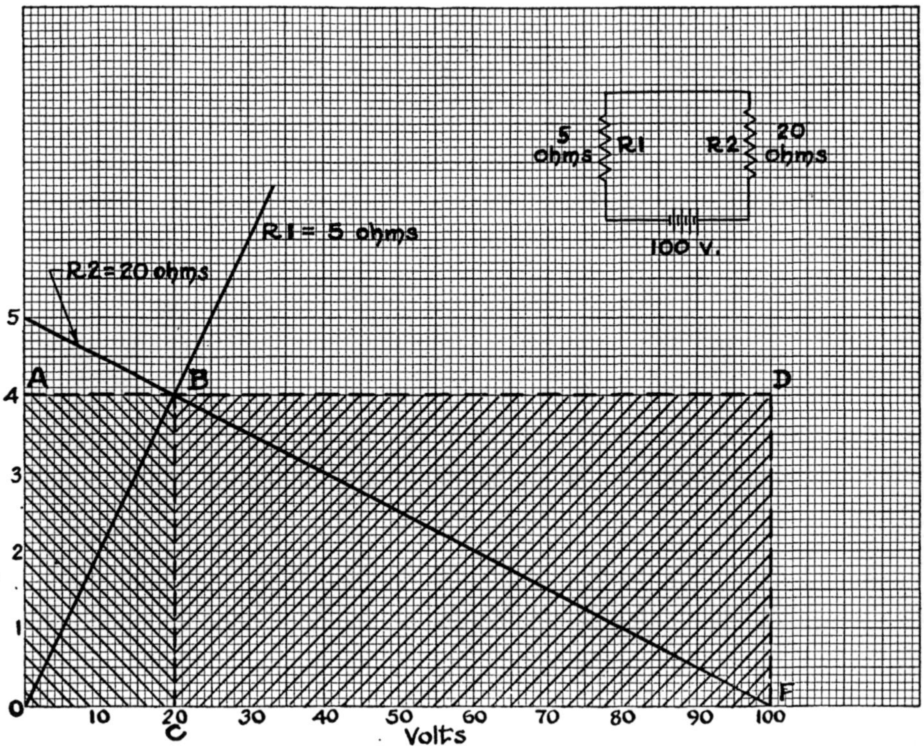

:G. 14-1. The area OABC represents the power dissipated in the 5-ohm resistor and area BDF represents the power dissipated in the 20-ohm resistor. The power supplied by the ıttery is represented by the area OADF.

Other Boundaries

Another boundary position which determines the range of operation of the load line is that member of the plate family for which the instantaneous grid voltage is zero ($e_c = 0$). This is because we do not want the grid ever to draw grid current, in consequence of which we must not allow the grid to become positive. This restriction excludes certain types of power amplifiers, but we do not consider such an omission detrimental, since our object is only to explain certain fundamental phenomena which will enable you to better understand more advanced texts. In Fig. 14-2 this $e_c = 0$ line of the plate family is drawn heavier than the other members of the family.

As we have seen in an earlier chapter, operation of the tube on the curved lower portions of the plate family results in distortion of the output wave. If we wish to limit the amount of distortion, we may draw a horizontal line across the plate family corresponding to some particular value of plate current. In Fig. 14-2 this line corresponds to a plate current of 15 milliamperes, a value which was chosen rather arbitrarily. In practice, the value of plate current for which this line should be drawn is dependent upon several factors, such as the type of vacuum tube employed, the permissible percentage distortion, and the power output desired.

We now have three boundaries, namely, the plate dissipation curve, the zero instantaneous grid voltage curve, and a minimum plate-current line. The operating portion of any load line must lie within (or touching) these boundary lines.

Power Output

The power output in watts that the tube can deliver to the load is dependent upon the magnitude of the load resistor. Let us draw a load line for a resistance $R_b = 2600$ ohms and using a plate supply voltage $E_b = 390$ volts, as has been done in Fig. 14-3. This load line falls below the plate dissipation curve as is required. In our previous discussion we stated that the minimum plate current should be 15 milliamperes, and since the load line intersects the $e_c = 80$ volt curve of the plate family at 16 milliamperes, we shall adopt this 16 milliampere value as the minimum value of plate current. The other end of the working range of the load line terminates on the $e_c = 0$ curve of the plate family.

Since we have chosen —80 volts as the maximum negative swing, our quiescent value of grid voltage is determined at —40 volts, for

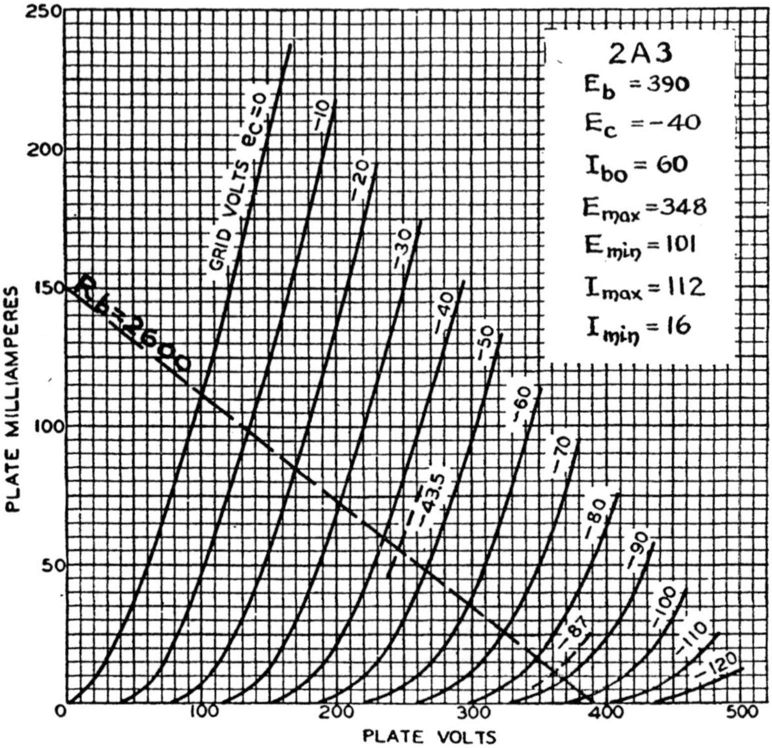

FIG. 14-3. The load line is used to determine the power output of a tube. Here the load line is for a resistor of 2600 ohms, the usable portion of which lies within the boundaries specified in Fig. 14-2.

this will enable the entire range from $e_c = 0$ to $e_c = 80$ to be covered. The intersection of the load line with the $e_c = 40$-volt curve gives a quiescent plate current $I_{bo} = 60$ milliamperes.

It can be shown that the power output of a tube is given by the expression

$$P_o = \frac{(E_{max} - E_{min})\ (I_{max} - I_{min})}{8}$$

where

$$\begin{aligned} P_o &= \text{power output (watts)} \\ E_{max} &= \text{maximum plate voltage} \\ E_{min} &= \text{minimum plate voltage} \\ I_{max} &= \text{maximum plate current} \\ I_{min} &= \text{minimum plate current} \end{aligned}$$

From Fig. 14-3 we observe that the maximum plate voltage and minimum plate current occur where the load line intersects the $e_c = -80$-volt curve of the plate family, so that $E_{max} = 348$ volts and $I_{min} = 16$ milliamperes. The minimum value of plate voltage and the maximum value of plate current occurs at the intersection of the load line with the $e_c = 0$ curve of the plate family, so that $E_{min} = 101$ volts and $I_{max} = 112$ milliamperes. Substituting these values in the power output equation, we obtain

$$P_o = \frac{(348 - 101)\ (0.112 - 0.016)}{8}$$

$$= \frac{247 \times 0.096}{8}$$

$$= 2.96 \text{ watts}$$

Later, we shall try several different values of load resistor in order to compare their influence on the power output.

Distortion

For a triode the distortion produced is predominantly second-harmonic distortion. The percentage second-harmonic distortion, represented as $\%\ H_2$, is given by the expression

$$\%\ H_2 = \frac{I_{max} + I_{min} - 2\ I_{bo}}{2\ (I_{max} - I_{min})} \times 100$$

From Fig. 14-3 we have already determined the maximum plate current $I_{max} = 112$ milliamperes, the minimum plate current $I_{min} = 16$ milliamperes, and the quiescent plate current $I_{bo} = 60$ milliamperes. Inserting these values in the distortion equation, we obtain

$$\%\ H_2 = \frac{112 + 16 - 2\ (60)}{2\ (112 - 16)} \times 100 = \frac{800}{192} = 4.2\%$$

As we shall see, the choice of the load resistor is a factor that influences the magnitude of the distortion.

Power Output vs. Plate Dissipation

We are now in a position to compare the effect of the load resistor on the power output and the plate dissipation. But first let us review our findings in connection with Fig. 14-1, and reinterpret the significance of this figure.

Let us consider, in Fig. 14-1, that resistor *R1* represents one of the curves of the plate family, in particular that one corresponding tc

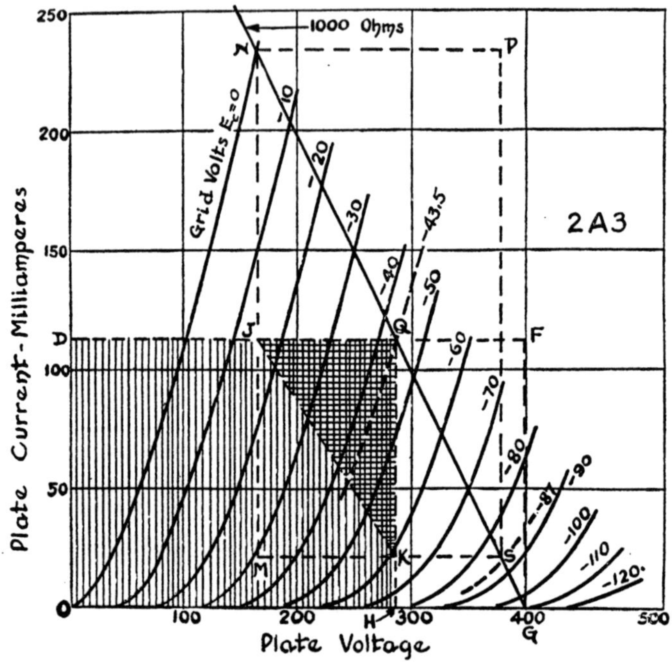

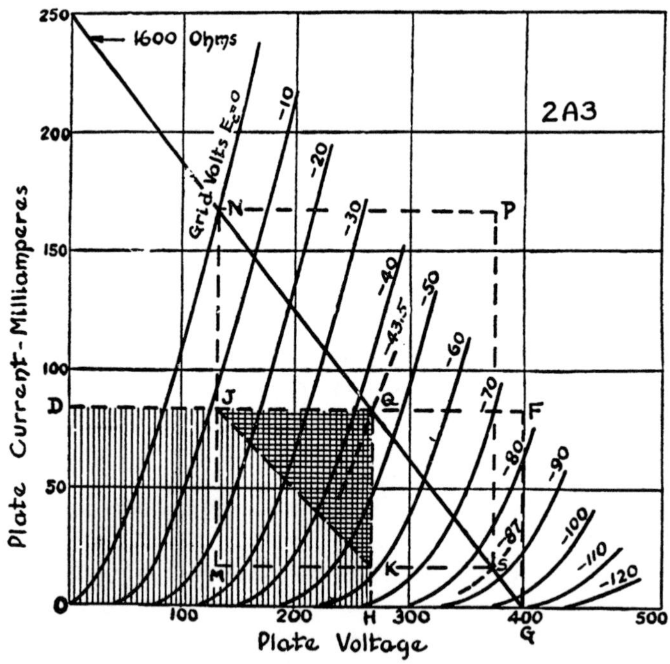

FIGS. 14-4 (above), 14-5 (below). In these two and the two graphs on the next page, different load lines are drawn on the plate family of a 2A3 for different values of loads as indicated, in order to indicate the power output and plate dissipation.

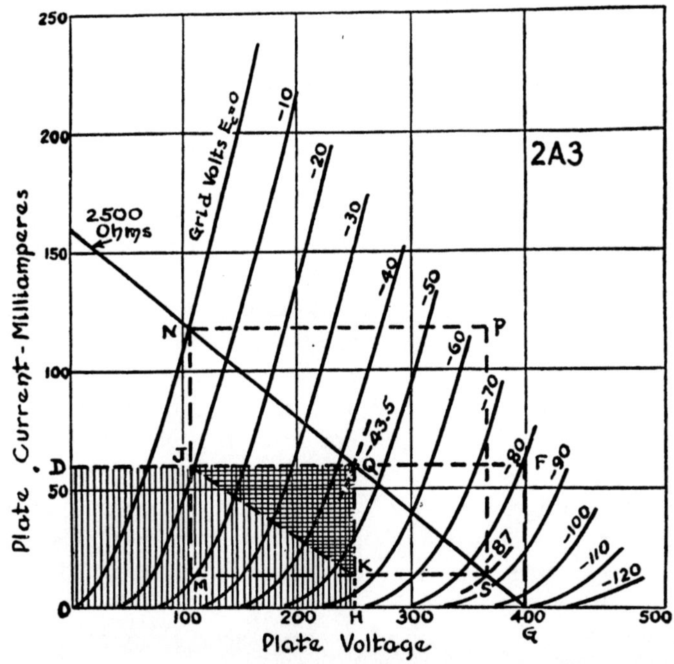

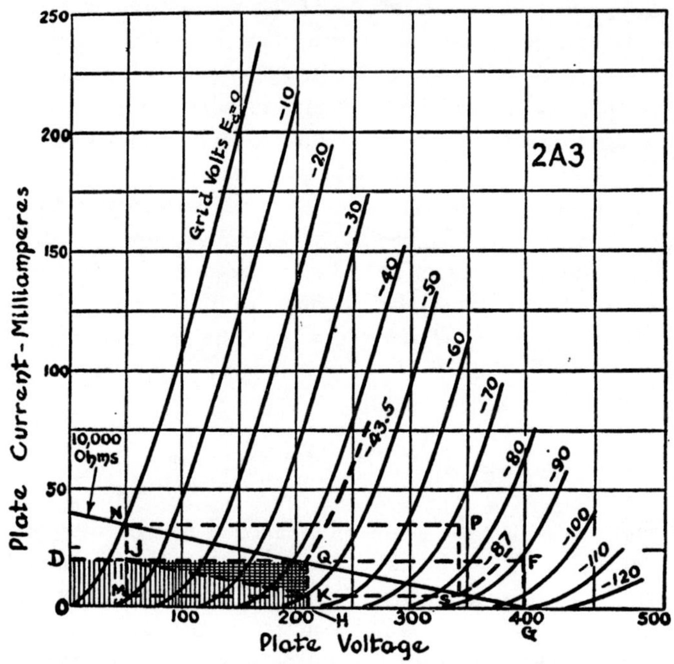

FIGS. 14-6 (above), 14-7 (below). In these four sets of curves, observe how the proportion of power output area (QJK) and the plate dissipation area (ODJKH) changes as the load resistance is changed.

$e_c = 0$. We shall further imagine that resistor *R2* represents a load line that has been drawn across the plate family. We now have the equivalent of a vacuum tube with a load resistance in the plate circuit.

The total power supplied by the battery is then represented by area *OADF*. Area *CBDF* corresponds to the power dissipated in resistor *R2*, which corresponds to the load resistor. Area *OABC* represents the power dissipated in resistor *R1*, which corresponds therefore to the power dissipated by the plate of the tube (plate dissipation), when $e_c = 0$.

In the light of the foregoing, let us examine Fig. 14-4, which illustrates a plate family of the 2A3, a load line for a 1000-ohm load resistor, and a plate-supply voltage of 400 volts. Consider first the condition when no excitation voltage is applied to the grid, namely, when the grid is being operated at its quiescent operating point *Q* ($E_c = -43.5$ volts). Then, area *ODFG* represents the total power delivered by the plate-supply voltage, area *HQFG* represents the power dissipated in the load resistor as heat due to the d-c flow of quiescent plate current I_{bo}, and area *ODQH* represents the power dissipated at the plate as heat (plate dissipation).

Now, let an excitation voltage be applied to the grid, so that the instantaneous grid voltage now covers the range from *N* to *S* on the load line. In addition to the d-c current flow there is now superimposed an a-c flow in the plate circuit. As in the previous case, area *ODFG* represents the total power delivered by the plate-supply voltage, and area *HQFG* represents the power dissipated as heat in the load resistor due to the d-c flow of quiescent plate current I_{bo}.

Point *S* corresponds to E_{max} and point *M* corresponds to E_{min}, so that line *MS* represents $(E_{max} - E_{min})$. Similarly, point *N* corresponds to I_{max} and point *M* corresponds to I_{min}, so that the line *MN* represents $(I_{max} - I_{min})$. Accordingly the product of *MS* and *MN*, namely, $(E_{max} - E_{min})\ (I_{max} - I_{min})$, corresponds to the area *MNPS*. We have already seen that the useful a-c power output P_o appearing in the load resistor is given by the expression

$$P_o = \frac{(E_{max} - E_{min})\ (I_{max} - I_{min})}{8}$$

Of the eight sections into which area *MNPS* can be divided, let us choose area *JQK*. This area *JQK* then represents the output power P_o that appears in the load resistor. A careful examination of the figure will disclose that area *JQK* is not exactly equal to one-eighth of area *MNPS*. This discrepancy comes about because the tube characteristics have a certain amount of curvature, and are not ideally straight as was assumed in the mathematical derivation of the formula. This fact, however, is of no essential importance in our discussion.

When the tube was operated at its quiescent condition, we saw that area *ODQH* represented the plate dissipation; now, however, it is the smaller area *ODJKH* that represents the plate dissipation. This brings out an important point: *the plate dissipation decreases in the same proportion that the useful a-c power in the load increases.* This shows that the maximum burden is imposed upon the tube when it is in its quiescent condition. It is interesting to note that the power tube in a radio receiver undergoes the least strain when the volume is at a maximum, and the most severe burden when operating at a low (or no) level of volume.

Fig. 14-4 should be compared with Figs. 14-5, 14-6, and 14-7. The plate-supply voltage and the quiescent grid voltage is the same in each of the figures, but the load line is different in each case. It will be observed that as the load resistance is increased, the power output (area *QJK*) and the plate dissipation (area *ODJKH*) both decrease. Some sort of compromise must therefore be made in order to obtain a reasonable power output while at the same time operating the tube within its allowable plate dissipation limit.

The condition illustrated in Fig. 14-4 and 14-5 results in a plate dissipation at the quiescent point considerably in excess of the allowable maximum of 15 watts. In the case of Fig. 14-4, the plate dissipation P_p for the quiescent condition, when the plate voltage $E_{bo} = 286$ and the plate current $I_{bo} = 0.113$ ampere is $P_p = E_{bo} \times I_{bo} = 286 \times 0.113 = 32.3$ watts. Similarly, for Fig. 14-5, the plate dissipation for the quiescent condition is $P_p = E_{bo}I_{bo} = 268 \times 0.084 = 22.5$ watts. The value of load resistor for both of these cases is too small and results in excessive plate dissipation.

The condition, of which Fig. 14-7 is illustrative, gives a plate dissipation $P_p = 210 \times 0.020 = 4.2$ watts. Although this value is well below the allowable 15 watts, the power output P_o is very low, as is indicated by the smallness of area *JQK*. Here the load resistor is excessively large.

Let us now re-examine Fig. 14-6. For the quiescent condition, the plate voltage $E_{bo} = 250$ (point *Q*) and the plate current $I_{bo} = 0.060$ ampere (point *D*), so that the plate dissipation is $P_p = E_{bo}I_{bo} = 250 \times 0.060 = 15$ watts. Consequently, a load resistor of 2500 ohms results in a reasonable power output and a suitable plate dissipation. Let us calculate the power output. From Fig. 14-6, we observe that $E_{max} = 365$

volts, $E_{min} = 104$ volts, $I_{max} = 0.118$ ampere, and $I_{min} = 0.013$ ampere. Accordingly, the power output P_o is

$$P_o = \frac{(E_{max} - E_{min})\ (I_{max} - I_{min})}{8}$$

$$= \frac{(365 - 104)\ (0.118 - 0.013)}{8}$$

$$= \frac{261 \times 0.105}{8} = 3.4 \text{ watts}$$

In order to find the percentage second-harmonic distortion $\%\ H_2$, in addition to the I_{max} and I_{min} values, we need the quiescent value of the plate current I_{bo}, and an examination of Fig. 14-6 shows that this is 0.060 ampere. Then

$$\%\ H_2 = \frac{I_{max} + I_{min} - 2\ I_{bo}}{2\ (I_{max} - I_{min})} \times 100$$

$$= \frac{0.118 + 0.013 - 2(0.060)}{2\ (0.118 - 0.013)} \times 100 = 5.2\%$$

Thus we see that the employment of a 2500-ohm load resistor results in a power output of 3.4 watts having 5.2% second-harmonic distortion. If this amount of distortion is permissible, then we have found a suitable load resistor. On the other hand, if this amount of distortion is too great for the purpose in hand, then other load lines will have to be tried until a suitable load is found.

We have shown several of the more elementary factors that must be considered in the design of a power amplifier. In actual practice a similar "trial and error" technique is employed, including the consideration of a number of factors which we have not concerned ourselves with in this book, for, as we stated in the beginning of this chapter, they were beyond the scope of this text.

The design of a pentode power amplifier or of a beam-power tube amplifier involves even more factors than were required for a triode. Accordingly, we shall discuss these types only briefly, and shall leave you to consult more advanced texts for additional details.

Pentode Power Amplifier

The principal reason that a pentode is used as a power amplifier is that it has a much greater *power sensitivity* than a triode, that is, the ratio of a-c power output to a-c input grid voltage is greater. For example, the 2A3 triode requires a 45-volt peak signal on the grid in order to deliver 3.5 watts to its load circuit. The 6F6 pentode, on the other hand, will deliver 3.2 watts to its load when a 15.5-volt peak signal is applied to its control-grid.

The principal objection to the use of a pentode as a power amplifier is the large amount of distortion that is produced. This is illustrated in Fig. 14-8, which shows the power output obtainable from a 6F6 for

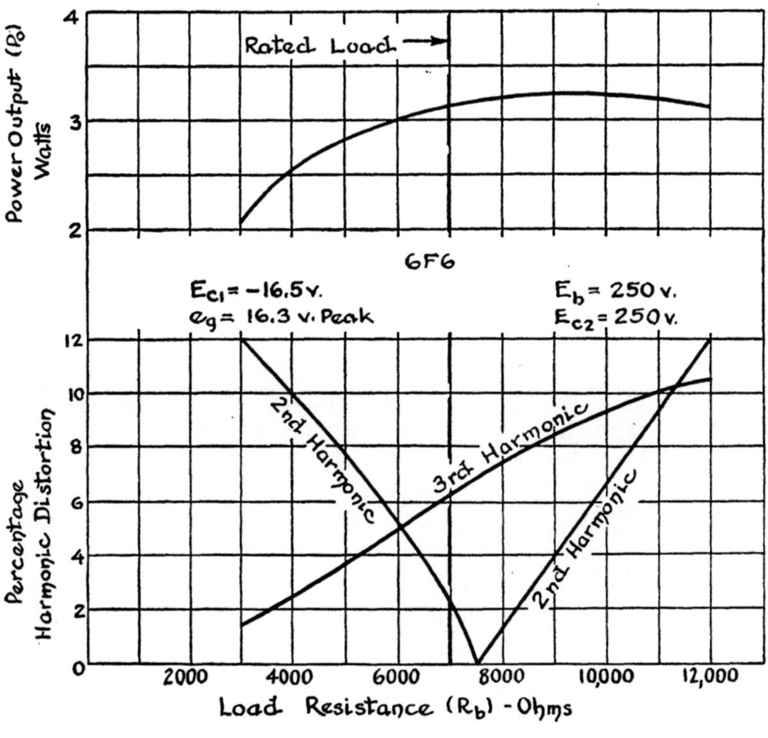

FIG. 14-8. The upper curve shows how the power output of a 6F6 tube varies with the value of the load resistance and the lower curves show the amounts of second and third harmonics which are produced for various values of the load resistance.

various values of load resistor, and also the amount of second- and third-harmonic distortion that is produced. This figure shows that a definite compromise must be made in the choice of a load resistor in order to obtain a suitable power output without excessive distortion.

In order to get a better idea of why this distortion occurs, let us examine the plate family of a 6F6 pentode, as shown in Fig. 14-9.

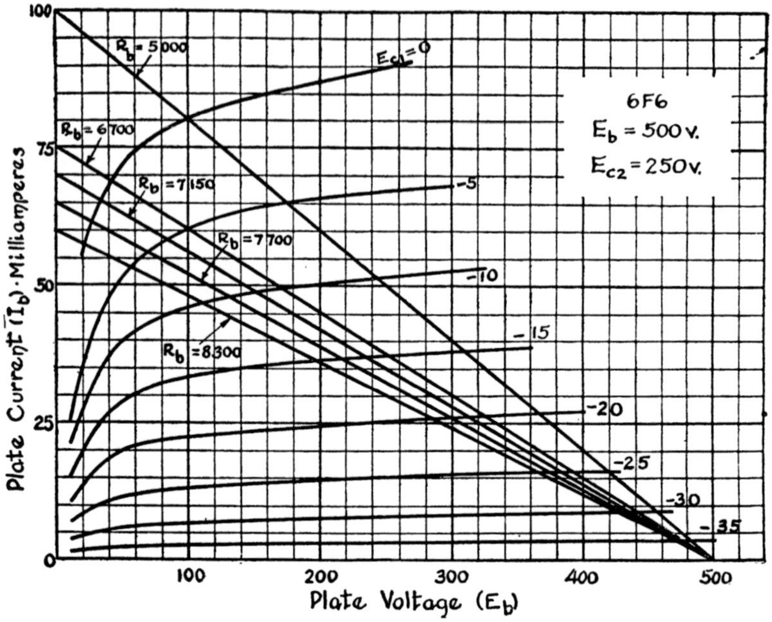

FIG. 14-9. The load lines drawn on this plate family of the 6F6 tube were used to obtain the dynamic transfer characteristics in Fig. 14-10.

Five load lines have been drawn on the plate family, the plate-supply voltage being maintained constant at 500 volts. The points of intersection of each load line with the various curves of the plate family are replotted to give the dynamic transfer characteristic, as shown in Fig. 14-10.

An examination of Fig. 14-10 reveals that the dynamic transfer characteristic has a decided curvature at its lower portion, corresponding to the highly negative control-grid voltages. As we have already seen earlier in this book, such curvature results in a flattening of the lower portion of the output plate-current wave. This accounts then for part of the distortion.

Further examination of Fig. 14-10 shows that the upper portion of each of the dynamic transfer characteristics is also curved, the larger the value of the load resistor the greater the curvature. The dynamic transfer characteristic corresponding to a load resistor of 5000 ohms is less curved in its top portion than the other characteristics, but such a low value of load resistance results in an inadequate power output. On the other hand, too high a value of load resistor would result in an

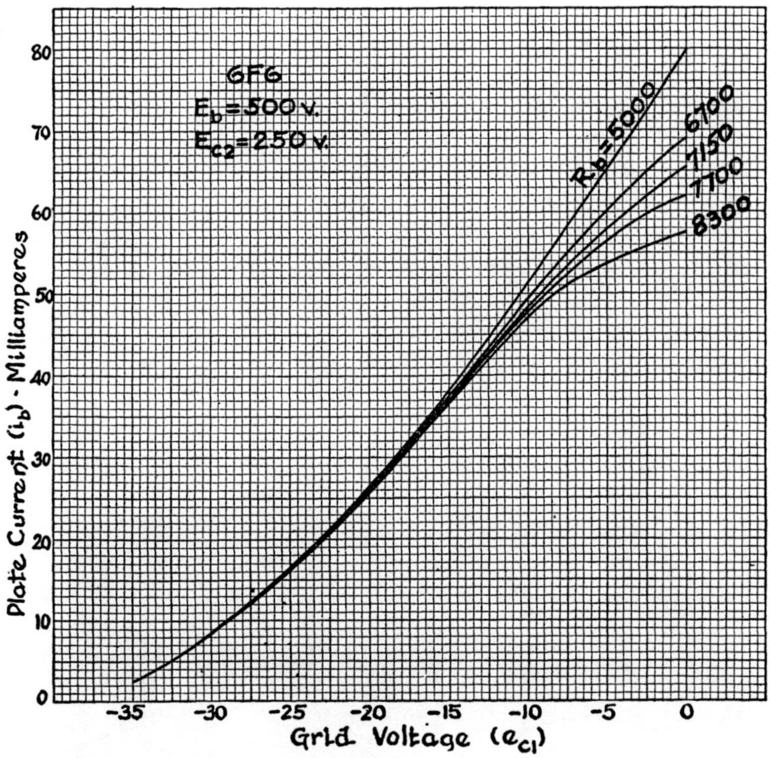

FIG. 14-10. This group of dynamic transfer characteristics of the 6F6 is used to select the proper load resistor.

excessive amount of distortion, as is indicated by the excessive curvature of the dynamic transfer characteristic corresponding to 8300 ohms. Consequently this value of load resistor, as well as those somewhat higher in magnitude, are not employed, despite the fact that they would result in a maximum of power output.

An effective compromise between power output and distortion is obtained by using a load resistor of 7000 ohms. As can be seen from Fig. 14-8, this will result in a power output of 3.2 watts, with 2.2% second-harmonic and 6.2% third-harmonic distortion.

A pentode power tube is generally operated with its screen-grid voltage equal to its plate voltage. It should be kept in mind that in using a pentode as a power amplifier, not only must the plate dissipation be kept within its maximum allowable rating, but also its screen dissipation.

Beam Power Tube

The *beam power tube* may be either a tetrode or a pentode wherein the electron stream is concentrated into a well-defined beam in the inter-electrode space between the control-grid and the screen-grid, thus increasing the power handling properties of the tube and providing a comparatively high operating efficiency. This is accomplished by the arrangement of the turns of wire comprising the control-grid and screen-grid and the introduction of two beam-confining electrodes. The purpose of this electrode arrangement is to reduce the secondary emission between the plate and the screen-grid without using a suppressor-grid as in the pentode.

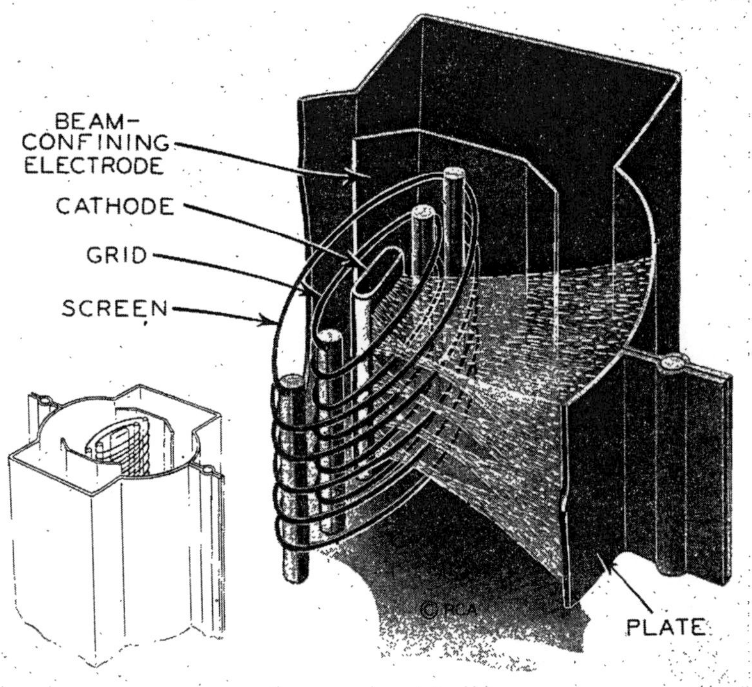

Courtesy RCA Mfg. Co.

FIG. 14-11. Arrangement of the electrodes of a beam power tube. The paths of the electrons are indicated by the light dotted portions between the cathode and the plate.

The electrodes are located in the beam power tube as illustrated in Fig. 14-11. The turns of wire comprising the screen-grid are immediately behind corresponding turns of the control-grid; therefore, most of the electrons traveling through the spaces between the wires of the control-grid continue on to the plate only slightly influenced for the most part by the screen-grid, even though this grid is positively charged. In the ordinary tetrode where the corresponding turns of wire

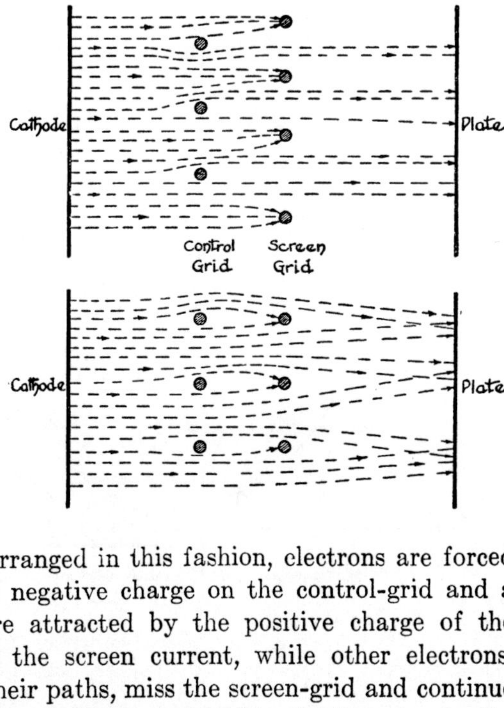

Figs. 14-12 (above), 14-13 (below). When the control-grid wires are not in line with the screen-grid wires, as in Fig. 14-12, some electrons are attracted to the screen-grid wires and do not reach the plate. When the wires of the two grids are lined up, as in Fig. 14-13, more electrons reach the plate than in the former case.

of the two grids are not arranged in this fashion, electrons are forced out of their paths by the negative charge on the control-grid and a large number of these are attracted by the positive charge of the screen-grid, thus forming the screen current, while other electrons, although deviating from their paths, miss the screen-grid and continue on to the plate. This action is illustrated in Fig. 14-12, wherein the lines indicate the paths of a few of the electrons.

Because of the alignment of the corresponding turns of the two grids in the beam power tube, a much larger percentage of the electrons go on to the plate than in the former case, because even though the electrons are attracted from their normal paths by the positive potential of the screen-grid, this deviation is insufficient to cause many of them to hit the screen-grid wires. In other words, an electron "shadow" is cast, as illustrated in Fig. 14-13, thus causing a reduction in screen current and a corresponding increase in plate current.

The pair of beam-confining electrodes, one located at each end of the grid structure as shown in Fig. 14-11, further influences the paths of the electrons from the time they pass the screen-grid until they strike the plate. These electrodes are connected inside the tube to the cathode and so are at zero potential. Because of this difference in potential between these elements and the screen-grid and the shape of these electrodes, the electrons coming from the screen are directed to a certain portion of the plate, as indicated in the top view of the tube elements in Fig. 14-14. As a result of this concentration of the electrons

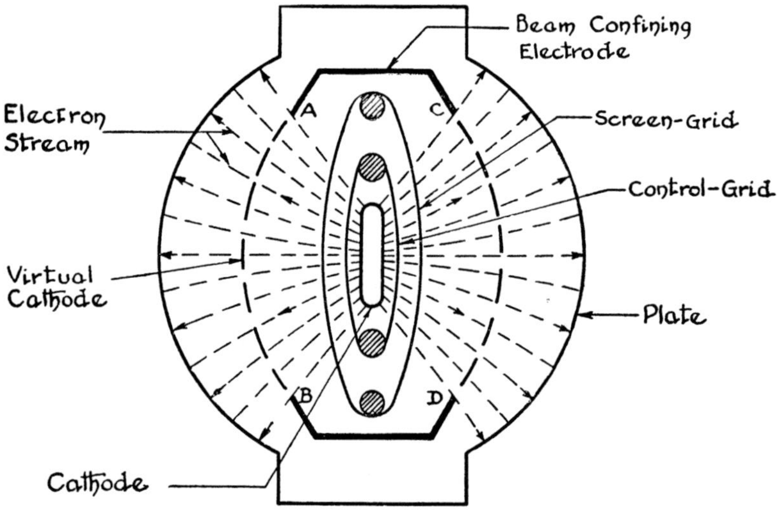

FIG. 14-14. Electrons are kept within certain bounds marked AB and CD by the beam-confining electrodes at each end of the grid structure of the beam power tube.

into a beam and their formation by the alignment of the two grids into layers or "sheets," a plane parallel to the plate and approximately indicated by the dotted line in Fig. 14-14, is at zero potential; this is called the *virtual cathode.* The effect of this virtual cathode is to turn back to the plate the secondary electrons freed from the plate in just the same way the suppressor in a pentode does. As you can see from Fig. 14-14, the electron sheets are kept within the regions indicated by AB and CD by the beam-confining electrodes; hence, because these are also at zero potential, any stray electrons that might emerge from the ends of the grid structure, will be suppressed and prevented from returning to the screen-grid. All these preventive measures tend to reduce distortion.

The relationship between the plate voltage and the plate current, is shown in Fig. 14-15. These curves were taken with the cathode tem-

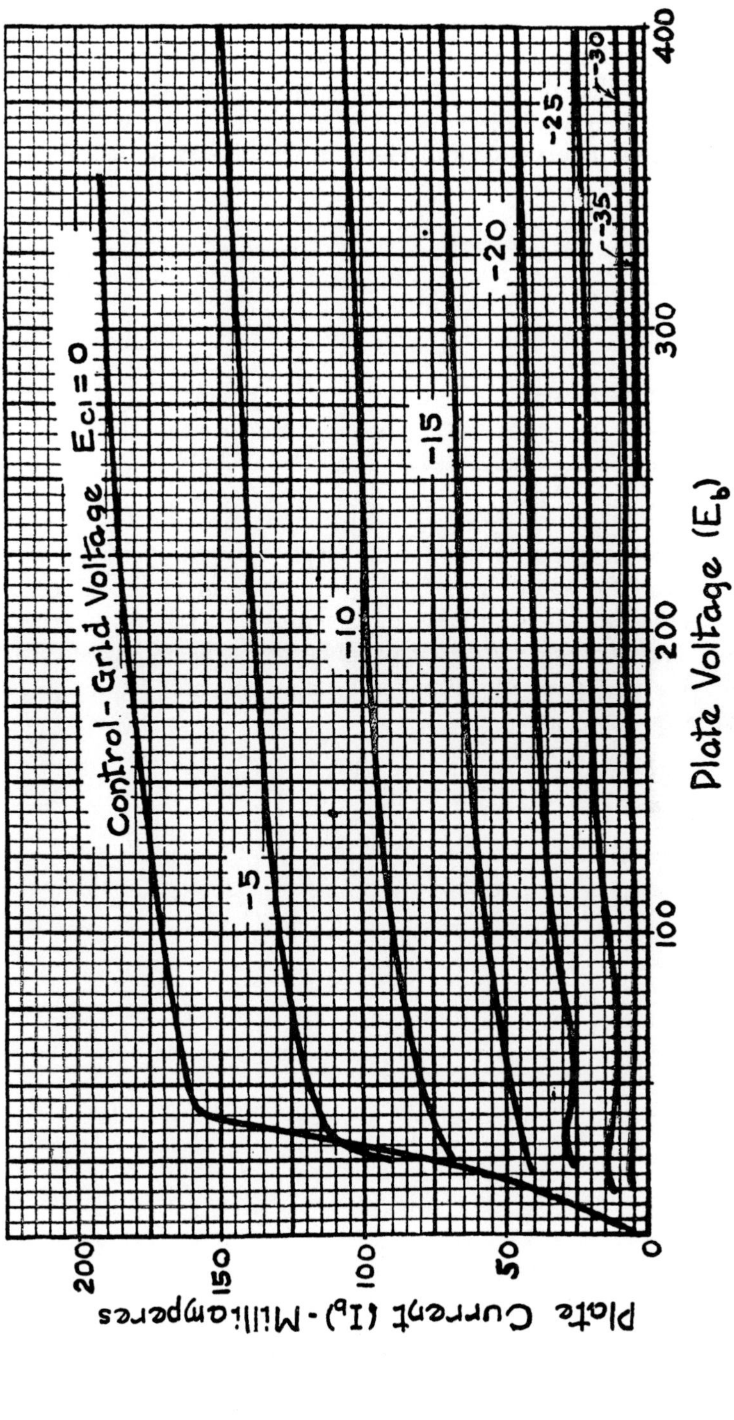

FIG. 14-15. Plate voltage—plate current characteristic curves of a beam power tube.

perature and the screen-grid voltage constant, the latter being held at 250 volts; the voltage E_{c1} on the control-grid, is marked at each curve. You will notice that the knees of each of these curves is much sharper than the knee of a pentode characteristic; this difference is illustrated in the two general curves of Fig. 14-16, in which the solid curve is for

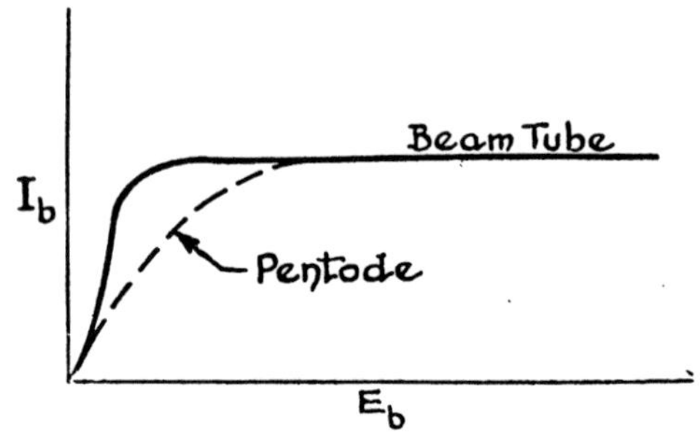

FIG. 14-16. The sharpness of the knee of the beam power plate-voltage characteristic curve is contrasted here with a similar curve of a pentode (dotted curve). The sharpness enables the beam power tube to handle greater variations of voltage and larger amounts of power with little distortion.

the beam power tube and the dotted curve is for the pentode, which coincides with the solid curve for higher values of plate voltage E_b. The difference in the sharpness of the knees of the two curves is marked. This sharpness makes it possible for the beam power tube to handle large variations of voltage and larger amounts of power with relatively low distortion.

Chapter 15

MISCELLANEOUS VACUUM TUBES

In the foregoing chapters we have explained the principles underlying the functioning of vacuum tubes from the simple diode to the more complex pentode. Static and dynamic characteristic curves have been presented of typical tubes of each of the basic classifications, so that you may understand their operating properties. You should know that these typical tubes were chosen from scores of triodes, tetrodes, and pentodes merely as examples, the thought being that with an understanding of the functioning and operating properties of typical tubes, you can discover for yourself what you require about any tube from a handbook or manual furnished by most tube manufacturers.

More than likely you have heard or read about some types of vacuum tubes which have not been mentioned thus far in this book. It is a physical impossibility to cover every device that goes under the general name of "vacuum tube," but a few have been selected for inclusion in this concluding chapter which we felt would give you at least an acquaintance with their general operating principles. No attempt has been made to go into these descriptions to any great extent as they would be beyond the scope of this book, but with the knowledge of electronic action which you have gained from the preceding pages and these brief explanations, you will have a foundation upon which you can build.

Before starting these descriptions of vacuum tubes that are found both inside and outside the radio field, it seems advisable to tell you something about the physical make-up of the tubes which you will frequently encounter, as well as the meanings of the numbers and letters by which the different tubes are designated.

Vacuum Tube and Tube Socket Construction

The earliest type of vacuum tube that was commercially available for a receiving set was a triode having a glass envelope and a plastic four-pronged base, two prongs being for the filament connection and one each for the control-grid and the plate. One or both filament prongs were larger in diameter than the grid and plate prongs in order to make certain that the tube would be inserted in the plastic tube socket correctly. When the indirectly-heated type of cathode was introduced, this necessitated the addition of more prongs on the tube base, as one more prong was required for the cathode connection. A tube of this kind is the type 76 triode which has five prongs so spaced that it can be inserted into a corresponding five-hole tube socket in one way only.

Upon the advent of the tetrode another terminal was necessary, for the screen-grid was connected to a base prong while the control-grid connection was made in the form of a cylindrical metal cap on the top of the glass envelope. Connection to this cap was made by a close-fitting spring terminal soldered on the end of a wire. The type 22 tetrode was an example of this type of construction. When the indirectly-heated type of cathode was introduced, this necessitated the addition of another prong on the tube base for the cathode terminal. An example of this was the type 24A tetrode.

Still another prong was needed when the suppressor grid was added to the four electrodes of the tetrode; thus the pentode had six prongs on the tube base, an example being the 2A5. This was not universal practice, however, for in some pentodes the suppressor is connected internally to the cathode and so only five base prongs are required in addition to the grid cap on the top of the tube. The type 75 is such a pentode and so uses a socket having five holes.

The socket for the type 59 pentode has seven holes to accommodate that number of prongs on the tube base, two being for the heater connections, and one each for the cathode, control-grid, screen-grid, suppressor-grid, and plate. The two heater prongs are larger in diameter than the other prongs on the six- and seven-prong bases and the holes in the sockets are made correspondingly large so that the tubes can only be inserted in the sockets in the one way, thus eliminating the chances of getting wrong voltages on elements that would ruin the tube.

The seven-prong tubes are manufactured in two sizes; the type 59 is a larger tube than the 6A7 and its prongs are arranged in a larger circle than the prong circle of the 6A7. One socket is necessary, however, to accommodate both tube sizes.

When metal tubes were introduced, since the metal envelope was sometimes employed as a shield, the metal envelope required a terminal for external connection; this together with the heater, cathode, the several grids, and plate, in the case of a pentode, complicated matters to a great extent when the products of the various manufacturers were marketed. As no standardization had been adopted, it was impossible to determine which terminals on the bases were connected to which electrodes without consulting the manufacturer's layout charts. Finally, this confusion was eliminated to a certain degree with the advent of the *octal base.* Primarily, it was intended to have the same element of any type of tube connected to the same prong and if any of the prongs were not used, they were to be left off the base or else no connection was to be made to them. Fig. 15-1 shows the base connections of a

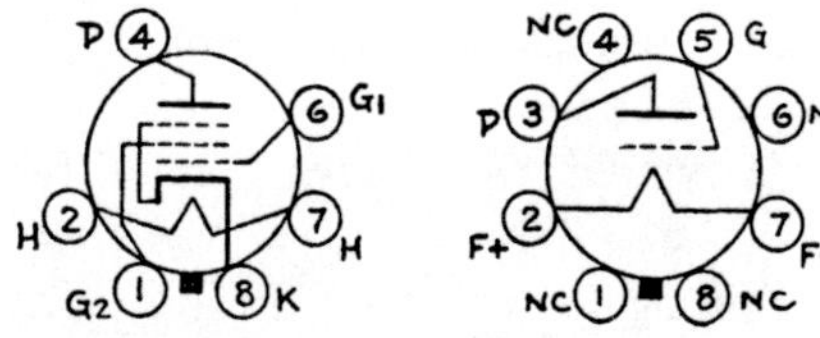

FIGS. 15-1 (left), 15-2 (right). The base of a pentode tube having 6 prongs, those numbered 3 and 5 being eliminated. Fig. 15-2 shows the base of a triode using 4 prongs, the others not being connected.

pentode wherein prongs 3 and 5 are eliminated from the base and Fig. 15-2 shows all the eight prongs on the base, but numbers 1, 4, 6, and 8 are unused, being marked "NC" meaning no connection.

It was mentioned above that it was intended to have the same elements of a tube connected to the same prongs no matter whether the tube was a triode, tetrode, or what. This system of numbering is illustrated in Fig. 15-3, which is the *bottom* view of the base. The post in the center of the base has a ridge on it, which acts as a key so that the prongs can be inserted in the holes in the socket in only one way. You will notice that the numbering of the prongs is clockwise, with prong 1 to the left of the key. The key, in the tube socket diagram Fig. 15-1 has a base of this type, and is shown as a small projection at the bottom.

In case of metal tubes, the metal envelope, which functions as a shield if desired, is connected to prong 1. The two terminals of the

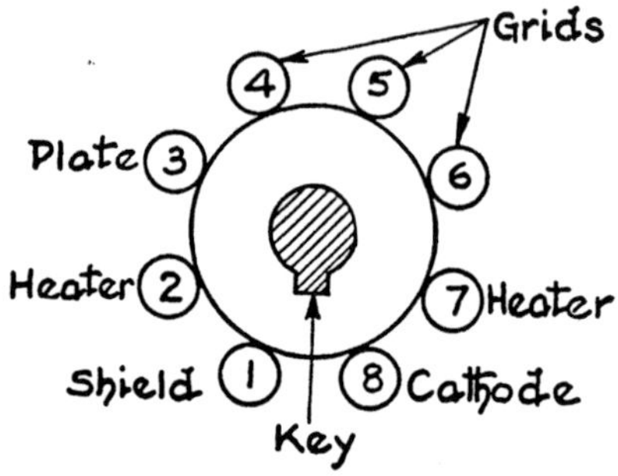

FIG. 15-3 (left). Bottom view of an octal tube base showing prongs to which elements are sometimes connected.

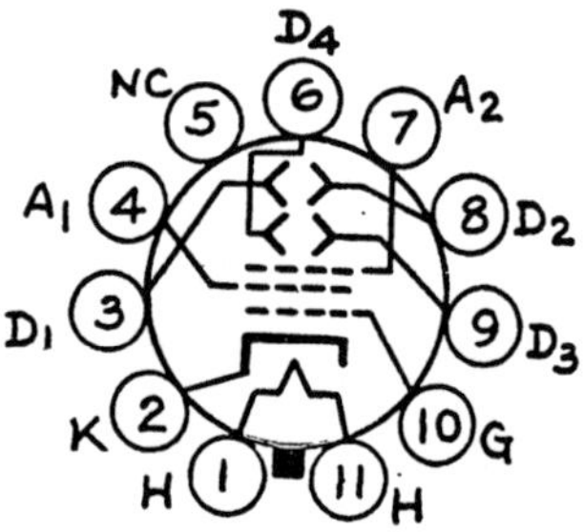

FIG. 15-4 (right). Diagram of an 11-prong tube base used with a cathode-ray tube.

filament or those of the heater are connected to prongs 2 and 7 and the plate or anode is connected to prong 3. Prongs 4, 5, and 6 are for the various grids, there being no set system for connecting these; in other words, the control grid is sometimes brought out to any one of the three prongs mentioned, although frequently to pin 5. Prong 8 is connected to the cathode.

This system was followed until the grid cap on the top of the tube envelope was abandoned and the control grid was brought down to one of the prongs on the base. The alternating current flowing in the widely separated leads to prongs 2 and 7 caused trouble; so one of the heater leads was sometimes changed over to prong 8. Other changes were necessitated, and so at the present time the system has become more or less disarranged. To be on the safe side, it is strongly suggested that a tube chart be consulted as to what prongs are connected to which electrodes of a tube.

As you can readily understand, eight prongs would be insufficient to accommodate all the connections necessary for a cathode-ray tube which has ten or more electrodes and which will be discussed later on in this chapter; hence, larger bases were produced with 11 and in some instances, as many as 14 prongs. In Fig. 15-4 is shown the connections to the deflecting plates and A1 and A2 are the prongs for the first and second anodes. In some instances, connections to the deflecting plates are made to cap terminals on the glass envelope of the tube, as is done in the case of some control-grid connections of tetrodes or pentodes. Of course, such bases as these are much larger in diameter than those of tubes used in the usual radio receiver, the former being about 2¼ inches in diameter and the latter about an inch smaller.

The sockets into which these various bases fit, as we have mentioned previously, are designed in one way or another to prevent the tube from being inserted unless the prongs are lined up correctly. The octal base has a keyed cylindrical post in the middle of the ring of prongs that fits into a corresponding keyed hole in the socket.

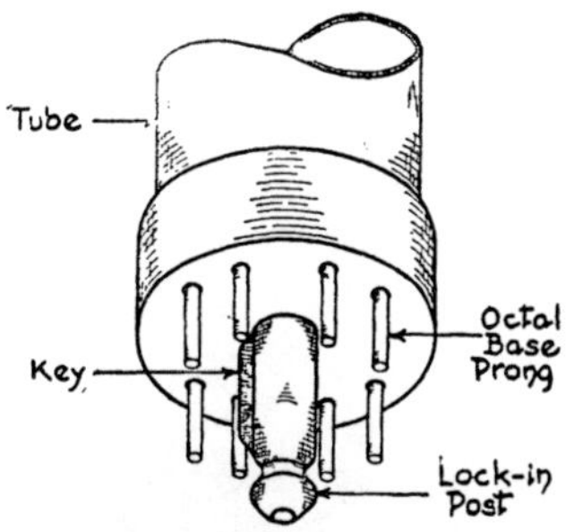

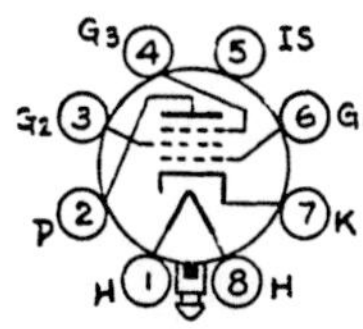

FIGS. 15-5 (left), 15-6 (right). The construction of the lock-in type base is shown on the left and the small projection at the bottom of the socket symbol indicates a lock-in base.

One of the newer types of tube bases is the *lock-in* type, a bottom view of the base being shown in Fig. 15-5. When a tube with such a base is inserted in its special socket, it has to be worked out by tilting the tube from one side to the other and then pulling straight out after being loosened, instead of just a straight pull as in the octal type. The tube base socket diagram for this type of tube sometimes differs from others in that there are two small projections (one inside the other) at the bottom of the symbol as shown in Fig. 15-6.

Another development that employed the eight-pronged octal base was the so-called *single-ended tube*. In order to do away with the grid cap on the top of the tube envelope, the internal structure of the tube was changed so that the control grid is brought out to one of the prongs on the base. Such construction permits all wiring to be done on the under side of the chassis inasmuch as no connection has to be brought out to the cap on the top of the envelope, and of course, this also makes for more rigid and permanent wiring between components of the equipment.

As we mentioned above, the earliest vacuum tubes consisted of a plastic base with metal pins or prongs, and on this base were mounted the tube elements enclosed within a glass envelope. As the elements of the tube were enlarged to accommodate greater amounts of power, the envelopes had to be enlarged correspondingly to permit the proper dissipation of the additional heat that was generated. In order to keep the overall dimensions of the tubes within reason and still have sufficient heat dissipation, metal envelopes were tried and found successful. It was found that metal being a better conductor of heat energy than glass, the metal envelope could be made smaller than a glass envelope for otherwise identical tubes. This difference in dimensions is shown

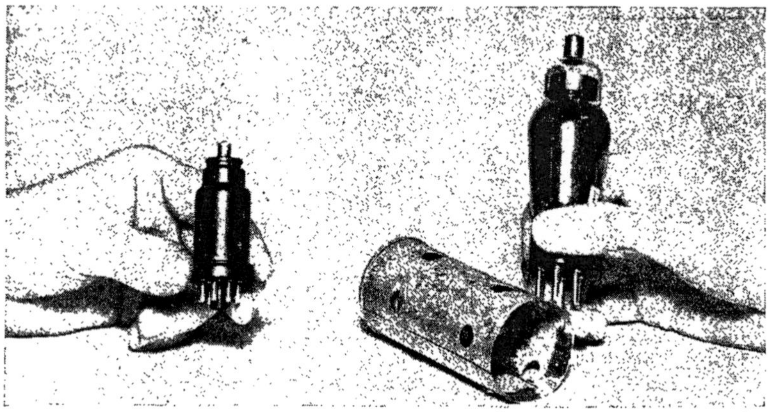

Courtesy of General Electric Co.

Fig. 15-7. The metal tube on the left can be made smaller than a glass tube of the same type because of the greater heat dissipating properties of the former. The metal shield is used to shield the glass tube against stray fields.

in Fig. 15-7. Another important factor is that the metal envelope acts as a shield against fields which would disturb the proper operation of the tube; this eliminated the metal shield that is also shown in Fig. 15-7, which is necessary when glass tubes are used in circuits when the frequency is high.

The internal construction of a metal tube is shown in Fig. 15-8. This is actually a combination of metal and glass, because the latter is used in several places for insulation and also the exhaust tube, through which the tube is evacuated, is of glass. One term in this illustration may be unfamiliar to you: the *getter*, item 11. This is a strip of magnesium, barium, or some other chemical which when heated to a temperature between 800° to 1000° F., will combine with any gases left within the tube and so make the vacuum that much better.

Another development is the special small and compact seven-prong miniature glass tube. One such tube is the 1L4 pentode, shown in Fig. 15-9, and its seven prongs are so spaced that the tube can be inserted into the socket in one way only.

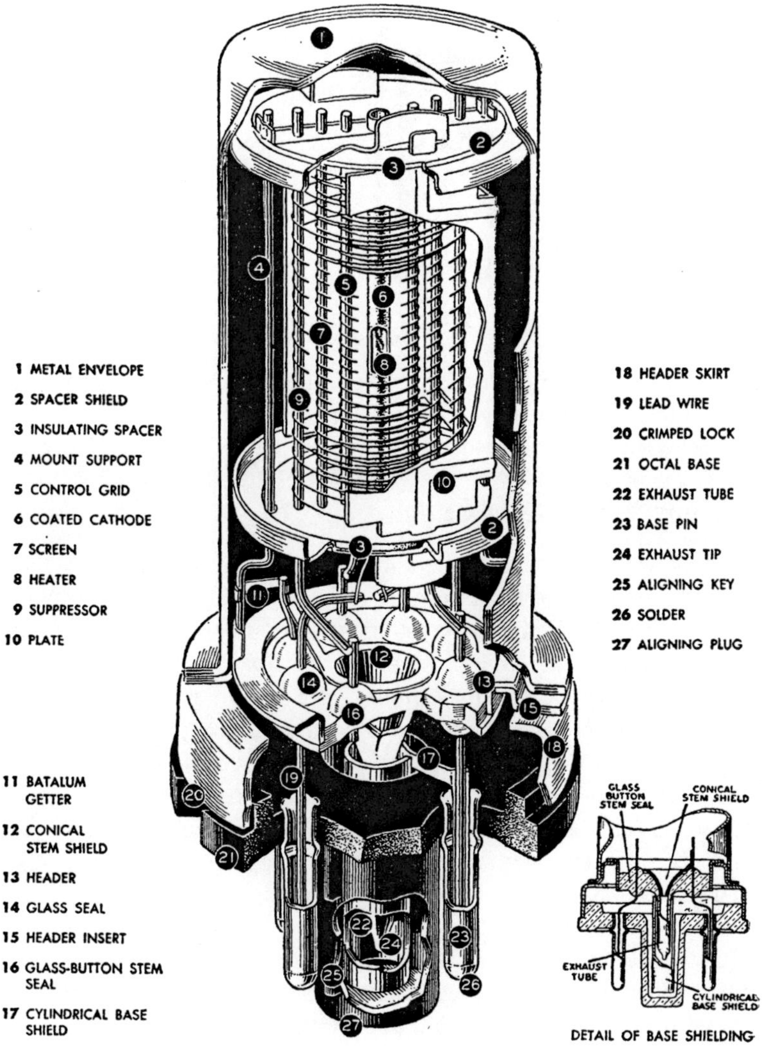

Courtesy RCA Mfg. Co.

FIG. 15-8. Constructional details of a metal pentode.

These popular miniatures are mostly high-frequency tubes wherein the time of travel of the electron between cathode and plate, (electron transit time) and the inter-electrode capacitance between the elements

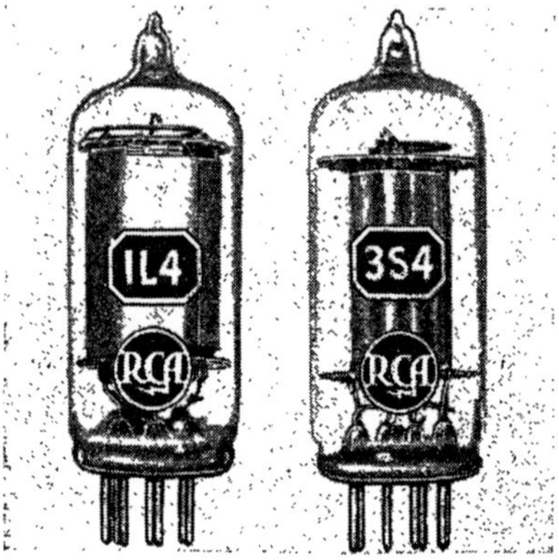

FIG. 15-9. Two types of miniature glass tubes. Because of their small size a great many tubes of this type can be used at radio frequencies, but some of this type are designed to be used at audio frequencies.

Courtesy RCA Mfg. Co.

are kept to a minimum. The type 6AL5 is such a tube; whereas the 3S4 pentode miniature, shown in Fig. 15-9, is one of several which are not intended for high-frequency use.

Vacuum Tube Designations

Every vacuum tube about which you will ever read or which you will use, is identified by a number or by a combination of numbers and letters. Unfortunately, the system of identification has not been uniform throughout the years. In the search for a system which will embrace all the types of tubes manufactured and in the changes necessitated by the transfer from one system to another, a certain amount of confusion exists. The following paragraphs outline the systems that have been used, but because of the numerous exceptions to almost all the cases, for absolute certainty regarding the functioning of a given tube, its operating voltages, etc., it is suggested that you consult a vacuum-tube handbook to find specific data on any tube in which you might be interested.

In the early 1920's, different tube manufacturers identified their own products by different letters prefixing the type number of the tube; for instance, the 2-volt triode manufactured by RCA was called WD-11 while the same type tube distributed by the Cunningham Co. was called C-11. In some cases manufacturers distinguished their tubes by giving them a certain number as well as an individual letter or letters, as UX-201A, UX-222, UY-227, etc. made by RCA and Cunningham's

CX-301A; CX-322, C-327; these were respectively identical tubes. Here the UX and UY identified different types of bases. Eventually the distinguishing letters and numbers were dropped, the designations becoming '01A, '22, '27, etc.

Such numbers had nothing at all to do with the tube function, filament voltage, etc., and the only way to determine anything about a tube was to look it up in a tube handbook. It was plain that some better system was needed, so in 1933 a systematic method of designations was developed.

The type number of a tube was divided into four parts, the last part not always being employed. These are:

(1) A number of one or more digits, designating the filament or heater voltage.

(2) One or more letters to designate the type or function of the tube.

(3) A number designating the number of useful elements in the tube.

(4) One or more letters designating the size or construction of the tube.

Let us take as an example, the designation of the power amplifier triode, 6A3. The first number shows that the filament voltage is 6.3 volts, the A indicates an amplifier, and the 3 stands for the three useful elements: the filament, control-grid, and the plate. (Notice that the fourth part of the designation is missing.) Another tube, a power amplifier pentode, the 6G6-G has the same heater voltage, namely 6.3 volts, the first G indicates an amplifier, six useful elements (the heater, the cathode, control-grid, screen-grid and suppressor-grid, and the plate), and the final G indicates a glass envelope of a certain size and base construction.

This type of identification was satisfactory until the metal-envelope tubes were put on the market. The metal envelope or *shell,* as it is commonly called, was brought out to a terminal on the base of the tube, as it can be used for shielding. Thus a metal triode of the indirect-heater type was designated as 6F5, for instance; this had the shell terminal in addition to the heater, cathode, control-grid, and plate. When a triode of the same characteristics was produced but with a glass envelope instead of a metal shell, the same designation 6F5 was used with the addition of the letter G. It must be noted, however, that this system of indicating a glass envelope is not universal, because the 6G5 is a glass tube of the electron-ray type and has five elements, yet its number is not followed by the letter G.

Other advances in tube design continued after the introduction of the metal tubes, and these necessitated other designations; these changes were the lock-in construction, the so-called single-ended tubes that eliminated the terminal on the top of the tube, and then the reduction in the size of the tube on many types.

In order to distinguish between the lock-in type of base and the standard bases that were then available, lock-in tubes having filament or heater voltages of 6.3 and 12.6 volts, were numbered 7 and 14 respectively, as the 7A5, a power amplifier pentode and the 14A5, a beam power amplifier. Inasmuch as the voltage designations of the 1-volt and 2-volt tubes already have a certain meaning, which will be explained in a moment, another designation had to be used. This was the addition of the letter L to the letter group in the type number; for example, 1LA4 is a power amplifier pentode with a lock-in base. It is unfortunate, but this system was not followed rigidly and so some exceptions to these rules have to be considered.

All the lock-in type of tubes have all the terminals on the base; in order to take advantage of this feature, the so-called *single-ended* tube was developed, thus eliminating the cap on the top of the tube. As single-ended tubes were put out that were similar in characteristics to existing tubes and as these were well-known by their type numbers, it was decided to add the letter S to the old type numbers. For example, the 6SJ7 is similar to type 6J7, but the former has all the connections at the base. This notation does not always hold true, for the 6SA7 is not similar to the 6A7.

The reduction in the size of the glass envelope was the next development that necessitated a change in the type numbering system. As the characteristics of the tubes were unchanged even though the tube itself was smaller, the letter T was added to the type numbers and letters. For example, the 6R7GT is the same electrically as the 6R7 and the 6R7G. In order to reduce the number of tubes, the decision was made to mark these tubes 6J7GT/G, which means that such a tube can be used to replace either the 6J7G or the 6J7GT.

Now let us consider the four groups of numbers and letters as listed above that make up the tube type designations. Below you will find the numbers that indicate the filament or heater voltages together with the various exceptions.

Number	*Voltage of Heater or Filament*
0	Requires no heater supply—cold cathode (Note 1)
1	Between 0.1 and 2.1 volts (Note 2)
2	Between 2.1 and 2.9 volts (Note 3)
3	Between 3.0 and 3.9

Number	*Voltage of Heater or Filament*
5	Between 5.0 and 5.9
6	Between 6.0 and 6.9
7	Between 7.0 and 7.9 (Note 4)
12	12.6 volts (Note 5)
14	12.6 volts
20 and above	Actual value of voltage

Note 1. Exception: The 01A from the old numbering system.

Note 2. Exception: The 1V from the old numbering system.

Note 3. The following tubes have filament voltages of 1.4 for series connection and 2.8 for parallel connection: 3A4, 3A5, 3A8GT, 3B5GT, 3C5GT, 3LF4, 3Q4, 3Q5GT, and 3S4. The 3LE4 has a filament voltage of 2.8.

Note 4. This is a "nominal" rating; the filament or heater voltage is 6.3.

Note 5. The 12A5 has a heater voltage of 6.3 for series connection and 12.6 for parallel connection.

It should be noted that many tubes have merely numbers, such as 1231 and 1851, that bear no relation to the filament or heater voltage. In the case of cathode-ray tubes, the first number indicates the diameter in inches of the screen; for example, the 3BP1 has a 3-inch screen, the 5AP4 has a 5-inch screen, etc.

The first letter group following the voltage designation was intended to indicate the function of the tube. The letters from U to Z inclusive are used for rectifiers, but the following types are exceptions to this: 6U5, 6U6GT, 6U7G, 6V6G, 6V7G, 6W7G, 6Y6G, 6Y7G, 6Z7G, 7V7, 7W7, and 14W7; these are electron-ray indicators or amplifiers of some type.

As explained above, the letter S preceding another letter, indicates a single-ended tube, and the letter L preceding another letter indicates a lock-in type base with the exceptions as noted above. Other letters of the alphabet before U are used for the most part to designate amplifiers of different types, but are employed more or less indiscriminately and with very few exceptions they may represent almost any type of amplifier. One or two exceptions should be noted: the letter H is generally used to indicate a diode; the letter P is used to indicate a cathode-ray tube. In certain instances combinations such as AB, AC, AD, AE, etc. are used, but this was done because the single letters of the alphabet had all been used.

The third portion of the type number are numbers from 1 up to 8

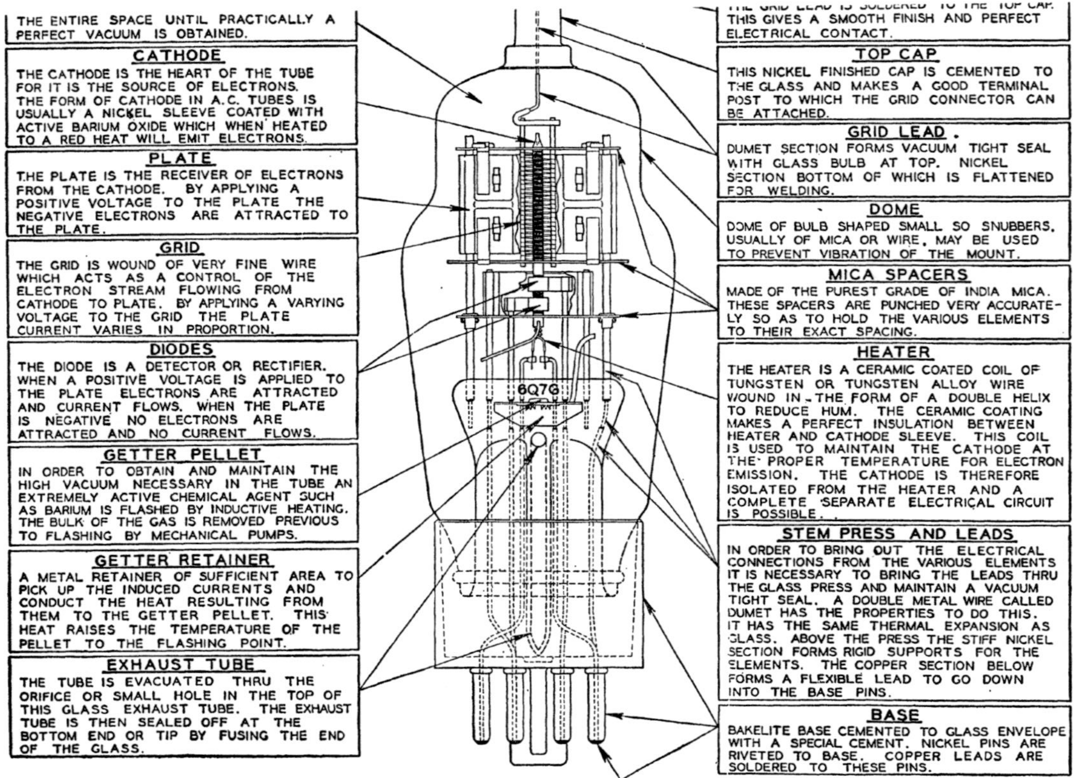

Courtesy of Ken-Rad Div. General Electric Co.

Constructional details of a duo-diode triode indirectly-heated tube.

that indicate the number of useful elements that are brought out to terminals on the base of the tube or as explained above. In some instances, the shielding in a metal tube is counted as an element, but this does not always hold.

The final group of letters has been explained above, as far as glass-envelope designations are concerned. Other letters than G and GT are sometimes employed but very seldom encountered. For example, the letter E is added at the end of the type numbering to designate a tube used for export.

Multi-Purpose Tubes

In order to gain compactness and to reduce costs, several combination tubes have been developed in which are incorporated the elements of more than one tube. One of the first of these tubes was the full-wave rectifier in which two diodes changed the a-c input into a pulsating direct current that can be utilized as the operating voltage on tubes. While this type of tube with its two sets of elements functions as a single tube, the other tubes outlined below, are really *multi-purpose*

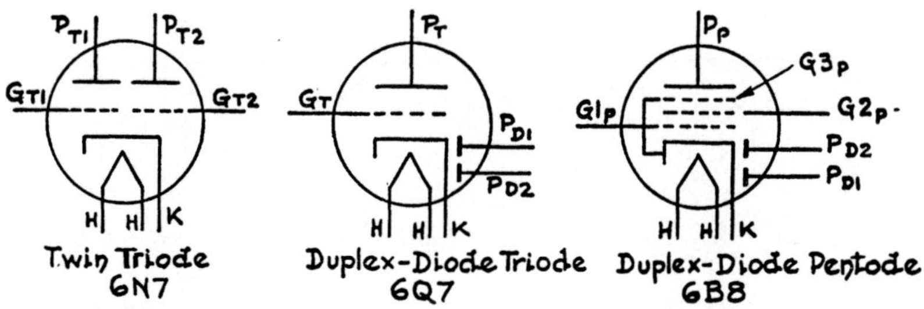

FIGS. 15-10, 15-11, 15-12 (left to right). Schematic symbols of three multi-purpose vacuum tubes.

tubes in that more than one function is performed by the electrons liberated from the same cathode, in most cases.

One of the simplest of these multi-purpose tubes is the twin triode, the 6N7, the schematic symbol of which is shown in Fig. 15-10. Here you see really two separate triodes *(K,* G_{T1}, P_{T1}, and *K,* G_{T2}, P_{T2}*)*, each having the same characteristics, which can be so connected that they function either separately, or together if the two grids and the two plates are connected respectively in parallel. Of course, when the triodes are connected in a circuit separately, each functions according to its individual characteristics; when the triodes are connected in parallel (i.e. the two grids are connected together to the input, the two plates to the output, and the cathodes connected together) then the plate resistance of the combination is halved, the mutual conductance is doubled, but the amplification factor remains the same as for a single triode.

Many of these types of tubes can not be connected to function as a single unit inasmuch as the whole idea is to have a tube wherein more than a single function occurs. In the duplex-diode triode, type 6Q7, shown in Fig. 15-11 three different functions can be obtained: each of the diode plates P_{D1} and P_{D2} will perform separately as well as the triode K, G_T, and P_T. A similar tube is the duplex-diode pentode, the 6B8 shown in Fig. 15-12. Notice that even though the suppressor grid $G3_P$ of the pentode portion of the tube is connected internally to the cathode, this does not interfere with the separate functioning of the two diodes P_{D1} and P_{D2}. A triple function tube, the diode triode pentode, the 6M8GT, is shown in Fig. 15-13. Here only a single diode plate P_D is incorporated. K, G_T and P_T are the triode section, and K, $G1_P$, $G2_P$, $G3_P$ and P_P are the pentode section.

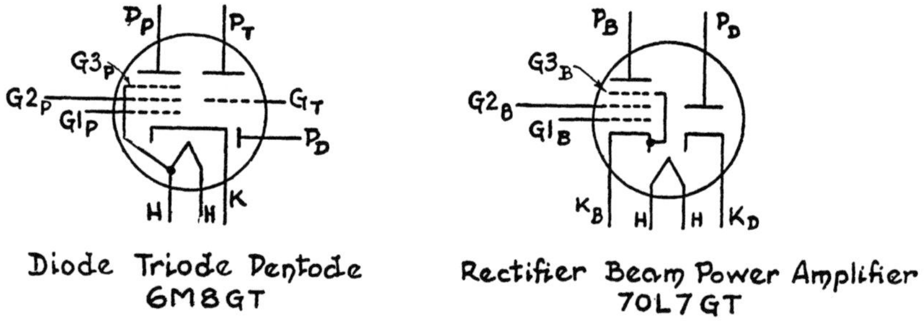

FIGS. 15-13 (left), 15-14 (right). On the left is the symbol of one type of a triple function tube, and that on the right is the schematic of one type of dual purpose tube.

In certain circuits it is desirable to have the heaters of the tubes connected in series; in order to avoid using automatic current control devices, (ballast tubes) and at the same time to obtain the proper voltages throughout the circuit, heater circuits were designed in certain tubes to function at a higher voltage than is usually encountered. Such a multi-purpose tube is shown in Fig. 15-14. This is a rectifier—beam power amplifier, the 70L7-GT, which, as the first number indicates, is designed to operate with a heater voltage of 70. Notice that this tube has two cathodes K_B for the beam power amplifier section and K_D for the rectifier, whereas the most of the other tubes have but one. This tube functions as a half-wave rectifier and an output beam amplifier.

In all these multi-purpose tubes the separate sets of elements function in exactly the same way as if each had been in an envelope of its own, with the exception, of course, of a tube like the twin-triode, of Fig. 15-10, when the elements are connected in parallel. It should be noted that a connection like this is rarely used as this defeats the main purpose of this type of tube.

Variable-Mu or Supercontrol Tube

The transconductance of a tube depends upon the physical arrangement of the electrodes and upon their construction among other things. For instance, the closer the control-grid is placed to the cathode, the more effective it is in controlling the electrons passing through it on their way to the plate.

Now what would be the effect if the turns of wire of the control-grid of a pentode were unequally spaced throughout its length, i.e. less space between the turns of wire at the ends of the grid and more space between them in the middle, as shown in Fig. 15-15? To illustrate the

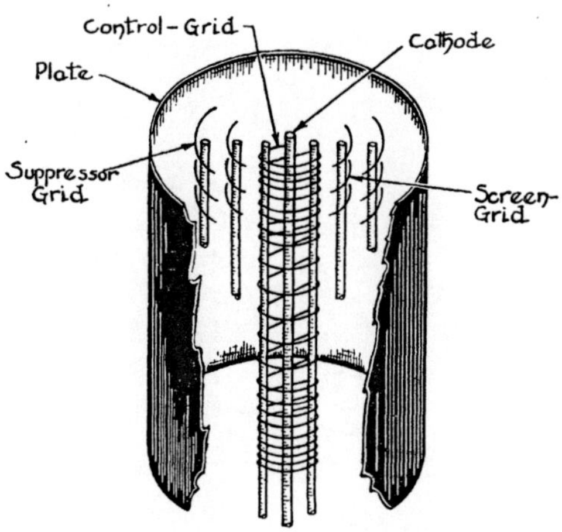

FIG. 15-15. The control-grid of the variable mu (variable mutual conductance) tube has unevenly spaced turns, those at the middle being more widely separated than the turns at the ends.

effect this would have on the grid voltage—plate current curve, let us assume that we have a single-stage two-tube amplifier, such as that shown in Fig. 15-16, wherein the tube T_1 has a high transconductance, and the tube T_2, which is connected in parallel with the first one, has a low transconductance. The plate circuit of each tube is connected to a separate primary winding of a three-winding transformer, the secondary being common, which means that any current flowing in either of the primaries will induce a corresponding voltage in the secondary. The corresponding $e_c - i_b$ curves for these two tubes are indicated in Fig. 15-17.

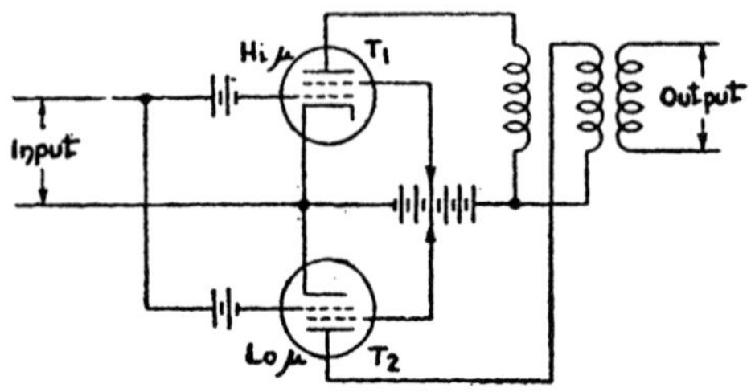

FIG. 15-16. Tube T_1 is a high-mu tube and T_2 is a low-mu tube. The combination results in a characteristic that is different from either of the individual tubes.

First assume that a low-voltage signal is applied to the input circuit of the combination. The value e_1 of this signal being low, will result in a relatively high output from T_1 and a low output from T_2 of Fig. 15-16. These theoretical values can be seen by points of intersection of the vertical line from e_1 and the two curves in Fig. 15-17. When

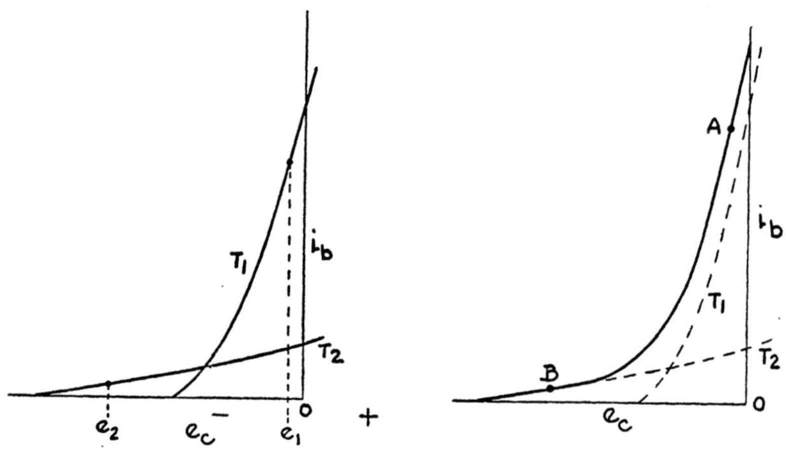

FIGS. 15-17 (left), 15-18 (right). The curves marked T_1 and T_2 are the dynamic transfer characteristics of tubes T_1 and T_2 respectively of Fig. 15-16. The curve on the right is a combination of the two curves on the left and is the characteristic of a variable-mu amplifier.

the value of the signal is raised to some point e_2, then only the tube having the low transconductance will have any output, as e_2 is beyond the cutoff point of T_1. In other words, when the grid bias is low, most of the amplification in the output is due to T_1, but as the grid bias is raised (i.e. made more negative) then T_1 functions less and less.

Inasmuch as these two tubes have a common input and a common output and as they function as one tube, we can combine the two characteristic curves; this has been done in Fig. 15-18. The common characteristic is drawn solid and the dotted curves, which are repeated here from Fig. 15-17, are included to assist you to visualize the combination. Having such a characteristic curve, the question arises, "Is a single tube capable of varying its transconductance with a change in grid bias so as to have a curve such as this?" The answer is in the affirmative, for if the spacing of the windings of the control grid be similar to those of Fig. 15-15, for example, if the grid's turns are close together at the top and bottom and widely separated in the middle, then a curve like the solid one in Fig. 15-18 will be realized. Such a tube is called a *variable-mu tube* in order to indicate that it has a variable mutual conductance (transconductance), the *mu* being derived from the first two letters of the word mutual.

Now let us see how a tube like this functions. If a signal that is weak, that is one with a low value of voltage, is impressed on the input circuit of this tube, the operating point will be some point on the characteristic curve such as *A* in Fig. 15-18. Inasmuch as the grid bias is only a few volts, the charge on the control grid is relatively low with very little hinderance to the passage of the electrons resulting. If a strong signal is applied to the grid circuit, the operating point may now be some point *B* on the characteristic curve of Fig. 15-18. Here the grid bias on the tube is much greater than formerly and if the turns of the control grid were equally spaced throughout its length, no electrons would be able to get past it to the plate, i.e. the tube would be biased beyond the cutoff point. Because of the variable spacing of the control grid turns, this does not occur. The charge at the two ends of the control grid, where the turns are relatively close together, is such that the electrons are prevented from passing through the turns, but in the mid-section, some electrons do get through because of the wider spacing of the turns and the smaller field that is there effective. Hence, even though the grid bias is very much greater than in the former case, some plate current does flow, as may be seen from Fig. 15-18.

You can get an idea of the relative functioning of a variable-mu (remote cutoff) tube as compared to a sharp cutoff pentode if you will compare the two $e_c - i_b$ curves shown in Fig. 15-19. In both cases, the plate voltage E_b was 250 volts, the voltage E_{c2} on the screen-grid was 100 volts, and zero volts on the suppressor E_{c3}. You will note that the sharp cutoff pentode curve, marked 6SJ7, is much steeper than that for

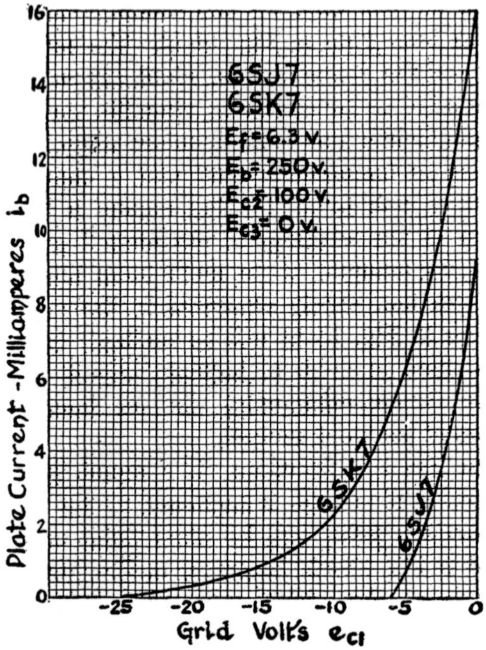

FIG. 15-19. The grid voltage—plate current curve marked 6SJ7 is for a sharp-cutoff pentode and that marked 6SK7 is for a remote-cutoff (variable mu) tube. These show that the 6SK7 can be operated at a grid voltage that is about five times more negative than the other.

the variable-mu, the 6SK7; also that the 6SK7 extends nearly five times further to the left, i.e. to −25 volts. This means that the 6SK7 can be operated at a point that is somewhere between 15 to 20 volts more negative than the 6SJ7. It is because the 6SJ7 has a plate-current cutoff for a small value of negative grid voltage (−6 volts) that it is called a *sharp-cutoff* tube, and since plate-current cutoff for the 6SK7 requires a large value of negative grid voltage (−25 volts) it is called a *remote-cutoff* tube or sometimes a *supercontrol* tube.

The plate family of characteristics for the 6SK7 is shown in Fig. 15-20. In general it may be stated that sharp-cutoff tubes are employed in audio-frequency amplifiers, while remote-cutoff tubes are utilized in radio-frequency amplifiers. In an audio-frequency amplifier, the quiescent grid bias is located at approximately the center of the linear portion of the characteristic of the tube in order to minimize distortion. The characteristic of a remote cutoff tube is extremely curved and this type of tube is therefore not suitable for an audio-frequency amplifier because the resulting distortion would be excessive. On the other hand, in a radio-frequency amplifier the quiescent grid bias usually is different for each of the various radio stations that are

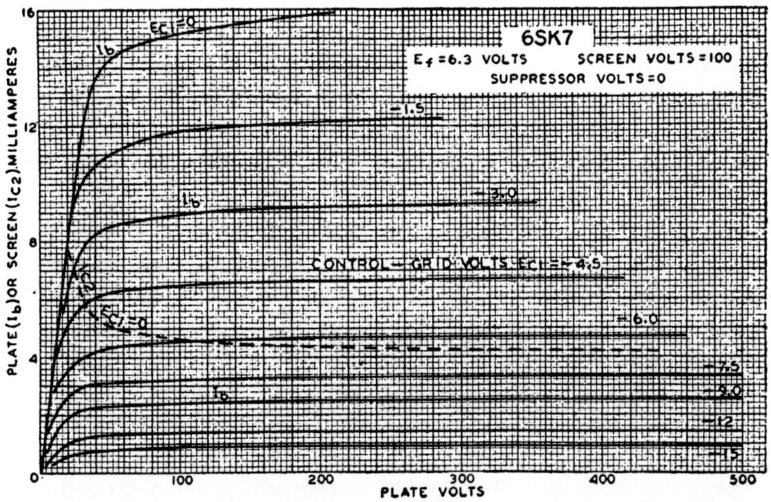

Courtesy RCA Mfg. Co.

FIG. 15-20. The plate family of the 6SK7.

received, and the curvature of the characteristic of a remote cutoff tube is an advantage, since the transconductance varies throughout the length of the characteristic. In this case, any distortion that is produced as a result of the curvature of the characteristic is unimportant, since it is eliminated by the associated tuned circuit.

Acorn Tubes

The triodes and pentodes about which you have read in previous chapters have been designed to operate at frequencies encountered in the ordinary receivers such as you have in your home. As you are doubtless aware, frequencies are in use today that are measured in millions of cycles (megacycles) instead of kilocycles (thousands of cycles) and for such ultrahigh frequencies the ordinary vacuum tube becomes practically useless. Hence triodes and pentodes that will function efficiently above say 200 megacycles have had to be developed. While a comprehensive explanation of the operation of these tubes is far beyond the scope of this book, yet it is felt that some mention should be made to acquaint you with these small *acorn* triodes and pentodes.

Throughout this text you have read about the electrons traveling from the cathode to the plate but did you happen to think of the length of time that is consumed by these inter-electrode travelers in making the trip? The time taken by an electron to go from the cathode to the plate can be neglected when considering the ordinary triode or pentode used in a conventional receiver or amplifier, because this time is extremely short compared to the time of one cycle of the frequency at which the apparatus is working. For example, at voltages encountered in ordinary tubes, it takes about one thousandth of a microsecond (1/1000 of a millionth of a second) for electrons to go from the cathode to the plate. Now at 1000 kc (1.0 mc), this represents only 1/1000 of a cycle, which is negligible. This is not so true at 200 mc, where this transit time represents 0.2 cycle. When the frequency of the signal voltages approaches the neighborhood of 200 mc or higher, then the time consumed by the electron going from cathode to plate, must be considered. You can readily see that if the signal voltage is of such a high frequency that it reverses the charge on the control grid before the electron has time to move beyond its influence, then there is bound to be an unwanted change in the behavior of the electron.

One way to overcome this handicap of the electron *transit time* problem, is to reduce the distance between the electrodes involved. It was found that if the physical dimensions of a vacuum tube were divided by some number *n,* such as 2, 3, 4, etc., the plate current, transconductance, amplification factor, and plate resistance will be prac-

Courtesy RCA Mfg. Co.

FIG. 15-21. The 954 acorn tube on the left is a pentode and the 955 is a triode. Their size can be estimated by comparing the tubes to the golf ball.

tically unaffected, but the transit time and the capacities of the tube elements will be reduced according to the number used as a divisor. However, the mechanical problems involved in the manufacture of the tubes make it unfeasible to reduce the dimensions of the tubes to too great an extent. In Fig. 15-21 are illustrated the acorn pentode (954) and triode (955), which you can see are smaller than a golf ball. The similarity of the triode to an acorn, in shape and size, is the reason these are called *acorn tubes*.

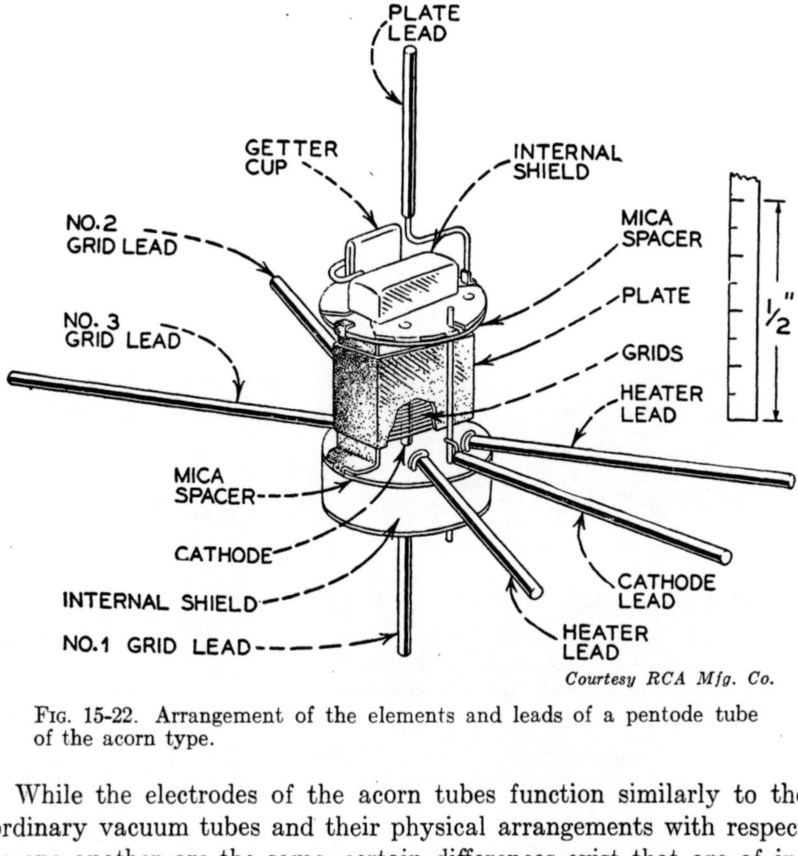

Courtesy RCA Mfg. Co.

FIG. 15-22. Arrangement of the elements and leads of a pentode tube of the acorn type.

While the electrodes of the acorn tubes function similarly to the ordinary vacuum tubes and their physical arrangements with respect to one another are the same, certain differences exist that are of interest. The cylindrical cathode is about one-half the length of a common pin and about the same diameter as a pin. The control grid is elliptical instead of round, as are the other grids in the pentode. The spaces between all the electrodes along the short axis of the ellipses are quite small where the emission from the cathode is most effective. Unlike ordinary tubes, the acorn triode and pentode tubes do not have the usual pronged base; instead mica discs called *spacers* keep the electrodes fixed in their position. These are indicated in Fig. 15-22.

The assembly is placed within a glass envelope consisting of two parts which are sealed together at the level of the lower mica spacer. Through this glass seal of the triode are brought out the five leads that are therefore in the same plane. In the pentode the plate lead is brought out through the top of the envelope and the control (No. 1) grid lead through the bottom. The arrangement of the elements of the pentode acorn tube is illustrated in Fig. 15-22. In passing, we want to call your attention to the location of the leads to the control (No. 1) grid and plate: remember that the plate is generally shown at the top of the elements in a vacuum tube symbol and that is where its lead is on a pentode acorn tube. The top and bottom of these pentodes look almost alike, but you can identify the top by the type number of the tube etched on the glass envelope. The triode acorn tube has a flattened top and identification of the top and bottom presents no difficulties.

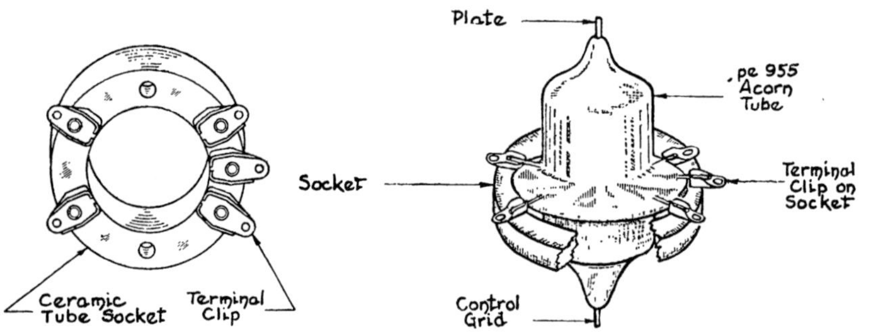

FIGS. 15-23 (left), 15-24 (right). The ceramic socket for an acorn tube, showing the terminal clips. Connections to the control-grid and plate of the pentode are made directly to the leads from the tube instead of to clips on the socket.

A special ceramic tube socket, as shown in Fig. 15-23, is used for acorn tubes. As illustrated in Fig. 15-24, the tube is inserted in the large hole in the center of the socket and is held securely in place by its five prongs which fit into the terminal clips on the socket.

Cathode-Ray Tubes

In the vacuum tubes described up to this point, the first consideration has been the *number* of electrons that pass from the cathode across to the plate: the plate current and how it is employed. In the cathode-ray tube, the *direction* the electrons take when they are emitted from the cathode is the all-important factor. Of course, *how many* electrons go in a certain direction does have some bearing, but the main thing is *where* the electrons go, not how many.

While the cathode-ray tube certainly comes under the general classification of vacuum tubes, inasmuch as it is evacuated to a high degree, and has a cathode, a grid, and an anode, yet the electrons emitted from the cathode do an entirely different job from what they do in a triode or pentode. In these tubes classed as amplifiers, detectors, etc., the main concern is how many electrons get into the plate circuit and how they can be put to work there. In the cathode-ray tube, the emitted electrons remain within the confines of the glass envelope and expend their kinetic energy in another way: the mechanical energy of their motion is transformed into light energy. The cathode-ray tube, although a vacuum tube in the strict sense of the word, has nothing to do with the amplification of voltage; instead it provides a visual means of indicating the wave-forms of alternating voltages and currents.

A cathode-ray tube consists of an electron source comprised of a heater and cathode, as shown in Fig. 15-25. Just beyond the cathode

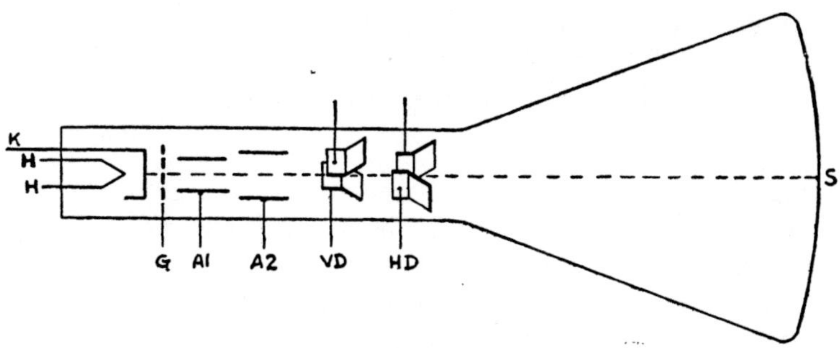

FIG. 15-25. The fundamental elements of a cathode-ray tube.

K, which is maintained at emitting temperature by the heater *H,* is a series of electrodes *G, A1,* and *A2,* which concentrate the electrons into a narrow beam and increase their velocity; these electrodes are the grid, and the anodes respectively. Between these electrodes and the large end of the tube there is located means for deflecting the electron stream both horizontally and vertically. These are two pairs of plates, *VD* and *HD* in Fig. 15-25. On the inner surface of the large end of the tube, which is only slightly rounded, is a chemical coating *S,* called the *screen,* which has the property of fluorescing at the point where the high-speed electrons strike it. Thus the electron stream striking the screen, is made visible as a small spot of light. By means of the deflecting electrodes, this spot can be moved over the screen in any direction and at such a speed that it appears as a continuous line of light.

The Electron Gun

Let us consider briefly the first-mentioned electrodes in the preceding paragraph, these electrodes in combination being called the *electron gun*. The grid G is cylindrical and almost surrounds the cathode K, as shown in Fig. 15-26. The grid is maintained at a negative potential

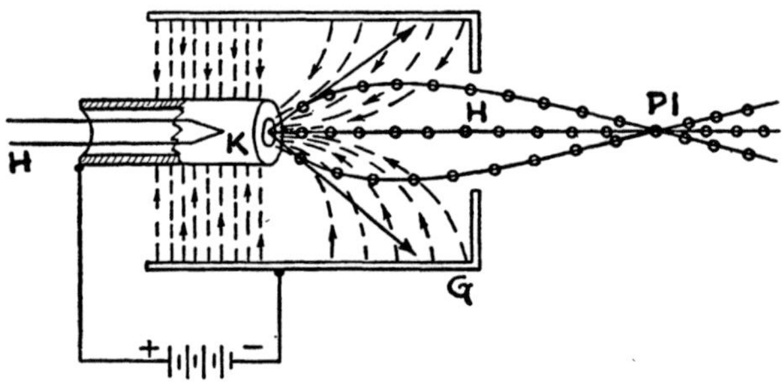

FIG. 15-26. The elements of the electron gun of the cathode-ray tube. The arrows on the dashed lines indicate the direction in which electrons are urged to move by the field, the resulting path of the electrons being indicated by the series of dots.

with respect to the cathode and this means that electrons coming from the end of the cathode are repelled by the negative field (indicated by the dotted arrows) set up by the grid so that most of them pass through the hole H in the end of the grid and to focus at point *P1*. The electron rays then start to diverge. This is controlled by other elements as shown in Fig. 15-27.

The purpose of anodes 1 and 2 is to accelerate the electrons forward, and also to cause the electron beam to become focused on the fluorescent screen. Although both anodes 1 and 2 are at a positive potential with respect to the cathode, the potential of anode 1 is lower than that of anode 2. Therefore an electric field (indicated by the dotted arrows) exists between the two anodes, this field being quite concentrated on the inner surface of the small-diameter first anode, and more spaced and less intense on the large-diameter second anode.

When electrons enter the anode 1 cylinder they are constrained to move along the direction of the lines of force of the electric field, and therefore towards the axis of the anode. The amount of this deflection is dependent upon the intensity of the electric field and the forward velocity of the electrons. When the electrons enter anode 2, they are again constrained to move along the lines of the electric field, which is now in a direction to tend to force the electrons away from the axis

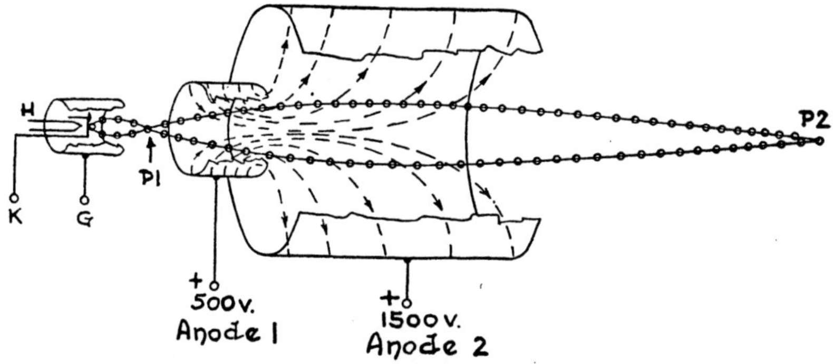

FIG. 15-27. The function of anodes 1 and 2 is to increase the velocity of the electrons and to focus them in the paths indicated by the series of dots that outlines the limits of the electron stream.

of the anode. This force on the electrons away from the axis in the interior of anode 2 is very much less than the force on the electrons toward the axis in the interior of anode 1 because of the difference in their respective field strengths. Therefore the overall tendency is for the electrons to converge towards the axis and so they focus on a point *P2* on the fluorescent screen.

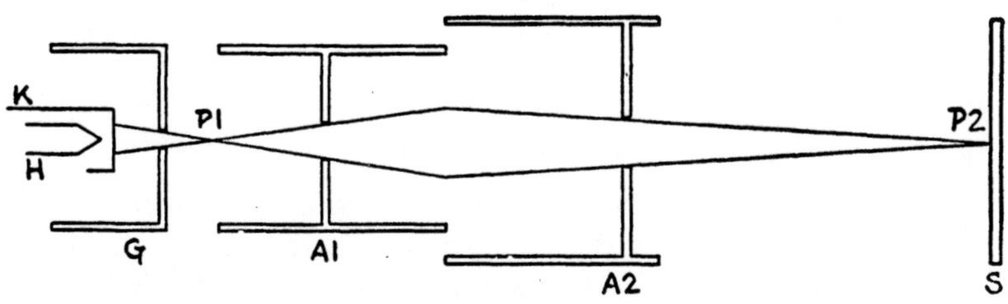

FIG. 15-28. The electron stream converges at P1 just beyond the grid, then diverges, and finally is concentrated to a small spot P2 on the fluorescent screen of the tube.

The combined action of the various focusing effects of the grid and the two anodes is illustrated in Fig. 15-28.

The Fluorescent Screen

The fluorescent screen *S* is the destination of the electron beam after it passes through the deflecting plates, which will be explained in a moment. As mentioned above, when the fast-moving electrons of the beam strike the coating on the inner surface of the glass envelope, most of their kinetic energy is released to the atoms of the chemical coating and the interaction results in the release of some light energy. This is called *fluorescence.* Some screens continue to glow longer than others, depending on the chemical coating; tubes provided with magnesium or calcium screens have a fast decay time lasting only a few thousanths of a second; *normal* or medium screens are formed of a zinc compound and they have a decay period of several hundredths of a second; the *slow* screens, which are formed of sulphides, glow for several seconds. Most of the screens glow with a greenish light where the electron stream strikes, but the large cathode-ray tubes used in television work respond with a light that is nearly white.

It might be well in passing to differentiate between two terms that are often found to be confusing: *fluorescence* and *phosphorescence.* It has been found that when certain materials are subjected to the bombardment of ultraviolet light, X-rays, streams of electrons, and light that is visible to the human eye, they glow with a color that is distinctive to the material. If the glow ceases almost immediately or within a few seconds after the bombardment of the material has stopped, then the glow is said to be *fluorescent.* On the other hand, if the glow persists for a long period, up to several hours, after the stoppage of the bombardment, then the glow is said to be *phosphorescent.* This latter phenomenon is illustrated in the material that is used in painting the hands and figures of watches and clocks, as they continue to glow for a relatively long period after they have been exposed to artificial light or sunlight.

Deflecting the Electron Stream

Two different means are employed for deflecting the electron stream in order to obtain the visual representations of alternating voltage and current waveforms; these are *electrostatic* and *electromagnetic* deflection. Inasmuch as the former of these is much more widely used, we shall consider this only.

In Fig. 15-29 we have reproduced the two pairs of deflecting plates indicated by *VD* and *HD* in Fig. 15-25, eliminating from consideration for the time the electrodes comprising the electron gun. Here we shall assume that the electron stream is properly focused so that it can be seen on the screen of the tube as a small spot of light. *VD1* and *VD2* represent the pair of deflection plates that while they are horizontal themselves, are nevertheless responsible for the *vertical* movements of the spot on the screen; *HD1* and *HD2* represent the *horizontal-deflec-*

tion plates, which are perpendicular physically to *VD1* and *VD2* and, as their name implies, they deflect the stream and therefore the spot on the screen, horizontally. These directions are shown in the two circles on the right.

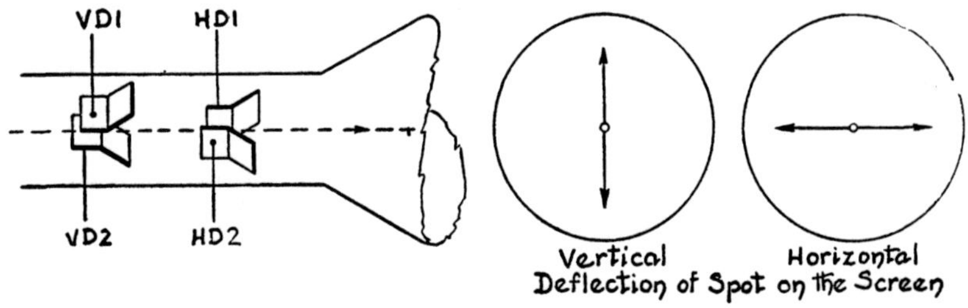

FIG. 15-29. VD1 and VD2 represent the plates that deflect the electron stream vertically, and HD1 and HD2 the plates that deflect the electron stream horizontally.

Now what causes these movements of the spot on the screen? You will recall in an earlier chapter that it was explained how a negative test charge tends to travel in an electrostatic field towards the plate on which is impressed a positive charge and how the attracting force varies with the amount of the voltage applied between the two plates. Here we have almost similar conditions, the exception being that the negative test charges (the electrons in the beam emitted by the cathode) are traveling at a high velocity through the electrostatic field and at right angles to it. Of course, the faster the electrons are traveling, the greater must be the force exerted on them to cause them to deviate through a given angle; that is, in order to make an electron change its course by a certain angle when it is traveling at a certain velocity requires less force when the velocity of the electron is low since it is under the influence of the field for a longer time.

Refer now to Fig. 15-30, which represents the pair of vertical-deflection plates, *VD1* and *VD2*. Assume that the upper plate is connected to the positive terminal of a battery, and *VD2*, the lower plate, is connected to the negative terminal. An electrostatic field is thus established that will force electrons passing through this field upwards towards *VD1*. As the electrons in the beam are traveling at such a high velocity, the force of the field acts upon them for a very short time, namely, while they are going from one end of the plates to the other; therefore the stream is deflected upwards a certain amount and the luminous spot on the screen rises from its normal position at the center of the screen, as indicated on the right of Fig. 15-30.

When the battery connections are reversed, as shown in Fig. 15-31, then by the same reasoning as outlined in the preceding paragraph, the spot on the screen will travel downwards from its normal position, as indicated in the circle of Fig. 15-31.

We have not considered here any specific amount of voltage impressed on the plates. Let us assume that we have a voltage difference

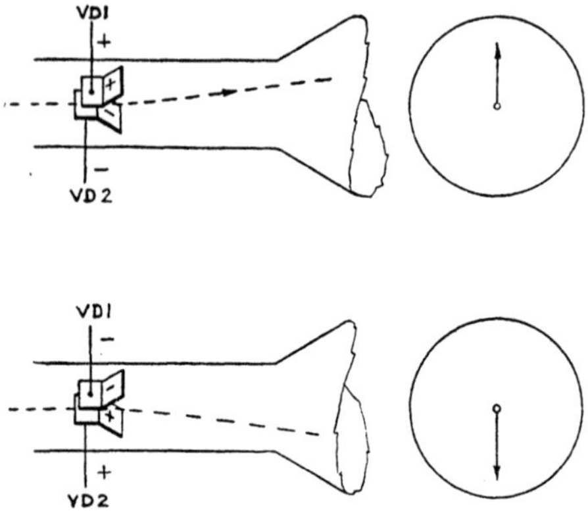

FIGS. 15-30 (above), 15-31 (below). When the upper vertical-deflection plate is positively charged, Fig. 15-30, the spot will move upwards on the screen, and when the polarities are reversed, Fig. 15-31, the spot will move downward.

between the two plates of approximately 100 volts; the spot will then move either upwards or downwards depending upon which plate is positively charged, about 1 inch. If this voltage be doubled, then the spot will move twice the distance. Thus you can see that here we have a means of determining the amount of the voltage that is impressed across the two plates. It should be noted in passing that this figure of about 100 volts per inch of deflection varies for different cathode-ray tubes.

Exactly the same reasoning that has been followed for the vertical-deflection plates can be applied to the plates which move the electron stream horizontally. If it is assumed that no voltage is applied to the vertical plates and a d-c voltage is applied to the plates *HD1* and *HD2* of Fig. 15-29, then the electron stream will be deflected to one side or the other of its normal position in the center of screen, as indicated in Fig. 15-32. If a positive voltage is applied to *HD1,* then the other plate *HD2* will be negative with respect to the former and the electron stream and consequently the spot on the screen, will be deflected to the right. If the polarity of the voltage is reversed, then the electron stream will be deflected in the opposite direction and the spot on the screen will move to the left.

From the foregoing it should not be difficult for you to foretell what happens when different voltages are applied to the two pairs of deflection plates. Let us suppose that a positive voltage is applied to the upper plate *VD1* and to the right-hand plate *HD1*. The result will be that the electron stream will be attracted upwards by *VD1* and at the same time it will be attracted to the right by the force of the field by *HD1*. The result of these two attracting forces will be that the spot on the screen will move at an angle upwards and to the right, as indi-

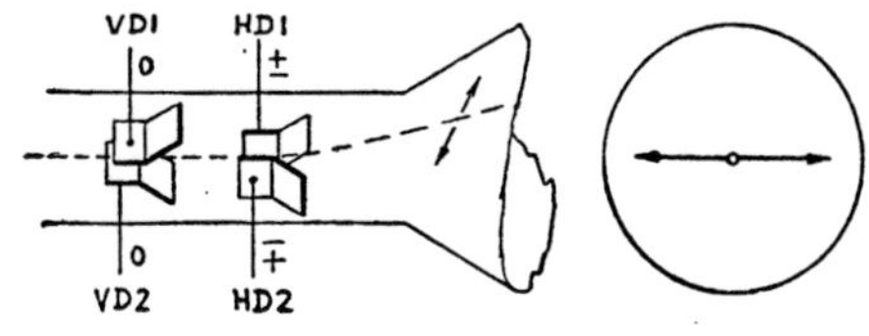

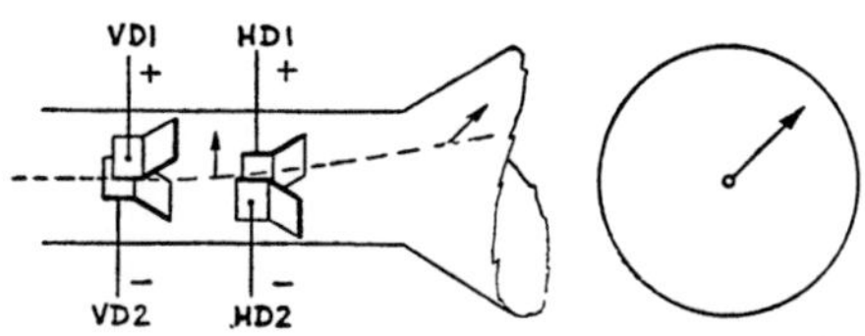

FIGS. 15-32 (above), 15-33 (below). Assuming no voltages applied to the vertical-deflection plates, the spot will move horizontally to the left or right in the center of the screen depending on the polarities of the plates HD1 and HD2. When equal positive voltages are applied to VD1 and HD1 (top and right-hand plates respectively) the spot goes towards the upper right side of the screen.

cated in the circle of Fig. 15-33. If the two voltages had been unequal but with the polarities the same, then the spot would have moved at some other angle, depending on which voltage was the greater.

Similar reasoning can be applied to any combination of voltages that may be applied to the plates. The spot can be moved anywhere on the screen, depending, of course, upon the polarity of the voltages and their relative strengths.

A-C Voltages on Deflecting Plates

Thus far we have been dealing with direct-current voltages applied to the plates. From all that you have read in the foregoing chapters, you can understand that if direct-current voltages can be employed, so also can alternating-current voltages. It is by the use of a-c voltages impressed on the two sets of deflection plates, that we can see the exact shape of the waves, for example, we can tell whether they are pure sine waves or whether they are distorted in some manner.

In order to illustrate how a-c voltages function on the deflecting plates of the cathode-ray tube, we shall still employ d-c voltages. In Fig 15-34 are two voltage-dividing devices connected across batteries, each voltage divider being tapped at certain points and a rotary switch being connected to one of each of the pairs of plates as indicated. The voltage divider that is connected to *VD1* is so divided that each tap will apply a voltage to the vertical-deflection plates that is approximately in proportion to the various values illustrated in Fig. 10-7 in Chapter 10 for a sine wave. In other words, as the movable arm travels around the taps, voltages will be applied to the vertical-deflection plates that starting at zero will increase to a positive peak (point 2 in Fig. 15-34), return to zero (point 4), then to a negative peak (point 6),

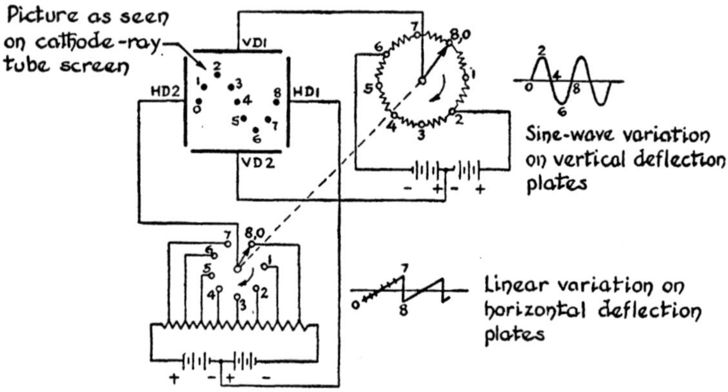

FIG. 15-34. A linear voltage is applied to the horizontal plates and this moves the spot from left to right across the screen; a sine-wave voltage is simultaneously applied to the vertical-deflection plates, thereby resulting in a cycle of the sine wave on the screen.

and finish up again at zero (point 8). A linear voltage divider of a similar type is connected to the horizontal-deflection plates and is so arranged that each division will apply equal voltages in steps, starting at the negative maximum (point 0), through the various points to a maximum positive (point 7), and ending with a sudden drop to the maximum negative (point 8). This series of voltages, if applied alone, would move the spot across the screen from left to right in equal steps and then return it to the starting point at the left. Furthermore, it is assumed that the movable arms of these two voltage dividers are connected mechanically to the same shaft so that they can be rotated together and are so arranged that the sine-wave voltage divider starts with a zero voltage and the linear voltage divider at the tap where the voltage is maximum negative, (point 0).

At the beginning the spot is at position 0 at the extreme left of the screen, due to the linear voltage divider on the horizontal-deflection plates and in the middle of the screen vertically, and due to the voltage on the vertical-deflection plates. When each movable arm is moved simultaneously, the spot on the screen moves to the right and upwards to the position 1. As the arms are moved from tap to tap, the spot progresses to the right-hand side of the screen and up and down approximately according to sine-wave variation, thus tracing on the screen one cycle of a sine wave. This method of tracing is very slow and the entire shape of the wave can not be exactly determined. If the movable arms were rapidly driven by a motor, then the spots would merge into a continuous line and we could see the waveform as if it were drawn in light on the screen.

This can be done electronically. One of the requirements is that a voltage be applied to the horizontal-deflection plates that will attract the electron stream steadily and so move the spot horizontally across the screen smoothly; then when it is at the right-hand side of the screen, the polarity will reverse and quickly return the spot to its starting point. Such a voltage is available and from its wave-shape, as

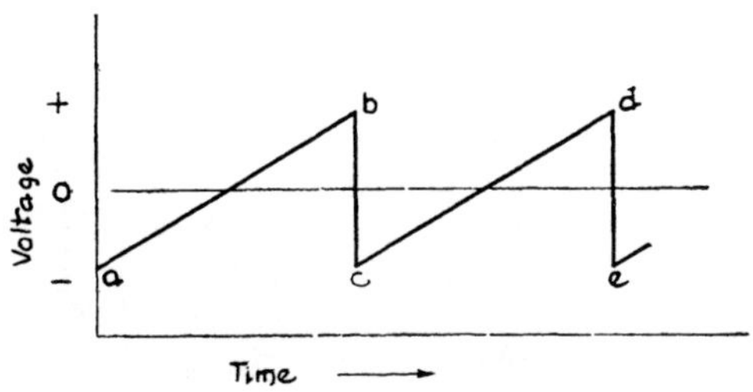

FIG. 15-35. The saw-tooth voltage applied to the horizontal-deflection plates increases linearly from a minus value, through zero, to a positive peak (from a to b) and then drops quickly to its negative peak (point c), thus moving the spot steadily from left to right and then quickly returning it to the left.

shown in Fig. 15-35, it is called a *sawtooth* wave. The point *a* is negative and as the time increases, the voltage rises through zero to a maximum positive value, point *b*. The voltage then drops through zero to a negative value *c*, which is the same as at point *a*. When such a voltage is applied to the horizontal-deflection plates of a cathode-ray tube and no voltage is applied to the vertical-deflection plates, the spot will move across the middle of the screen from left to right in a horizontal straight line. At the right-hand side, the spot will fly back to its starting position, due to the sudden drop in voltage, as shown in the almost vertical slope of the wave.

Let us assume that we apply an alternating voltage on the vertical-deflection plates that is a sine wave, as indicated in Fig. 15-36. Further, we will apply a sawtooth voltage from a linear-sweep generator to the horizontal-deflection plates. Now if each of these voltages has the same frequency, we have exactly the same result that was explained in connection with Fig. 15-34; that is, we have one cycle of a sine wave traced on the screen of the tube, as shown in Fig. 15-37. If the frequency of the sine wave be doubled and that of the sawtooth wave be left unchanged, then you can see that one cycle of the sine wave will be completed when the spot has only gone one-half the distance across the screen and by the time the spot has traveled all the way to the right-hand side of the screen, the second cycle of the sine wave will be completed. In other words, we now have two cycles of the sine wave appearing on the screen at the same time, as shown in Fig. 15-38.

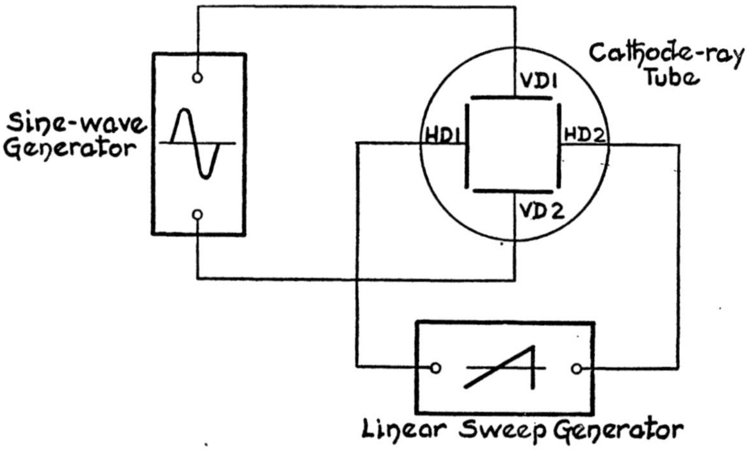

FIG. 15-36. When the sine-wave voltage and the linear sweep voltage have the same frequency, then a single cycle of the sine wave will be traced on the screen, as shown in Fig. 15-37.

For the sake of simplicity, we have made little reference in the preceding paragraphs to the return of the spot from the right side of the screen to the left. This, of course, must be taken into consideration,

FIGS. 15-37 (left), 15-38 (right). A single cycle and two cycles of a sine-wave voltage as they appear on a cathode-ray tube screen. The light horizontal line in each represents the return trace.

but as the time consumed by this return of the spot is so extremely short, it is neglected for the most part. We have indicated it in Figs. 15-37 and 15-38, as a faint line that is almost horizontal. In most of the modern *oscilloscopes* (as one instrument employing the cathode-ray tube is called) a voltage is applied to the grid of the electron gun, so that the spot is blanked out during the brief period of its *flyback* from the right side of the screen to the left.

Space does not permit further elaboration of the subject of cathode-ray tubes, but if you desire further and more detailed information on the functioning of this important tube and how it is used, you are referred to the author's book on this subject, entitled "The Cathode-Ray Tube at Work."

Visual Indicator Tubes

A *visual indicator* tube, which is sometimes called a *tuning indicator*, is essentially a special type of cathode-ray tube. Although it is frequently used on test equipment as an indicator, its most common use is as a tuning indicator that shows when a receiver is properly tuned in on a radio station.

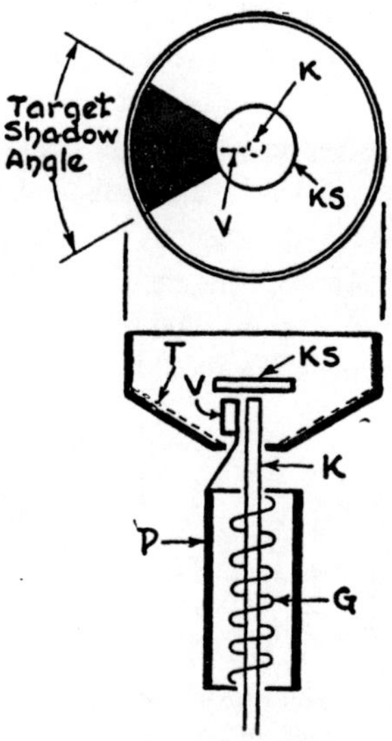

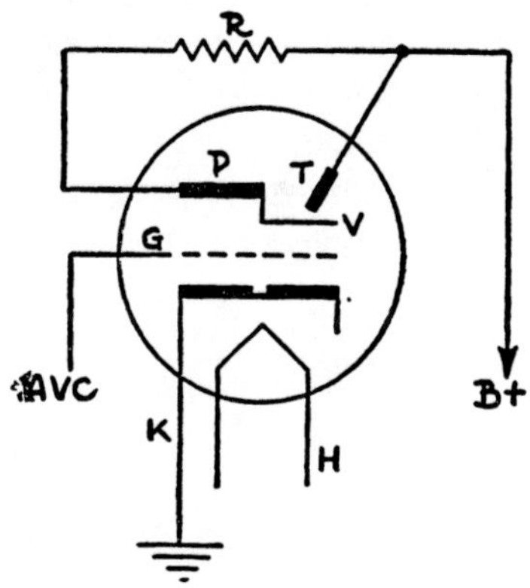

FIGS. 15-39 (left), 15-40 (right). The arrangements of the elements of a visual-indicator tube are shown on the left, and the way this tube is connected in a circuit is shown on the right.

As is shown in Fig. 15-39, a visual indicator tube consists of a triode *P,G,* and *K* and a modified cathode-ray tube, *K,V,KS,* and *T* in the same glass envelope. The cathode extends beyond the triode plate section and also serves as a source of electrons for the cathode-ray section. A cathode shield *KS* is positioned in such a manner as to prevent the red-hot cathode *K* from being seen from the top of the tube. A thin metal ray-control vane *V*, which functions to control the opening of the shadow angle, is located between one side of the cathode *K* and the fluorescent target *T*, and is electrically connected to the triode plate *P*. The fluorescent target is inclined at an angle to the axis of the cathode.

The target *T* is connected directly to the power-supply voltage, as shown in Fig. 15-40, and a resistance *R* is connected between the target and the triode plate *P*. The plate and the ray-control vane *V* are consequently always at a voltage that is less than that of the fluorescent target by an amount equal to the voltage drop across the plate-target resistor *R*. This voltage drop depends on the magnitude of the plate current, which, in turn, is dependent upon the value of the grid voltage.

Suppose that a sufficiently high negative voltage is applied to the grid *G* to drive the plate current beyond its cutoff value. Since no plate current flows, there is no voltage drop across the plate-target resistor. The triode plate and ray-control vane are therefore at the same potential as the target. Now electrons are being emitted by the hot cathode in all directions and since the fluorescent target is positive with respect to the cathode, the target attracts these electrons. As a result of this electron bombardment the entire fluorescent target becomes illuminated, with the exception of a narrow line resulting from the mechanical construction of the ray-control vane. The appearance of the target when the triode grid is highly negative, is shown in Fig. 15-41.

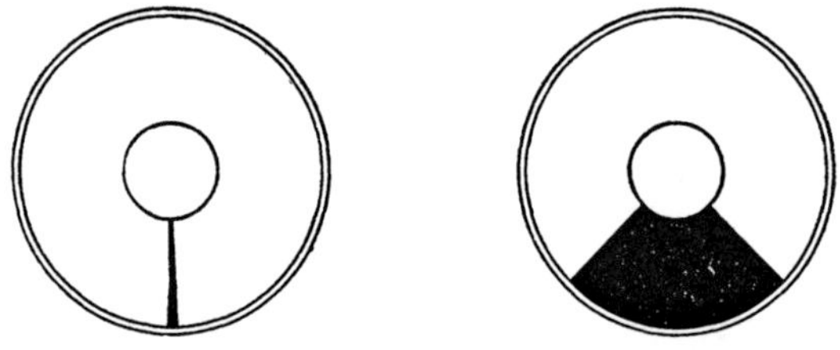

FIGS. 15-41 (left), 15-42 (right). When the triode grid of the indicator tube is highly negative, the shadow on the target is narrow (left) and this shadow widens as the voltage approaches zero.

Now assume that the grid is no longer negative, but instead has zero volts applied. Plate current will flow and a large voltage drop appears across the plate-target resistor. The positive voltage of the triode plate and therefore of the ray-control vane is now considerably lower than the positive voltage of the target by an amount equal to the voltage drop across the plate-target resistor. Although the ray-control vane is positive with respect to the cathode, this vane is now highly negative with respect to the target. The electrostatic field between the ray-control vane and the target is therefore of such character as to result in a repulsion of the electrons emitted by the cathode in a direction away from the ray-control vane. Consequently, no electrons will strike the fluorescent target in the vicinity of the ray-control vane, so that this portion of the target will be dark. The appearance of the target for zero volts on the triode grid is shown in Fig. 15-42.

The negative control voltage applied to the triode grid (or the lack of it for the zero-voltage condition) may be obtained from a circuit in a radio receiver, this circuit being known as the *automatic volume control* or *avc circuit.* In a circuit of this type a highly negative voltage is developed when a radio station is accurately tuned in, and if this negative voltage is applied to the grid of a visual indicator tube, it will be seen that the shaded sector will reduce to a narrow line as indicated in Fig. 15-41. On the other hand, if the receiver is not accurately tuned to the radio station, then very little (or no) negative voltage will be developed by the avc circuit and the angular sector of the visual indicator tube will appear as shown in Fig. 15-42. The angle of opening

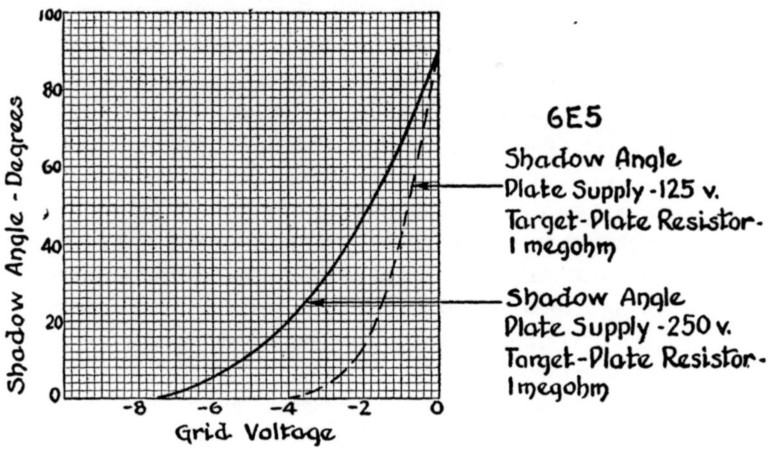

FIG. 15-43. These curves show that the shadow angle of a 6E5 visual indicator tube decreases in width the more negative the grid of the tube is made.

of the sector for a 6E5 tube for various values of control-grid voltage and for two different values of plate-supply voltage, is shown in Fig. 15-43.

Gas-Filled Tubes

You will recall that in Chapter 4 it was stated that the electrons coming from the cathode formed a negatively charged blanket around that electrode, and impeded to a certain extent the passage of other electrons through this space charge. Now if some means were available to do away with this retarding effect, it would be possible to increase the amount of current going from the cathode to the plate. Such a means is found in the gas-filled tubes.

May we refresh your memory about ionization before explaining the functioning of these tubes. In Chapter 1 it was stated that when an electrically balanced atom was struck by a high-speed electron, an electron attached to the atom might be dislodged from its orbit, and if the kinetic energy transfer were sufficiently great, the electron would leave the atom and go off into space. This subtraction of an electron from the atom leaves the atom positively charged and this action is called ionization. This *positive ion,* as it is now called, is electrically unbalanced. Therefore if a stray electron comes within range, the electron is attracted to the ion and takes the place of the electron which was removed, thus restoring the electrical balance.

In the vacuum tubes which were discussed in the preceding chapters, every effort was made by the manufacturer to have a very high degree of vacuum so that the electrons going from the cathode to the plate will encounter a minimum of opposition—that is to say, that the electrons will strike against as few gas atoms as possible. Hence, if the tube has not been evacuated to the proper degree, too many electrons will hit gas atoms and the functioning of the tube will not follow its normal characteristics. On the other hand, if some gas atoms, which meet certain physical requirements, are introduced into the tube, the electrons being emitted from the cathode will meet considerable opposition but the overall effect will be an increased current.

Let us see how this comes about. Suppose that we follow one electron from the cathode of a diode in which some gas molecules have been introduced. The electron acquires a high velocity in its flight from cathode to plate. Before it travels very far, it strikes an atom of a gas molecule and, if the kinetic energy transfer is sufficient, an electron is knocked away from the influence of the atom and goes off into space, as well as the original electron which continues on its way with a slightly diminished velocity. These freed electrons immediately come under the influence of the positive field set up by the positively charged anode and are attracted by it. In their course towards the anode or plate, they may strike other gas atoms and ionize them in the same way as described above, or they may continue on to the anode with no further hinderance. Thus more electrons arrive at the anode than are emitted from the cathode.

What happens to the positive ions? Inasmuch as the cathode is more negative than any other part of the system, the positive ions tend to travel in that direction. Before they reach the vicinity of the cathode,

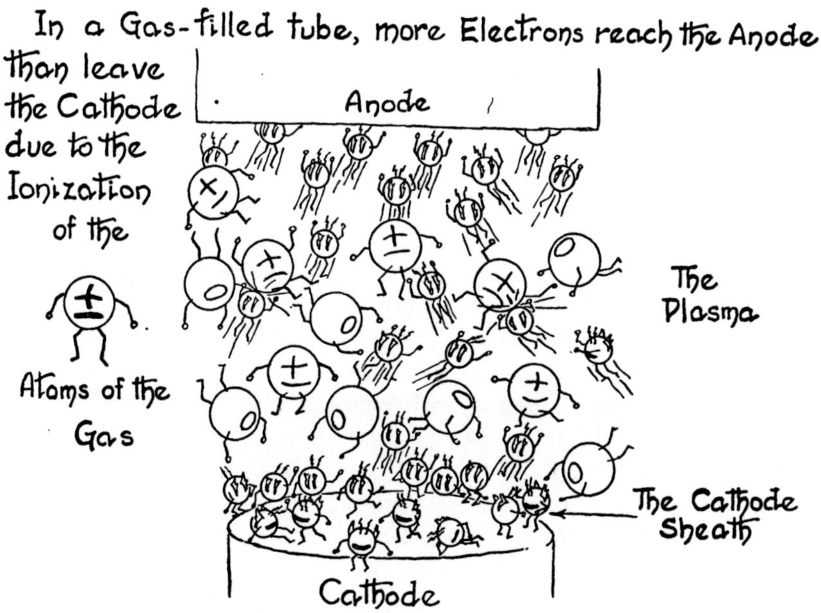

traveling much more slowly than any of the electrons, they have to pass through the negatively charged space charge, and as they are electrically unbalanced in a positive sense, they attract the nearest electrons. As the electrons in the space charge have comparatively little kinetic energy, the ions gather their lost electrons in their slow progress. These electrons, of course, are eliminated from the space charge, the effect of which is thereby reduced gradually in accordance with the number of ionized gas atoms that are restored to their original electrical balance. Thus you can see that the retarding effect of the space charge, which is called the *cathode sheath* in gas-filled tubes, is reduced approximately in proportion to the number of gas atoms that are introduced into the tube. This means that more electrons eventually reach the plate, not only those electrons that are emitted from the cathode but also those which are freed during the process of ionization. The result is that the tube is capable of passing a relatively great amount of current.

During the time that the gas atoms are being ionized, most of the space between the cathode and the anode is filled with free electrons on their way to the anode, gas atoms, and positive ions, the latter slowly going towards the cathode. This group of electrons, atoms, and ions is called the *plasma.*

In a preceding paragraph it was stated that when an electron had a high velocity it would ionize an atom if they collided. How is this high velocity imparted to the electrons? By the application of a potential of the proper value between the cathode and anode of the tube. If this potential is about 8 or 9 volts when the tube contains a certain amount of argon or xenon gas, nothing more happens than in an ordinary diode except that the electrons are impeded to a certain extent by the gas atoms on their way to the anode, but the electrons are not traveling quickly enough (they do not have sufficient kinetic energy) to cause ionization. When the voltage is raised to between 10 and 15 volts (depending again on the gas and some other factors) then the force in the field set up between the anode and the cathode is enough to cause the electrons to move with sufficient velocity (kinetic energy) so that when one of them collides with an atom, it gives up some of its energy to one or more of the electrons of the atom, and that electron leaves the atom. Thus, you see that the velocity of the electrons is dependent on the voltage impressed across the electrodes.

It might well occur to you that with an increase in the electron flow of this nature by the introduction of gas into a tube, it would be a simple matter to employ gas in a triode so that the plate current would be increased without any other changes in the functioning of the tube. Unfortunately, this is impossible, inasmuch as the electrons entering and leaving the plasma have a far different effect than the electrons that go directly from the cathode to the anode in a completely evacuated tube.

Let us assume that we have a triode which is filled with a certain amount of gas; here we will find that the control grid only exerts a control on the electrons over a certain limited range of voltage, not over a wide range as in the case of a triode vacuum tube. When a negative voltage of 10 or 12 volts is applied to the grid of our *thyratron,* as a gas-filled triode is called, the field set up between it and the cathode tends to retard the electron flow from the cathode towards the anode, upon which is impressed a constant positive voltage. This is just like the action in a triode. When the grid voltage is made less negative, more and more electrons are released as it were by the grid action and at a certain voltage, the electrons attain sufficient velocity to start the ionizing action, and the plasma is established. From this point on the grid has a negligible effect on the electron flow and can not stop it once it has started. The way to stop the electron flow from the cathode to the anode, once it has started, is to remove the anode voltage. In other words, the thyratron is a means of causing an action to start by impressing a certain voltage on the grid, and then causing the action to stop by removing the anode voltage.

Although the electrodes of the thyratron are given the same names as those in the triode, the construction of the tube itself is much different in many respects. The heater is enclosed within the cylindrical cathode, which in turn is almost completely surrounded by the control grid. Instead of being made of wire, the grid is a solid cylinder which extends below the bottom of the cathode and above the anode, as shown in Fig. 15-44. Between the cathode and the anode is a perforated disc

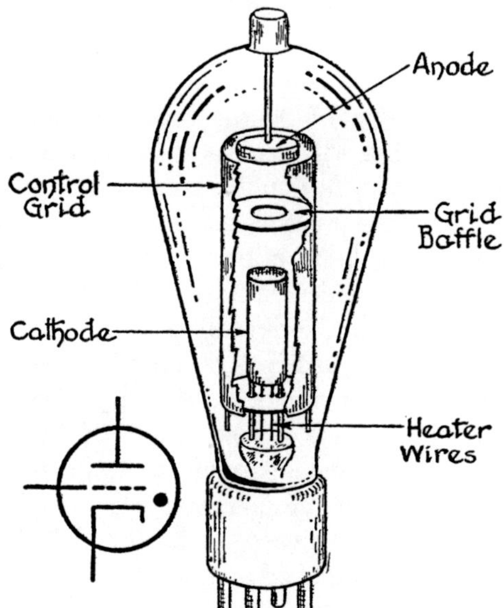

FIGS. 15-44 (right), 15-45 (left). The arrangement of the elements of a thyratron tube are shown on the right. The dot in the schematic symbol at the left indicates that the tube is gas-filled.

which is part of the grid structure and is called the *grid baffle.* These electrodes are connected to prongs on the base of the tube in the usual manner. The anode consists of a thicker disc than the grid baffle and is connected to the external circuit by a terminal cap on the top of the glass envelope. The grid extends above the anode and below the cathode in order to shield the so-called *discharge path* from charges that may accumulate on the glass walls of the envelope. The opening in the grid baffle is made large enough so that the field set up by the anode can easily penetrate into the space between the baffle and the cathode and so influence the electrons emitted from the cathode.

The construction just described is for a so-called *negative-control* tube, that is to say, a tube that triggers the ionization at a negative grid potential. By making some changes in the construction of the grid, it is possible to hold back the ionization until the potential on the control grid has become positive.

The schematic symbol for a gas-tube is similar to that for a vacuum tube, except that a dot is placed in some convenient position, as indicated in Fig. 15-45.

Photoelectric Cells

In Chapter 2 it was mentioned that electrons could be liberated from a cathode by three different means, one of the three being by *light energy*. It is this form of radiant energy that sets off the electrons in their flight from the cathode to the anode in *photoelectric cells,* which are also called *phototubes* and *photocells.*

Before explaining the fundamentals underlying the functioning of photocells it will necessary for you to know certain facts about light energy. As you doubtless are aware the light energy that affects our eyes so that we can see is, according to scientists of one school of thought, in the form of electromagnetic radiant energy that is emitted from an incandescent body like the sun, electric-lamp filament, or a candle flame. These radiations are similar to radio waves, heat waves, X-rays, and cosmic rays, differing only in their frequency, or if you prefer, their wavelength. Broadly speaking, visible light energy to the physicist consists of that portion of the electromagnetic spectrum which lies between the approximate frequency limits of 430 million megacycles and 859 million megacycles, which in terms of wavelength would be that band between 0.000070 and 0.000035 centimeter with violet at the low end and red at the longer wavelength. Thus you see that the color of the light that induces the sensation of sight is merely a matter of wavelength, or if you like, of frequency.

The nature of light has been a controversial subject for centuries: some physicists held that the radiant energy is in the form of waves and have performed elaborate experiments that substantiate their views; others said that light is not a continuous wave motion but is in the form of a series of bursts of energy and they also have experimental proof. We bring this matter to your attention here because in explaining the functioning of photocells, we shall have to refer to both wavelength and to these energy bursts.

When light energy falls on certain substances, the electrons within the substances are disturbed in their regular routine—some of them no longer whirl around their nucleii and it is the behavior of these disturbed electrons that gives rise to the photoelectric effect. This effect is not the same in all the substances in which the electrons are disturbed. For example, in one *photosensitive* material when the bursts of light energy, called *photons,* strike it, a change in potential within the substance occurs and this is known as the *photovoltaic* effect. In another substance, the resistance changes when photons strike its surface and this is called the *photo-conductive* effect. When photons strike another group of substances, some electrons are emitted from the substance and this is called the *photo-emissive* effect. And it is this last effect with which we are most concerned for its application is by far the most widely used.

The photocell itself is a simple affair compared with the construction of some of the complicated vacuum tubes that have been described earlier in this chapter. The cathode, which is coated with one of the

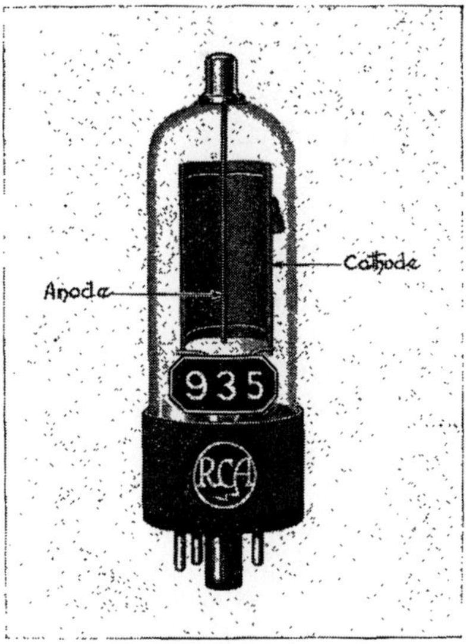

FIG. 15-46. One type of photocell has its anode, a rod, centrally located with respect to the curved cathode from which electrons are emitted when light impinges upon it.

Courtesy RCA Mfg. Co.

photosensitive materials mentioned above, is a section of a cylinder, while the anode is in the form of a rod, as shown in Fig. 15-46. Light energy enters the glass envelope and strikes the photosensitive surface.

In order that as many electrons as possible arrive at the anode when they are released from the cathode, a positive potential is impressed on the anode just the way it is in the case of a triode. A typical schematic for a photocell is shown in Fig. 15-47, in which the symbol for the cell appears. The anode is connected to the positive side of the battery E_b through a current meter and the cathode is the large semicircle connected to the negative side of the battery through the load resistor, R_b.

Now let us see how a photocell functions when it transforms light energy into electrical energy. Above the schematic of Fig. 15-47 has been drawn an electric lamp which is connected to a variable source of voltage so that the amount of light energy emitted by the filament of the lamp can be changed. In this illustration no voltage is impressed

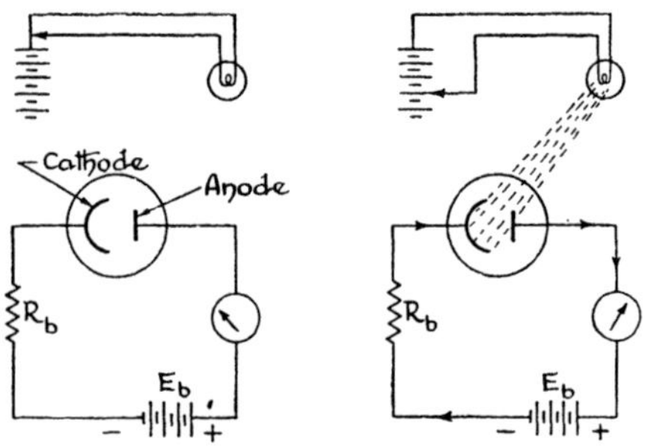

FIGS. 15-47 (left), 15-48 (right). When no light energy falls on the cathode of a photocell, no electrons flow through the anode circuit in which is connected the load R_b. In Fig. 15-48 light is indicated hitting the cathode of the photocell with a resulting current flow through the anode circuit.

across the terminals of the lamp filament, so no light energy is being emitted and no electron current is flowing in the photocell anode circuit. If a small amount of voltage is applied to the electric lamp so that the filament emits some light energy, then electrons will be emitted from the cathode of the photocell and as a positive potential is impressed on the anode, as shown in Fig. 15-48, the emitted electrons will travel across the inter-electrode space to the anode and so through the external circuit around to the cathode. The passage of these electrons will be indicated on the current meter, which here is a microammeter inasmuch as the amount of current is much less than in the case of a triode. The load resistor is included in the anode circuit for the same reason that a load is used in the output of a triode: the voltage drop developed across the load is impressed on the grid of an amplifier tube where it is amplified to usable proportions.

Two very important facts should be observed. Just as soon as light energy falls on the cathode of the photocell, electrons are emitted; in other words, the *time lag is practically non-existent.* The second fact is brought out when the amount of voltage applied to the electric lamp filament is increased: when more light energy falls on the cathode of the photocell more electrons are emitted from the cathode and the anode current is increased in proportion. Stated differently: *the number of electrons emitted per unit of time is proportional to the intensity of the light energy falling on the cathode.* This latter fact is of the utmost importance in the many uses to which the photocell is put in talking motion pictures, television, and many other industrial applications.

The question may arise in your mind as to the effect that the anode voltage may have on the electron flow. Substantially, it is the same as in the case of a triode: when the anode voltage E_b in Fig. 15-49 is increased more electrons are attracted to the anode from the cathode, with a resulting increase in the anode current. Of course, this holds true for a given intensity of the light energy falling on the cathode. In this respect, the light intensity can be compared with the control-grid voltage in the case of a triode. A family of anode voltage—anode current curves is shown in Fig. 15-49. Curve No. 1 is for the highest

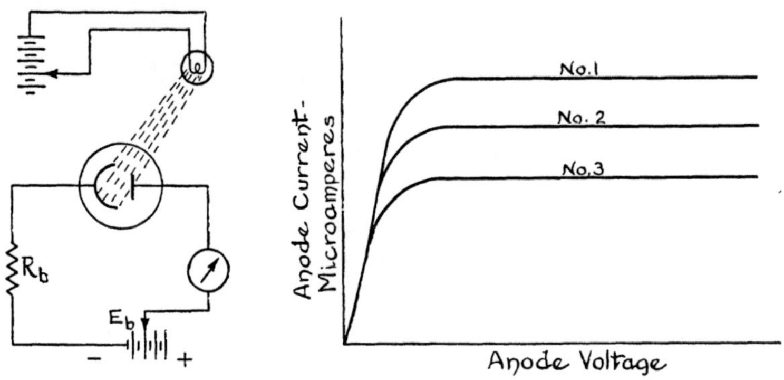

FIG. 15-49. If the anode voltage of a photocell is held constant and the amount of light falling on the cathode is varied, the anode current will change correspondingly. Curve No. 1 in the family of characteristic curves is for the greatest light intensity which is held constant while the anode voltage is increased; curve No. 2 is for less light intensity, and curve No. 3 is for the smallest amount of light intensity.

light intensity, curve No. 2 for a medium amount of light, and curve No. 3 for a small amount. You will recall that these curves are similar to the plate family of a pentode. These curves show that if the anode voltage is maintained constant, a change in the light intensity will cause the anode current to change in proportion to the amount of light. And bear in mind that these light changes cause an almost instantaneous change in the anode current.

Thus far we have assumed that the light source is of a constant frequency: it is the same color or, as it is called, *monochromatic.* The different chemicals with which the cathode surface is coated determine the frequency of the light at which the photocell starts to function; this is called the *threshold frequency.* In other words, a photocell might not start functioning until the light has changed from a deep red into a yellowish color and then it would continue to function until the frequency of the light was raised above some upper frequency limit.

APPENDIX

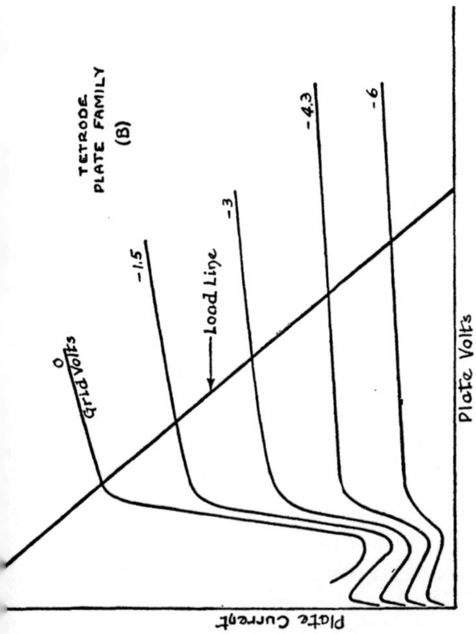

The general plate voltage—plate current characteristic curves with a typical load line on each for the three fundamental types of amplifier tubes; (A) the triode, (B) the tetrode, and (C) the pentode.

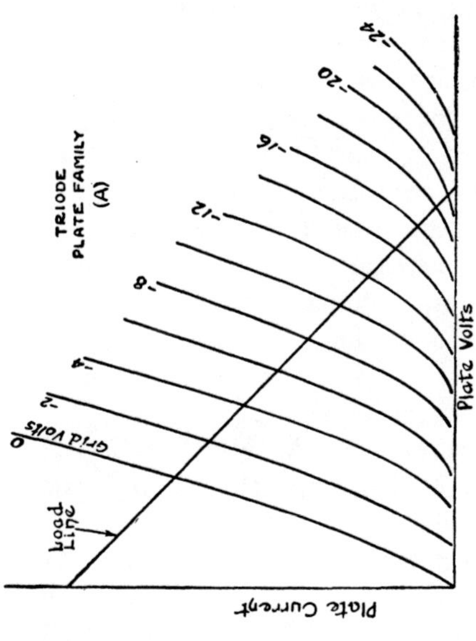

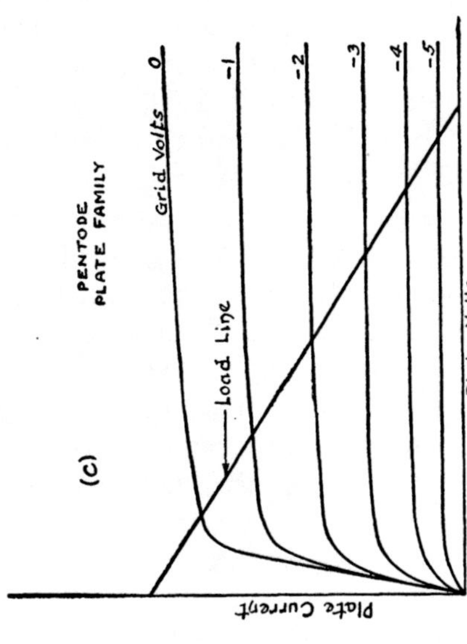

Letter Symbols

A LARGE NUMBER of letter-symbols are required to designate the various factors that are involved in vacuum-tube circuits. When first confronted with this vast array of symbols, you may feel that it would be a difficult task to memorize them. There are, however, several rules which so organize the symbols as to make them almost self-explanatory. The following descriptions will be better understood if you refer to Fig. 11-4.

(1) *Subscripts* indicate the circuit; for example, subscripts $_c$ and $_g$ refer to the grid circuit and subscripts $_b$ and $_p$ refer to the plate circuit.

(2) *Capital letters* refer to steady or direct-current values; for example, E_b indicates that the d-c voltage E is located in the plate circuit. Similarly, E_c represents the control-grid supply voltage.

(3) *Lower-case letters* refer to *instantaneous* values of varying quantities.

(3A) If the lower-case letter has the subscript $_b$ or $_c$, then it refers to the *total* instantaneous values of varying quantities; for example, e_c signifies the instantaneous total control-grid voltage, and i_b represents the instantaneous total plate current.

(3B) If the lower-case letter has the subscript $_g$ or $_p$, then it refers to the instantaneous values of the *alternating components* of varying quantities; for example, e_g signifies the instantaneous value of the alternating component of the grid voltage, and i_p represents the instantaneous value of the alternating component of the plate current.

(4) *Double subscripts* are employed in order to make a still further distinction; for example, E_{bo} represents the quiescent value of the plate voltage, while E_{ba} represents the average value of plate voltage. Also in the case of multi-grid tubes it is necessary to make a distinction among the various grids. The grids are numbered in sequence, grid number 1 being the nearest to the cathode. Thus, E_{c1} would represent the control-grid bias voltage, E_{c2} would signify the screen-grid voltage, and E_{c3} would represent the suppressor-grid voltage.

Although the letter-symbols used throughout this book are, for the most part, in accordandce with the standards for letter-symbols purposed by the Institute of Radio Engineers, it should be noted that these symbols are not universally employed. Other books may use different letter-symbols; for example, R_L is sometimes used instead of R_b to represent the plate load resistor.

List of Letter Symbols

A_v =	Voltage amplification (gain)
C_k =	Capacitance of cathode resistor bypass condenser
E_b =	Plate-supply voltage
e_b =	Instantaneous total plate voltage
E_{ba} =	Average value of plate voltage
E_{bo} =	Quiescent (no-signal) value of plate voltage
E_c (or E_{c1}) =	Control-grid voltage supply
e_c (or e_{c1}) =	Instantaneous total control-grid voltage
E_{c2} =	Quiescent value of screen-grid voltage
e_{c2} =	Instantaneous total screen-grid voltage
E_{c3} =	Quiescent (bias) suppressor-grid voltage
e_g =	Instantaneous value of alternating component of grid voltage
e_k =	Voltage across cathode resistor
E_{max} =	Maximum value of plate voltage
E_{min} =	Minimum value of plate voltage
e_p =	Instantaneous value of alternating component of plate voltage, measured relative to the time axis
e_{Rb} =	Instantaneous voltage across the plate load resistor
g_m =	Transconductance (mutual conductance)
$\% H_2$ =	Percentage second harmonic distortion
i_b =	Instantaneous total plate current
I_{ba} =	Average value of plate current
I_{bo} =	Quiescent (no-signal) value of plate current
i_c =	Instantaneous total grid current
i_{c2} =	Instantaneous total screen-grid current
I_{max} =	Maximum value of plate current
I_{min} =	Minimum value of plate current
i_p =	Instantaneous value of alternating component of plate current, measured relative to the time axis
P_o =	Power output (watts)
P_p =	Plate dissipation (watts)
R_b =	D-c resistance of the plate load resistor
R_c =	D-c resistance of the grid load resistor
R_k =	Resistance of cathode resistor
R_p =	D-c plate resistance
r_p =	A-c plate resistance
μ =	Amplification factor (constant)

CPSIA information can be obtained at www.ICGtesting.com
Printed in the USA
BVOW061709190412

288120BV00006B/2/P

9 781882 580422